THE ROUTLEDGE HANDBOOK OF MOBILE SOCIALITIES

This is a state-of-the-art survey of an emerging area of study in media, communication and cultural studies, mobility studies and mobile communications.

'Mobile socialities' demarcates a new area of research that captures people's various and contrary experiences of media in relation to their mobilities and socialities. The chapters in this volume are written by a range of international scholars offering a comprehensive overview and source of inspiration for a diverse range of topics on the contingent practices and finite resources of people and media on the move. The book demonstrates through empirical and theoretical research how mobile socialities is a generative concept for thinking through power, identity and the contexts of media in public and mediated spaces, work and everyday life, addressing a spectrum of mobile socialities and lived politics. The research and various cases make visible previously hidden or obscured social practices and allow us to rethink the meanings of mobility, digital media or the home in these examples of people living within the centre and peripheries of society.

The *Handbook* establishes mobile socialities as a new area of academic enquiry, ideal for advanced undergraduate students and scholars across the disciplines of media, communication and cultural studies, anthropology, cultural geography and sociology.

Annette Hill is Professor of Media and Communication at Lund University, Sweden.

Maren Hartmann is Professor for Communication and Media Sociology at the University of the Arts (UdK), Berlin.

Magnus Andersson is Associate Professor at the Department of Communication and Media at Lund University, Sweden.

THE ROUTLEDGE HANDBOOK OF MOBILE SOCIALITIES

Edited by Annette Hill, Maren Hartmann and Magnus Andersson

Routledge
Taylor & Francis Group
LONDON AND NEW YORK

First published 2021
by Routledge
2 Park Square, Milton Park, Abingdon, Oxon OX14 4RN

and by Routledge
52 Vanderbilt Avenue, New York, NY 10017

Routledge is an imprint of the Taylor & Francis Group, an informa business

British Library Cataloguing-in-Publication Data
A catalogue record for this book is available from the British Library

Library of Congress Cataloging-in-Publication Data
Names: Hill, Annette, editor. | Hartmann, Maren, editor. | Andersson, Magnus, editor.
Title: The Routledge handbook of mobile socialities / edited by Annette Hill, Maren Hartmann and Magnus Andersson.
Description: Milton Park, Abingdon, Oxon ; New York, NY : Routledge, 2021. | Includes bibliographical references and index.
Identifiers: LCCN 2020048698 (print) | LCCN 2020048699 (ebook) | ISBN 9780367543976 (hardback) | ISBN 9781003089872 (ebook)
Subjects: LCSH: Communication—Social aspects. | Mass media—Social aspects. | Mobile communication systems—Social aspects.
Classification: LCC HM1206 .R677 2021 (print) | LCC HM1206 (ebook) | DDC 302.23—dc23
LC record available at https://lccn.loc.gov/2020048698
LC ebook record available at https://lccn.loc.gov/2020048699

ISBN: 978-0-367-54397-6 (hbk)
ISBN: 978-0-367-54617-5 (pbk)
ISBN: 978-1-003-08987-2 (ebk)

Typeset in Bembo
by Apex CoVantage, LLC

CONTENTS

CONTRIBUTORS

Magnus Andersson is Associate Professor at the Department of Communication and Media at Lund University, Sweden. His research is within the field of media and cultural studies with a particular interest in questions related to mediation, media practices and spatial practices in the context of everyday life. He has conducted research projects on transnational migration, digitalization of worklife and within rural media studies.

Deborah Chambers is Professor of Media and Communication Studies at Newcastle University, UK. Her research centres on the changing roles, uses and social consequences of media technologies with a focus on mobile media, media in the home, social media and mediated intimacies. She has taught and researched in the UK and Australia, and is author of numerous articles and books including *Cultural Ideals of Home* (Routledge, 2020), *Changing Media, Homes and Households* (Routledge, 2016) and *Social Media and Personal Relationships* (Palgrave, 2013).

Shuhan Chen is a postdoctoral research associate at the University of Sheffield, UK. Her research explores digital media and culture in Asian countries and behaviour change for sustainable development in the UK. Her recent book (with Professor Peter Lunt) *Chinese Social Media: Face, Sociality and Civility* (Emerald, 2021, in press) discusses the modernity, culture and social changes in contemporary China through the lens of people's use of WeChat.

Jamie Coates is an anthropologist who examines the relationship between technology, mobility and imagination in urban Northeast Asia. He is currently investigating how forms of play, consumption and media use among Chinese people living in Japan are changing the ways interpersonal Chinese relations and Sino-Japanese relations are imagined. His recent publications include 'The Cruel Optimism of Mobility' in *positions: asia critique* 27(3), and the film 'Tokyo Pengyou', in the *Journal of Anthropological Films*.

Elisabetta Costa is Assistant Professor at the Department of Media and Journalism Studies at the University of Groningen, the Netherlands. She is a social anthropologist researching social media, personal relationships, politics and gender in Turkey, the Middle East and Italy. She is the author of *Social Media in Southeast Turkey* (UCL Press, 2016) and co-author of *How the*

World Changed Social Media (UCL Press, 2016). She has published in *Ethnos, Global Perspectives, the Middle East Journal of Culture and Communication, New Media & Society*, and *Oriente Moderno*.

Deniz Neriman Duru is Assistant Professor at the Department of Communication and Media at Lund University, Sweden. She has worked in different UK-, Danish- and EU-funded projects on migrants and refugees in Europe, and published articles and book chapters on conviviality, diversity, migrants' social media use, media ethnography, anthropology of Turkey and migration in Europe (in *South European Society and Politics, Southeast European and Black Sea Studies and Journal of Ethnic and Migration Studies*).

Leopoldina Fortunati is Professor of Sociology of Communication and Culture at the University of Udine, Italy and is a member of the Academia Europaea. She has conducted several research projects on the relation between mobility and digital media. Her most recent publication is: Ling R, Fortunati L, Goggin G, Lim S.S., Li Y. (eds). *The Oxford Handbook of Mobile Communication, Culture, and Information* (Oxford University Press, 2020). With Richard Ling and Gerard Goggin, she directs the series Studies in Mobile Communication at Oxford University Press. Her works have been published in eleven languages.

Maren Hartmann is Professor for Communication and Media Sociology at the University of the Arts in Berlin and is a member of the Academia Europaea. Her research focuses on media appropriation in everyday life, but also on cyberculture, the urban, mobile media and mobilities. She recently finished a research project on time and (mobile) media, and is now conducting one on homelessness and media use (DFG, 2019–2022). Her most recent book is the edited collection on *Mediated Time* (Palgrave, 2019).

Annette Hill is Professor of Media and Communication at Lund University and Visiting Professor at King's College London. Her research focuses on audiences and popular culture, with interests in media engagement, everyday life, genres, production studies and cultures of viewing. She is the author of eight books, and many articles and book chapters which address varieties of engagement with reality television, news and documentary, television drama, entertainment formats, live events and sports entertainment, film violence and media ethics.

Carlos Jimenez is Assistant Professor in the Department of Media, Film, and Journalism at the University of Denver, United States. His research broadly examines the role of media (mobile phones, social media, community radio and automation) in the everyday lives of low-wage immigrant workers. In California, he helped farmworkers build a community radio station. You can tune in to Radio Indígena www.mixteco.org/radio. His research in Denver currently focuses on the role of media technology in the everyday lives of day laborers.

Orvar Löfgren is Swedish Professor Emeritus of Ethnology at Lund University. Among his contributions to ethnological research, Löfgren has been innovative in his broad views of source material. He has studied the everyday conditions of consumption, leisure, travel, tourism, adventure tourism and cross-border transnational activities such as in the Öresund Region in connection with the Öresund Bridge. His research interests include the cultural analysis of everyday life, for example in the arenas of consumption and media use. He has a special interest in mundane activities that often are overlooked, as routines, waiting and daydreaming. Another central field is studies of national identity formations and transnational processes, and in the history of transnational mobility, as in tourism and travel.

Peter Lunt is Professor of Media and Communication at the University of Leicester, UK. The author of seven books and over one hundred papers on media audiences, regulation and social theory. His recent books include *Media Regulation* with Sonia Livingstone (Sage, 2012) and a book with Shuhan Chen on *Chinese Social Media* (Emerald, 2021, in press). He is currently working on a book for Polity on Goffman and the Media.

Shaun Moores is Emeritus Professor of Media and Cultural Studies, Faculty of Arts and Creative Industries, University of Sunderland, UK. He is the author or co-author of seven books, including *Media and Everyday Life in Modern Society* (Edinburgh University Press, 2000), *Media/Theory: Thinking about Media and Communications* (Routledge, 2005), *Media, Place and Mobility* (Palgrave Macmillan, 2012) and *Digital Orientations: Non-Media-Centric Media Studies and Non-Representational Theories of Practice* (Peter Lang, 2018).

David Morley is Professor of Communications, Goldsmiths College, University of London, UK and author of *Television, Audiences and Cultural Studies* (Routledge, 1992); *Home Territories: Media, Mobility and Identity* (Routledge, 2000); *Media, Modernity and Technology: The Geography of the New* (Routledge, 2006) and most recently *Communications and Mobility: The Migrant, the Mobile Phone and the Container Box* (Wiley-Blackwell, 2018). He is also the editor of *Stuart Hall: The Essential Essays* (Duke University Press, 2019). His work has been translated into eighteen languages.

Roger Norum works across social anthropology, media studies and the environmental humanities, asking how mobility and environment mediate human relationships to the world. He is currently carrying out research on ecologies of infrastructure, community and remoteness. Recent collaborative publications include *Tourism Geopolitics: Assemblages of Power, Mobility and the State* (Arizona Press, 2021) and *Migrantes* (Ediciones Ekaré, 2019). He has worked at the Universities of Leeds, Oxford and UCL, and is founding editor of Palgrave's Arctic Encounters book series.

Erika Polson is Associate Professor in the department of Media, Film and Journalism Studies at the University of Denver. Her research involves critical cultural studies of digital media and mobility in global contexts, and specifically on new way that status is accrued or projected through mobilities. She is author of *Privileged Mobilities: Professional Migration, Geo-social Media, and a New Global Middle Class* (Peter Lang, 2016), and co-editor of the *Routledge Companion to Media and Class* (Routledge, 2020).

Anna Lisa Ramella is a postdoctoral student in the DFG-funded Collaborative Research Center "Future Rural Africa" at the University of Cologne, Germany. She has researched mobile communities, such as musicians and railroaders in the United States and Mali before turning to future-making practices of migrants in the Kenyan Rift Valley. Her research interests include im/mobilizations, place-making, digital/media ethnography, audiovisual methods and practice theory. She is co-convenor of the European Association of Social Anthropologists networks Anthropology and Mobility (AnthroMob) and Collaboratory for Ethnographic Experimentation (#Colleex).

Anna Reading, PhD, is Professor of Culture and Creative Industries, Kings College, University of London, UK and Honorary Visiting Professor, Institute for Culture and Society at Western Sydney University, Australia. She has published *Polish Women, Solidarity and Feminism*

(Palgrave, 1992); *Communism, Capitalism and the Mass Media* (Sage, 1998) with Colin Sparks; *The Social Inheritance of the Holocaust: Gender, Culture and Memory* (Palgrave, 2002); *Gender and Memory in the Globital Age* (Palgrave, 2016); co-edited *The Media in Britain* (Macmillan, 1999); *Save As . . . Digital Memories* (Palgrave, 2009) and *Cultural Memories of Nonviolent Struggles: Powerful Times* (Palgrave, 2015). She is joint Managing Editor of *Media, Culture and Society*.

Lynn Schofield Clark is Chair and Professor of Media, Film and Journalism Studies at the University of Denver, and President of the Association of Internet Researchers. She is author or co-author of four books, most recently *Young People and the Future of News* (with Regina Marchi, Cambridge University Press, 2017) and editor or co-editor of three, including *The Routledge Companion to Media and Class* (second editor with Erika Polson and Radhika Gajjala, 2020). She is a collaborator on the YELL initiative (Youth Engaged in Leadership and Learning) that partners Black and brown young people with university students of diverse backgrounds to examine systemic racism and amplify youth voice through communication innovation.

Philipp Seuferling is a PhD student at the Department for Media and Communication, Södertörn University, Stockholm, Sweden. In his PhD project he researches media practices in refugee camps in Germany between 1945 and 2000, historically exploring the roles of media technologies in contexts of forced migration. His general research interests are media and migration, media history, memory studies and media technologies.

Jolynna Sinanan, PhD is Research Fellow in Digital Media and Ethnography at the University of Sydney, Australia. Her interdisciplinary research in anthropology, development and media focuses on digital media practices in relation to regionally comparative mobilities, family relationships, work and gender. Her books include *Digital Media Practices in Households* (Hjorth et al., Amsterdam University Press, 2020), *Social Media in Trinidad: Values and Visibility* (UCL Press, 2017), *Visualising Facebook* (Miller and Sinanan, UCL Press, 2017), *How the World Changed Social Media* (Miller et al., UCL Press, 2016) and *Webcam* (Miller and Sinanan, Polity, 2014).

INTRODUCTION

Mobile socialities

Annette Hill, Maren Hartmann and Magnus Andersson

Framing mobile socialities

Our *Handbook of Mobile Socialities* builds upon a collective of international scholars who are researching at the intersections of media, mobility and sociality. We use a new concept of mobile socialitites to capture the dynamic interplay between connected sites of analysis regarding media and communications (e.g. mobile phone), actual movement (e.g. transportation) and social relationships (e.g. friendship). Researchers explore the divergent vantage points of the interplay between mobilities and socialities, analysing material conditions, situated contexts and the ascribed meanings to people's movements, their media practices and social relations.

We argue that mobile socialities is not only concerned with connections across media, mobility and sociality, but also with the contrary reality of material and imaginary worlds. For example, travelling on the train and reading a crime novel can evoke feelings of being 'on the go' and of creating a still moment in the slipstream of life; or photographs, personal letters and text messages can nourish dreams and keep memories alive during a refugee's journey. Thus, this book is about *immobilities* as much as *mobilities*, the abilities and possibilities, or the barriers and setbacks in being mobile or immobile. And this book is about *anti-socialities* as much as *socialities*, the performance of civility and incivility in social encounters, or the disruptive citizenship of social activism, in being social or anti-social.

Thus, the term mobile socialities is about mobile and social practices which are accentuated by, and brought into relief through, the mobility of *subjects*, *objects* and *content*, set within a *plurality of socialities*. In such a way, this concept enables us to address varieties of media, mobile and social experiences: such experiences may be richly suggestive of the flow and stillness of media technologies and our lived realities, the power dynamics of emerging forms of 'the social' in mobile times and the challenging question of how we want to live and communicate both now and in the future.

We shall see how guiding questions in this book address three key areas. First, what are *emerging modes of mobile socialities* for people within different spheres of society and culture? Second, how do *economic*, *temporal* and *spatial* relations form part of mobile socialities? And third, what are the *power dynamics* of mobile socialities in media, culture and society? These questions enable the scholars contributing to this collection to explore, redefine and critically reflect on mobile socialities as *a generative concept* for thinking through issues of power and identity, materialities and imaginaries, in public and mediated spaces, work and everyday life.

A generative concept

We find the authors in this *Handbook* use the concept of mobile socialities to question earlier studies, build upon existing research and cultivate new questions regarding media, mobility and sociality. We will briefly address each of these uses of the concept in turn before expanding upon what we mean by a generative concept and how this can be useful for scholars working at the intersections of mobilities and socialities.

Questioning earlier works

First, to consider how the concept is used to question earlier works, we briefly address a core theme of the individual and their digital networks in previous research. For example 'networked individualism' (Rainie and Wellman, 2012) and 'networked self' (Papacharissi, 2010) to a certain degree have neglected socialities; these terms also tend to over-emphasize the individual. The term 'network sociality' (Wittel, 2001) does address sociality by focusing on the fleeting and informational nature of social relations supported by digital media. However, this concept stresses boundaries between online and offline environments, whereas in contemporary media, environments these boundaries are increasingly traversed, re-aligned or reconstituted. While Andreas Wittel uses network sociality in contrast to Gemeinschaft and emphasizes the integrative and disintegrative nature of current processes, in our view the fleeting nature of the online world is overdrawn in his argument.

Certainly, with regard to dominant paradigms such as Manuell Castells' concept of network society, authors in this *Handbook* consider more malleable notions of boundaries within and between the social. This argument is partly based on tracing historical roots of earlier forms of mobile socialities (as in Orvar Löfgren's [Chapter 2], David Morley's [Chapter 1] and Philipp Seuferling's [Chapter 19] contributions in this book). Mobile socialities also moves beyond understandings of global or networked societies towards more culturally nuanced work. In particular, concepts such as *deterritorialization, reterritorialization, spatial practices, mediascapes* and *complex connectivity* by authors such as Arjun Appadurai (1996), Ulf Hannerrz (1996), John Tomlinson (1999) or David Morley and Kevin Robins (1995) are seen as steps towards understanding the spatial dimensions of mediations and the myriad relationships between the global and the local.

Building upon existing research

The concept of mobile socialities is also used to build upon and take inspiration from existing studies. We briefly offer two examples arising from chapters in the book relating to media, transportation and cultural geography that illustrate how mobile socialities is a concept that builds on existing studies in the field. One source to build upon is the work of David Morley (e.g. 1991, 2000, 2017). In the beginning of his career, Morley was concerned with the place of ethnographic studies of media consumption in the analysis of the simultaneous dynamics of globalization, localization and domestication in contemporary culture. At that time, television was still very much at the heart of these concerns. Already in his book, *Home Territories*, this concern has been broadened by engaging with the question of home on several levels. Not only does Morley broaden the notion of 'home' to include Heimat and similar concepts, but he also adds the city and suburbia, mobility and migrancy, and even cybernetic geographies into his approach where media and space are closely connected.

The question of mobility and migrancy return in his latest book on *Communications and Mobility* (2017) with the telling subtitle 'The Migrant, the Mobile Phone, and the Container Box.' Morley places a clear emphasis on material objects, to follow his own call for a materialist, *non-media-centric* media studies (Morley, 2009). The additional focus on human subjects, like the migrant, offers a significant means of researching all aspects of the intersections between media, geography and transportation. A recurrent feature of Morley's work is his ability to draw together meta-theoretical structures and the daily chores of local people, a combination of the macro and micro perspectives in research that features across chapters in this collection.

Similarly, Doreen Massey is an important figure for many authors in this book, especially regarding her argument about an understanding of place as relational, which involves social interaction within and beyond the place per se (1994). In a similar vein as Morley, she advocates studies of local everyday life to shed light on the complexity of global structures. In her analysis, local practices are not more concrete than global structures; we should not conflate "geographical scale with processes of abstraction in thought" (1994: 129). Moreover, her work on power geometry highlights how "some people are more in charge" of mobility, initiating flow, whilst others are more on the receiving end, and "some are effectively imprisoned" by mobility (1994: 149). Thus, as well as acknowledging relations of power imposed from above, an understanding of mobilities from below helps us to make sense of "the life paths of and lived spaces of the displaced – the camps and shelters, streets and fields [that] illuminate the workings of power in the world" (Delaney, 2004: 848). These two brief examples signal how the concept of mobile socialities builds upon and extends work in mobilities, mobile communications and social interactions.

Cultivating new research

We find authors using the concept of mobile socialities to cultivate new questions on media, mobility and sociality. For example, the relationship between *immobilities* and *mobilities* is explored in Morley's reflections on the possible consequences of the pandemic for being in lockdown, stuck and socially distanced. In Anna Reading's Chapter 16 in this book, she addresses the question of breathing – both metaphorically as well as in real terms: without lungs that move, nothing else counts. We have recently seen how mobilities are equally dominant when they come to a halt, as they are at the time of writing this text: the COVID-19 crisis that has caught the world and stopped many movements; forced immobilities underline our unquestioned reliance on mobilities and the intertwined nature of these endeavours.

Another key question to ask concerns how to analyze 'mobility' in relation to questions of *power dynamics*, to emphasize that movement carries both meaning and power for people and things and to highlight the embodied practices and social significance of mobility (Cresswell, 2010). This is particularly strong in Part III of this book, where mobile socialities can be seen in diverse work settings: from the food courier to the Australian mining worker, the tension created by power dynamics and individual appropriation strategies is often at the forefront.

Yet another question relates to the *cultural and social imaginary*. While these frame all the socialities described in this book, they are particularly prominent in instances of attachment. The imaginaries are embedded in dating applications (apps) as well as appropriated in their use (see Deborah Chambers' Chapter 6), just as much as they are present in social media practices in southeast Turkey that negotiate quite different settings of intimacy (see Elisabetta Costa's Chapter 8 in this book). Urry first emphasized the important additions to questions of mobility in the form of imaginative, virtual and communicative travels (Urry, 2007). When Anna Lisa Ramella shows in her chapter on Kenyan fisherpeople and their modes of migrant labour how they are unable to return home (both temporarily and permanently) but nonetheless consider themselves as 'commuters,' constantly engaged in imaginative travels in relation to home, this stresses the multi-layered nature of mobilities (and socialities).

Perhaps most importantly for this *Handbook* the various chapters offer a means of framing mobile socialities as a generative concept; that is to say, the term energises researchers in creative and critical modes of enquiry, diassembling and reassembling the concept for particular cases, places and problems. Here, then, the protean characteristics of mobilities and socialities come to the fore as media and mediation, mobility and stillness, sociality and anti-sociality connect, disconnect and reassemble to enable new lines of thought on people and media on the move. For example, in research on digital travel, authors find that mobility is mediated, formatted and commercialized (e.g. Erika Polson's Chapter 10), indeed suggesting a co-option of mobility and sociality into internet celebrity culture and platform society.

Referring to the work of Stuart Hall (see Brunsdon's edited collection of Hall's writings on the media forthcoming in 2021), authors in this *Handbook* engage in deconstruction and reconstruction work in relation to mobile socialities. The terms deconstruction and reconstruction refer to Hall's interpretation of photographs of post-war Black Afro-Caribbean settlement in Britain. Tracing the contradictory meanings and imprints of the images, Hall identifies powerful discourses within the original context of the photographs; the weaving of these images into popular memory, and the various accentuations of meanings in his deconstruction and reconstruction work. We find that the authors in this book are deconstructing the concept of mobile socialities in their empirical and theoretical research, by looking at and reflecting on the disassembled parts of the concept such as freedom of movement, communicative rights or the mediation of travel, and then reconstructing the concept as multi-accentual in meaning and expectation. Above all, we find researchers in this book contradicting their expectations about what they might discover in their research, to signal the contrary realities of people and their media practices, movements and social relations.

Locating mobile socialities research

The theoretical underpinning of mobile socialities plays out within and across existing disciplines and traditions – media anthropology, cultural/human geography, mobility studies, mobile communications, micro-sociology, media and cultural studies. In this second section of our Introduction, we briefly map out key works in mobilities, mobile communications and sociality, drawing attention to key works which signpost the theoretical trajectory of the concept of mobile socialities to bring into relief the richness of intersections across these areas and the values of academic research that address the multidimensional nature of people and media on the move.

On the move: (Im)mobilities

In the first instance, we actively build on work that has developed under the so-called mobilities paradigm (Sheller and Urry, 2006, 2016; Cresswell, 2006). The primary concerns of the mobilities paradigm encompass "both the large-scale movements of people, objects, capital and information across the world, as well as the more local processes of daily transportation, movement through public space and the travel of material things within everyday life" (Hannam et al., 2006: 1). Thus, beyond the issue of movements, the mobilities paradigm addresses the social, cultural, economic and political implications of all of these movements (i.e. the meaning invested in movements), and the new theoretical approaches and methodological questions related to these movements. These mobilities are located "at the center of constellations of power, the creation of identities and the microgeographies of everyday life" (Cresswell, 2011: 551). Authors in this *Handbook* are specifically interested in approaching 'mobility' in relation to questions of power dynamics, to emphasize that movement carries both meaning and power for people and things, and to highlight the embodied practices and social significance of mobility (Cresswell, 2010: 160). Many authors are also particularly interested in processes that could best be described with the concept of motility, which refers to the – unevenly distributed – resources and abilities needed to be mobile (see also Hartmann, 2019).

From a 'mobilities turn' to a 'mobilities paradigm,' the argument has been that mobilities are becoming *the* dominant factor shaping social relations. This paradigm became more widely recognised in both sociology as well as geography in the early 21st century, and was soon followed by anthropology and interdisciplinary fields of research such as migration, tourism and gender studies (Adey et al., 2013). The mobilities paradigm – which connects movement with broader social questions while still emphasizing the meaning-making processes involved in and around movement – has roots in the transdisciplinary discussion on globalization in the 1990s. Media always had a prominent position in this debate. We thus argue that there always have been connections between media/communication studies and the mobilities paradigm, although they need to be brought to the forefront (see also Hartmann, 2013).

This connection also led Urry to argue against a "metaphysics of co-presence" and instead proclaim that "presence is thus intermittent, achieved, performed and always interdependent with other processes of connection and communication" (Urry, 2007: 47). In the course of his work, Urry coined the often-quoted definition of mobilities as: a) the corporeal travel of people; b) physical movement of objects; c) imaginative travel; d) virtual travel; and e) communicative travel (Urry, 2007: 47). Again, this underlines the intertwined nature of mobilities and media – a connection that needs to be drawn out in more depth and detail. For example, in this book, we combine imaginative, virtual and communicative travels, and our main interest is how forms of mediated mobility are related and entangled with movement of people and objects.

In the wake of the 'mobility turn,' there have been a number of studies of mobile transnational elites, sojourning the global nodes of the world (see for example Nowicka, 2006; Favell, 2008; Jansson, 2018; Polson, 2016; Andersson, 2013). From the beginning, privileged lifestyles of the Western world and new technology have been an important focus of mobilities research, with arguments about how digital systems and network capital are constitutive parts of mobile lives and lifestyles (Elliot and Urry, 2010). Yet the social differentiation of mobility is complex and encompasses many parameters; it is not just a question of who moves and who does not. It is also a question of power in relation to mobilities of a different kind. To grasp the full meaning of 'mobilities,' the studies of privileged urban people of the West can be complemented with studies of social groups who do not have the same degree of control over their mobility. Therefore, we add the motilities concept to the mix.

In this context, Cresswell and colleagues (2016) discuss a spectrum of employment-related geographical mobility and point out less noticeable mobile work such as the transnational flows of workforces within hospitality, construction and sex industries. In a similar way, we adopt a contextual approach to researching mobility 'from below.' This is exemplified by the service and social sectors that support mobile professionals and privileged white collar workers by providing this group of people with services like transport, food delivery, housekeeping and domestic renovation, labour that is discussed in several chapters. In such studies, labour and mobility are interlinked with someone else's immobility (Massey, 1994), connecting with 'mobility justice'

(Sheller, 2018), and our work is informed by unequal access to digital mobile media or labour, transportation and home life, which helps us to understand the material and subjective conditions of mobility from below. Mobility becomes a floating signifier: it is a positive sign and resource for the privileged and a negative sign of deprivation and burden for the poor. Additionally, COVID-19 is shifting the signification yet again.

Moving meaning: mobile communication

In the mobile communication field, this relatively recent area of study has focused primarily on digital mobile media, as opposed to older forms of mobile media such as books or transistor radios. Scholars in this field have tended to address the affordances and practices of mobile media technologies, researching the micro-level of mobile media use in everyday life, across diverse lifeworlds (e.g. Ling and Campell, 2011; Hjorth et al., 2012a; Licoppe, 2004). In Goggin's and Hjorth's introduction to their 2014 *Routledge Companion to Mobile Media*, they list many questions. Amongst these we find some that reflect our concerns:

> What were the implications of the mobile phone, its improvisatory possibilities in the redrawing of the public and private spheres, its support for new kinds of intimacies, friendship, and relationship, and its new affordances? What were the meanings attached to mobile phones in different places, with their particular power relations, social dynamics, and particular histories?
>
> (Goggin and Hjorth, 2014: 1–2)

It is particularly questions about relationships that we are addressing here, while stressing that presence is not just a geographical matter. With concepts such as 'connected presence' (Licoppe, 2004), 'absent presence' (Gergen, 2002), 'ambient co-presence' (Madianou, 2016) and the 'triadic communication relationship' (Höflich, 2005), researchers in the field of mobile communication have also managed to answer these questions to identify key aspects of mobile socialities. Hence, the definition of mobile socialities draws us to questions of boundaries: absence and presence (see also Milne, 2010; Hjorth, Wilken and Gu, 2012; Bourdon, 2020), online and offline (Jurgenson, 2012;) and the connection between and multiplicity of (virtual and non-virtual) actors (Miller et al., 2016).

Another concept we take from this early research is the idea of *tethering*. In early engagements with Japanese so-called keitai culture (keitai denwa, short keitai, translates as portable telephones), important key words appeared, defining keitai as "not so much about a new technical capability or freedom of motion but about a snug and intimate technosocial tethering, a personal device supporting communications that are a constant, lightweight, and mundane presence in everyday life" (Ito, 2005: 1). The notion of technosocial tethering is central to the concerns and enquiries presented in this *Handbook*. While it can imply a technical sharing of networks and resources, it also refers to the related imaginary built on close social relations, intimate communication, and the multiple experiences of belonging or exclusion. These practices (as well their counter-practices) have since spread widely, infusing our understanding of tethering (for example, see Hill's contribution in Chapter 18 of this volume).

Mobile communication scholars have more recently been engaging with the mobilities research field (see Hartmann, 2013). These enquiries have directly influenced our work. Their original questions ranged from questions about how smartphones give mobility to traditional media in terms of distribution and use in public spaces (Fortunati and Taipale, 2017: 561) to questions about how digital networks – here labelled as 'networks of personal technologies' – perform as providers of sociability (Fortunati and Taipale, 2017: 561; see also Fortunati, Chapter 4 in this book). Emily Keightley and Anna Reading similarly addressed what they call "mediated mobilities" which name "the broad emergent cluster of ways in which different kinds of mobilities . . . are themselves experienced and articulated through particular historically situated media ecologies" (2014: 286). Their emphasis is on methodological concerns and on two key features of mediated forms of mobility: multimodality and multi-scalarity (Keightley and Reading, 2014: 300). The multimodal is visible in most of our empirical contributions, where platforms and media forms are used across the range, while the multi-scalar is most visible in the way that WeChat is employed strategically for different audiences by young adult Chinese users (see Shuhan Chen and Peter Lunt, Chapter 9 in this book) or where equally strategically different forms of address are used for different people in chanelling socially acceptable desires in southern Turkey.

Worth mentioning in this context is also the research on locative media, which refers to media that feature location awareness. The combination of the mobile phone providing users with information about his/her current surroundings and the network company with spacial information about the user (Thielmann, 2010; Wilken and Goggin, 2015; Frith, 2015; Farman, 2020) is an important prerequisite for the labour of the food courier, for example (see Magnus Andersson, Chapter 12 in this book).

There remains a distinction between mobilities research and mobile communications which can hinder critical thinking about the more fundamental and underlying processes of (im) mobilities. For the present context, the research field of mobile communication in general focus too much on technology, its affordances and the micro-level of mobile media. While drawing on the insights of the field, we will to contextualize, socially and culturally, the mobile media – an approach of 'non-technology-centric mobile media studies,' in order to paraphrase David Morley's urging. Put differently, we consider mobile media and communication as embedded within larger processes of mobilities, incorporating the dialectics of micro and macro perspectives in the study of people's experiences of mobile, social and media contexts (Goggin and Hjorth, 2014).

The orchestration of social relations: socialities

Sociality is about the dynamic relations between individuals and society where there is a push and pull between societal structures and strategies, and individual relations and experiences. In this book, we draw mainly on the anthropological discussion of sociality and the micro-sociological understanding of social relations, and combine them with the question of mobilities and media.

In a problematization of sociolocal approaches to the social where there is an an implicit scale between 'individuality' and 'society,' Marilyn Strathern (1996: 64) defines sociality as the "the relational matrix which constitutes the life of persons." Such a definition emphasizes the processual, dynamic and open character of sociality, i.e. "the complex processes of relating" (Bruun et al., 2011: 13). Furthermore, socialities are always plural; it is a perspective where individuals are considered as active nodes in a set of relationships (Miller, 2012; Pink et al., 2016: ch. 6). Socialities can be scalable, an approach for describing individuals' ability to, via social media, regulate the scale of their social interaction (Borgerson and Miller, 2016: 525; Miller et al., 2016). The fundamental change to earlier forms of sociality is not the size (or scale) itself, but the fact that these different scales are accessible to many more people – plus their accessibility is combined and articulated in relation to mobility.

Another theoretical strand of sociality that also is present in the chapters of this *Handbook* is symbolic intereactionism (Blumer, 1969), ethnomethodology (Garfinkel, 1967) and the sociology of interaction (Goffman, 1961, 1963). These various traditions paid attention to social interaction, social conventions and the regulatory power of communication and appearance in the micro-contexts of everyday life. These perspectives from the middle of the 20th century have become very relevant in the light of the increased modes of interaction enabled by digital media. Christian Licoppe has been a pioneer when it comes to revitalising and adapting these theories of interactions to a multi-layered social world, where a range of mediated forms of communication has changed the conditions and modes for sociality (Licoppe, 2004; Arminen et al., 2016).

The fields of mobilities and socialities are both characterized by a theoretical and empirical openness; a curiosity of seeing (old) things in new ways. This openness forms part of our understanding of mobile socialitites; as a generative concept, it can be stretched and adapted in very different social contexts. What we have described here is a particular lens to make constellations and configurations of media, mobilities and socialities visible so that we can understand new and old patterns emerging at this juncture in time.

Connecting mobile socialities

Our journey towards the concept of mobile socialities started from a place of friendship. A mobility exchange took place between Sweden, Germany and the USA, and during this collaboration, a group of scholars connected over a united research endeavour. We found ourselves curious, and concerned, about issues of power relations, social relations and spatial relations in media and communications. It was the spaces in between media and culture, geography and transportation, social interactions and mobility, that drew us together.

We also found ourselves passionate about the inter-relations between our research specialisms: from domestication theories of the media and what the home means to people who are homeless, to the privileged and mediated lifestyles of mobile elites; from roaming audiences following media industry byways and highways and making their own pathways in the media landscape; to mobile media practices among youth and professional workers in the context of participation and everyday life. Another example of our mode of enquiry is the transient work of day labourers and their reliance on mobile communications which contrasts with the experience and practice of mobility as a mooring for food delivery workers waiting for the next order to come through. We began to understand that people's media, mobility and sociality are connected in their work, home and travel; indeed, their whole lives are intertwined with, and mediated by, mobility in multiple ways.

As we met and talked about our interests at the intersections of media, mobility and sociality, a number of sources inspired us: specific fields of study that we worked with and against led us to the formation of a generative concept, elaborated in the first section of this Introduction and within the chapters of the *Handbook*. Certainly, if the concept of mobile socialities originated from a place of friendship, then this has only been strengthened by the people who have participated in our network, both those writers in this book and those who are part of our larger collective. Indeed 'Handbook' is an appropriate term – the hand of friendship extends across our group of scholars; their different disciplines, methodologies and methods; and theoretical trajectories. We met at various events, came together to present our ideas, and offered creative and critical readings of our emergent work in mobile socialities. This collection of original research arises within our interests in the intersections of media, mobility and sociality. In this part of the Introduction, we pull together the threads that are woven into our collective interests, highlighting five patterns within the book. These patterns include various *modes* of media, mobility and sociality; sensitivity to *encounters* and *ambience* within mediated mobility and social interaction; the play off between the *material* and *imaginary* within media practices, movement and social relations; the significance of *contingency* and *contrariness* in people's experiences of (im) mobile (anti)sociality; and finally an *ethics of mutual respect* which draws together our work across sections and chapters in the book overall.

Modes

Authors' work in the *Handbook* is characterized by an attention to *modes* of media, mobility and sociality. 'Mode' is a porous word that means multiple styles, processes and practices. For example, mode means a manner, or style of something, say the mode of transportation of buses. It also means a system or technique, such as a mode in which dating applications present images. And mode can mean the practices and routines we draw upon in our experiences, for example a *mode* of thought or feeling. Thus, we can see how various 'modes' of mobile socialities signals *methods*, the infrastructures and systematic tools we apply within our research and experiences of mobility and sociality; *mood*, in particular different ways of being, atmospheres and affective and emotional processes; and *form*, referring to genres, formats, platforms and applications, or styles of aesthetics, movement or soundscapes.

David Morley, for example, revisits some of the core concerns of his earlier work to question the importance given to modes of communication technologies in relation to social change. He suggests that other kinds of mobilities and materialities shape the need that information and communication technologies began to answer. His work offers a series of focal points that he regards as vital to the analysis of media communication today. He starts from a historical perspective via an emphasis on the materiality of communication to question the inequalities in communication and other societal processes. Morley's chapter can be read as a set of guidelines for various modes of mobile socialities research – and many of his emphases are taken up in the chapters that follow.

For Leopoldina Fortunati, it is the methods and modes of research in sociality that underpins her argument; by using political economic analysis, she builds on earlier work (conducted with, above all, Sakari Taipale), which argued with and beyond Marx to analyze the work of socality and mobility in the reproductive sphere. In her current contribution, she integrates these aspects into her formulation of a modes of sociality, a concept that has attracted much academic debate and is of importance to future types of research in mobile socialities. Turning from modes of sociality to modes of media, Deborah Chambers, in her analysis of dating apps, argues for new possibilities for mobile socialities, since geolocational dating is both algorithmically steered and negotiated by users. She develops the trope of the geosocial flâneur and his/her moral dilemmas, as well as whole range of examples of algorithmic dating. Such techno-discursive practices enmeshed in current dating app conventions highlight a mode of *geo-enabled socialities*.

As a long-term ethnographer, Lynn Schofield Clark has a particular interest in modes of methods. Through long-term commitment to participatory research and community involvement, Schofield Clark reflects on the mode of anchoring narratives in digital spaces. Her argument is clearly shaped by ethnographical methods, narrative and storytelling, and engagement in civic society. That is indicated by her use of methods such as digital storytelling, photovoice, and media-rich Youth Participatory Action Research (YPAR) with young people. Schofield Clark highlights the potential pitfalls of such methods (such as losing sight of context), but also shows how empowering such methods and modes of narratives can be for young people and their potential for conversation and civic action.

Similar to Schofield Clark, Jamie Coates conducted long-term ethnographic fieldwork – in his case in the Sino-Japanese context. He is concerned with how material objects and social relations flow between these two spheres and the informal digital economies – Daigu – that stem from them. His question is how different modes of socialities (interpersonal 'flat' ones and national or transnational relations) interplay within both material and imaginary settings. We also find with Shuhan Chen and Peter Lunt an analysis of modes of sociality in an Asian context through their study of university students in Bejing. These young adults are mobile in more than one sense: geographically, socially as well as via mobile technologies. Through an interview study, they are able to show different modes of socialities emerging through the various types of mobile communications they use, where parents are treated rather differently to best friends, but also to elite circles that they want to advance to.

Encounters and ambience

The foregrounding of 'modes' of mobile socialities enables us to explore and connect strongly with social encounters and ambient spaces. This mood work and the micro-universes people construct and maintain in their everyday lives is a significant feature of many chapters in this book. This mood work needs close attention; it is what Orvar Löfgren calls 'micro-moves': he explores Swedish life between 1930 and 1960, i.e. a pre-digital life, a time when people learned new skills by using media (radio and gramophone, but also cinema and magazines). He labels this process as "everyday mood work" and shows how mobility and media were intertwined in new ways that led to novel combinations of moving bodies and roaming minds. He draws our attention to how the everyday emotional climate in a domestic setting is attuned to the affective structures of family values and social relations. Significantly, he identifies how understanding present mood work in the home is connected to the past and historical contexts, thus underscoring how the old and the new are integral to contemporary media practices. Löfgren's emphasis is on routines and rituals, but also on daydreams and micro-moves, a running theme in other chapters. We find this type of detailed research and attention to encounters and ambience crucial to many authors in this book; uncovering small details, mundane routines, and what we do to occupy our time in between jobs, or while stuck in a traffic jam, set within broader affective practices and environments.

In Shaun Moores' chapter, the kinds of movement he addresses are pre-reflective bodily movements. He uses a large range of theoretical reference points such as Merleau-Ponty, Heidegger and Ingold, but most importantly, he re-introduces the mostly unknown David Sudnow. Moores draws on Sudnow's concept of the "*community of co-movers*" – a core concept that combines the question of the social with his reflections on movements. Magnus Andersson turns to another example, that of the gig economy or platform economy, but rather than paying attention to the macro level political economics of such digital labour, he focuses on the ambience of the job and micro-moves with mobile phones. Andersson sheds light on how these workers oscillate between 'mobility-intense' delivery jobs, the frustration of immobile waiting for delivery jobs, and how the mobile phone is vital for both in terms of monitoring of labour and the mood of time in between jobs.

In many respects, mobile phone practices become ambient practices that blur the boundaries between online and offline, presence and absence, work and dwelling. In relation to Deniz Neriman Duru's (auto)ethnographical study of Workaway, she reveals the intimate sociality that emerges and thrives over time when you share a roof together to create conditions for mutual cultural understanding. She highlights how conviviality, the mood work and problem solving that is a feature of the Workaway experiences for traveller and host, does not come without attention to small details, an ethics of mutual respect and continual reflection on how to shape convivial experiences, either in the media platform, or through everyday encounters in the home.

Materialities and imaginaries

Materialities and *imaginaries* are two important aspects of work in this book, linking to previous themes on modes, encounters and ambience within mobile socialities. For example, the materiality of mobility and the resources and infrastructures people need to be mobile connect, and are in tension with, modes of the imaginary represented, articulated and experienced through media (see for example the work of Appadurai, 1996). Roland Barthes' (1979) writings are a useful point of reference for the material and imaginary; in his work, he underscores the potential significance of a cultural resource, what a cultural artefact, or social encounter, might be; and he compares this to what an artefact or encounter actually is, signalling the existing cultural and social meanings embedded in norms, values and practices at a given moment. Thus, we find materialities and imaginaries intertwined in analyses of a social imaginary for migrant workers, or a techno-tourism imaginary for travellers and the mediation of mobility, where the meanings and values attached to media devices, platforms and applications, or social encounters, and so forth, become enmeshed in the capabilities and potentialities of these same objects, places, people and their experiences. What is also key to a Barthes' inspired interpretation of the social and cultural imaginary is that this can be contrary; a dialectics is at work in opposing movements, or the contra positions taken by particular actors. Thus, the dialectics of materialities and imaginaries is a feature of work in this *Handbook*, as authors strive to unravel the entanglement of actors, material objects, and environments that shape cultural and socio-technical imaginaries.

The dialectics of the material and imaginary signals how significant it is to research material conditions, such as media artefacts, devices and objects we draw upon in our daily lives and work routines, and to explore the symbolic power of stories, for example in photographs and through social networks, since these nourish imaginaries of other lives, other futures. The play off between the material and imaginary comes to the fore in Erika Polson's analysis of 'hashtag sociality,' the social relations that appears on social media platforms in relation to '#solotravel.' Polson's study illustrates how a traditional phenomenon, travelling alone, emerges as a new cultural form via social media and the system of tagging to become an established set of values within a global middle-class culture. Her study underscore how socialities today traverse the already blurred boundary between the material and imaginary, real world spaces for travellers and digitally mediated spaces for tourism branding in the experience economy. In the case of Roger Norum's research, he considers what happens in the tourism industry under forced stillness due to a pandemic. He looks closer at two cases – the Faroe Islands and Santa Lucia – and the use of geolocative media and live streaming to create and shape presence and to ignite tourism imaginaries. Norum uses these cases to elaborate on the meaning of representation, temporality and sociality for ways we imagine and mythologise presence in the tourism industry.

Carlos Jimenez has studied precarious group of workers, people who often lack both paper and homes (i.e. paperless and homeless), by focusing on their everyday use of the mobile phone. Reflecting on the phone as a digital archive, he finds that it has meaning: as a CV, with pictures of work you are particularly proud of; as a personal and affective diary with pictures of previous and/or desired possessions; and as sociality. Jolynna Sinanan, a media anthropologist who, in this context, draws on two different case studies of mobility-related work within the tourism industry of Mount Everest and Australian mining. Sinanan's study highlights emotional geographies of mobile work and the meaning of the mobile media for the maintenance of relations with significant others. Sinanan analyses mobility and related communicative practices for the imaginaries and aspirations of the good life; how life, intimacy and social relations should be organized.

Contingency and contrariness

Closely connected to the previous thread is the significance of contingency and contrariness in people's experiences of media, mobility and sociality. A number of authors in the *Handbook* turn their attention not to movement, but to being stuck in one place, such as in the case of migrant camps; not to sociality, but to being deliberately anti-social such as a form of disruptive civic action within social movements for equality. Here, we find chapters turning the concept of mobile sociality into counter perspectives on media, movement and social relations; immobile mobility, anti-sociality, transported immobility – these contrary terms enable authors to start a process of breaking down and disassembling the concept of mobile socialities, as mentioned earlier in this Introduction. For example, Elisabetta Costa draws on her fieldwork in south-Eastern Turkey to analyze how young Muslim women, who are homebound and living in conservative family environments, use online communication to create and maintain social relationships that are separated from their offline lives, some of them romantic, others not. The separation between the two spheres is important to all Muslim women in order to avoid conflicts. Costa describes the women's agency as *immobile mobility*.

In Anna Reading's chapter, she takes us on a political journey to elaborate mobile socialities and the implicit normative values that come with it. Reading begins with the "I can't breathe" quote set against the backdrop of COVID-19 (and political strategies to control it), environmental catastrophies and demonstrations against structural racism (Black Lives Matter, or BLM). She subsequently deploys a metaphorical method using corporeal metabolization and the physiology of the lungs as 'tools' along the way. Throughout the chapter, she explores the interdependencies between stillness and anti-sociality. Reading ends with an argument about a respiratory ontology and calls for new ways of understanding and valorising neurodivergent socialities.

Maren Hartmann's chapter outlines not only the concept of mobile figures, but also explores their history and their changing social status over time. Hartmann focusses in detail on one such figure, the wandering craftsperson. This traditional figure, who still roams the streets today, offering to work in their craft while on the road for at least three years, signifies both a highly modern as well as an anti-modern streak. The figure of the wandering craftsperson offers insights into ways of maintaining and reproducing sociocultural traditions related to questions of socialities and mobilities. Annette Hill takes us on yet another journey, this time with the help of Barthes' concept of 'transported immobility' set within the rural forests of Sweden. Her chapter is a reminder about how we tend to take connectivity and affiliated socialities for granted all too easily. Hill sheds light on the question of what it means to live in a digital media blackout zone through analytical reflections intermingled with her own lived experiences. She thereby addresses the complex relationship between stillness and flow, between media connections and breakdown.

A further layer to these media practices is added by Philipp Seuferling in his analysis of refugee camps in Germany after the Second World War: the media practices he found represent centre and periphery, mobility and immobility, permanence and the temporary. Seuferling analyses the architecture, spatial design and administrative regime of the camps, the media repertoires of the camps and the media practices and protests: his research is contingent on the conditions and archival materials of the time under investigation, and yet he also finds a continuity of paradoxes for the study of the conditions of refugee camps today. In Anna Lisa Ramella's research on the transregional migrant community, she uses the mobile sociality concept to spotlight the complexity of living in a fishing village in Kenya. She underlines how these migrants are stuck in a place where fishing is not only an economic practice, but also an important social practice: they form a sociality of practice. At the same time, members of this migrant community dream of mobility, of going back to the place of their ancestors. Hence, imaginaries of home form an important part of their daily social practices, linking this thread on contingency and contrariness with the materialities and imaginaries of being a migrant worker.

Ethics of mutual respect

In recent writing, Charlotte Brunsdon (see her edited collection on Stuart Hall, 2021) commented on the underlying ethic of mutual respect within Hall's teaching and written work on media cultures. In thinking about an ethics of mutual respect, we pay attention to the power of listening and observation, to the care work involved in researching people and their everyday environments, where trust and integrity are key to these moments in the field, and to the relationships we form with participants over time (see also Sennett, 2002; Morito, 2012). The visible and invisible ethics within academic research emerge through the attention given to mutual respect of different disciplines, methodologies and theories which inform the work in this book.

There is mutual respect for people and their movements and media practices in the chapters in the collection: for example, in research involving individuals and communities, we find close understanding of people within their environments, building a framework of trust and mutual understanding between researcher and participants in their media, mobility and social practices. In other work involving participatory research, for example, or observations in the home and workplace, we find how qualitative and ethnographic methodologies require constant attention to an ethics of mutual respect, where listening and dialogue are so important to research work.

There is also mutual respect for interdisciplinary research. In this *Handbook*, we find scholars working in one area of expertise striving for better understanding of another discipline and looking for new methodologies and theories which can enhance their thinking. Such work is exemplified by a non-media-centric perspective, for attention given to quotidian media cultures. For example, scholars pay close attention to the environment, such as mood work in the domestic sphere, or the natural environment of forests and lakes for rural inhabitants. Chapters in this book detail the micro-moves we make on a daily basis, for example the movement of the hand and our mobile phones, or the movement of bicycling through a street protest; these small moves are often taken for granted but say big things about the frameworks, theories and methodologies which underpin our research. Inspiration from ethnomethodological traditions, from cultural studies, history and memory studies, signal how research at the intersections of different disciplines can enable fresh perspectives on media and communication, movement and social relations.

Scholars in this book argue against one-sided, or one-theory-fits-all, perspectives on global media and digital society; they choose to explore the material conditions and constraints on various modes of communication, labour conditions, infrastructure and social relations. Such attention to material and mediated environments, to unusual encounters and contrary realities, is a significant feature of this book. There is a curiosity in exploring the spaces in between movement and civic action, or being stuck somewhere and dreaming of home, which bring to light how people express their capabilities, hopes and dreams regarding media, mobility and sociality.

References

Adey, P., Bissel, D., Hannam, K., Merriman, P. and Sheller, M. (eds.) (2013) *The Routledge Handbook of Mobilities*. London: Routledge.

Andersson, M. (2013) "Cosmopolitan practices in social contexts: Swedish skilled migrants in the Netherlands", *Crossings: Journal of Migration and Culture*, 4(2): 155–173.

Appadurai, A. (1996) *Modernity at Large: Cultural Dimensions of Globalization*. Minneapolis: University of Minnesota Press.

Arminen, I., Licoppe, C. and Spagnolli, A. (2016) "Respectifying mediated interaction", *Research on Language and Social Interaction*, 49(4): 290–309.

Barthes, R. (1979) *The Eiffel Tower and other Stories*. London: Penguin.

Blumer, H. (1969) *Symbolic Interactionism*. Berkeley, CA: University of California Press.

Borgerson, J. and Miller, D. (2016) "Scalable sociality", *Consumption Markets & Culture*, 19(6): 520–533.

Bourdon, J. (2020) "From correspondence to computers: A theory of mediated presence in history", *Communication Theory*, 30: 64–83.

Brunsdon, C. (2021) *Writings on Media: A History of the Present*. Durham, NC: Duke University Press.

Bruun, M. H., Skrædderdal Jakobsen, G. and Krøijer, S. (2011) "Introduction: The concern for sociality – practicing equality and hierarchy in Denmark", *Social Analysis*, 55(2): 1–19.

Cresswell, T. (2006) *On the Move: Mobility in the Modern Western World*. New York: Routledge.

Cresswell, T. (2010) "Towards a politics of mobility", *Environment and Planning: Society and Space*, 28: 17–31.

Cresswell, T. (2011) "Mobilities I: Catching up", *Progress in Human Geography*, 35(4): 550–558.

Cresswell, T., Dorow, S. and Roseman, S. (2016) "Putting mobility theory to work: Conceptualizing employment-related geographical mobility", *Environment and Planning A*, 48(9): 1787–1803.

Delaney, D. (2004) "Tracing displacements: Or evictions in the nomosphere", *Environment and Planning A*, 22: 847–860.

Elliot, A. and Urry, J. (2010) *Mobile Lives*. London: Routledge.

Farman, J. (2020) *Mobile Interface Theory: Embodied Space and Locative Media*. 2nd edition. New York: Routledge.

Favell, A. (2008) *Eurostars and Eurocities: Free Movement and Mobility in an Integrating Europe*. Oxford: Wiley-Blackwell.

Fortunati, L. and Taipale, S. (2017) "Mobilities and the network of personal technologies: Refining the understanding of mobility structure", *Telematics and Informatics*, 34: 560–568.

Frith, J. (2015) *Smartphones as Locative Media*. Cambridge: Polity Press.

Garfinkel, H. (1967) *Studies in Ethnomethodology*. Cambridge: Polity Press.

Gergen, K. J. (2002) "The challenge of absent presence", in J. E. Katz and M. A. Aakhus (eds.) *Perpetual Contact: Mobile Communication, Private Talk, Public Performance*. Cambridge: Cambridge University Press.

Goffmann, E. (1961) *Encounters: Two Studies in the Sociology of Interaction*. Indianapolis: The Bobbs-Merrill Company.

Goffmann, E. (1963) *Behavior in Public Places*. New York: The Free Press.

Goggin, G. and Hjorth, L. (2014) *The Routledge Companion to Mobile Media*. London: Routledge.

Hall, S. (2021) *Writings on Media: A History of the Present*. Edited by C. Brunsdon. Durham, NC: Duke University Press.

Hannam, K., Sheller, M. and Urry, J. (2006) "Editorial: Mobilities, immobilities and moorings", *Mobilities*, 1(1): 1–22.

Hannerrz, U. (1996) *Transnational Connections*. London: Routledge.

Hartmann, M. (2013) "From domestication to mediated mobilism", *Mobile Media & Communication*, 1(1): 42–49.

Hartmann, M. (2019) "The normative framework of (mobile) time: Chrono-normativity, power-chronography and mobilities", in M Hartmann, E. Prommer, K. Deckner and S. O. Görland (eds.) *Mediated Time: Perspectives on Time in a Digital Age*. Basingstoke: Palgrave Macmillan.

Hjorth, L., Burgess, J. and Richardson, I. (eds.) (2012a) *Studying Mobile Media*. London: Routledge.

Hjorth, L., Wilken, R. and Gu, K. (2012b) "Ambient intimacy: A case study of the iPhone, presence and location-based social media in Shanghai, China", in L. Hjort, J. Burgess and I. Richardson (eds.) *Studying Mobile Media: Cultural Technologies, Mobile Communication and the iPhone*. London: Routledge.

Höflich, J. R. (2005) "An mehreren Orten zugleich. Mobile Kommunikation und soziale Arrangements", in J. Höflich and J. Gebhardt (eds.) *Mobile Kommunikation. Perspektiven und Forschungsfelder*. Frankfurt am Main: Peter Lang, 19–41.

Ito, M. (2005) "Introduction: Personal, portable, pedestrian", in M. Ito, M. Matsuda and D. Okabe (eds.) *Personal, Portable, Pedestrian: Mobile Phones in Japanese Life*. London, Cambridge/MA: MIT Press, 1–16.
Jansson, A. (2018) *Mediatization and Mobile Lives*. London: Routledge.
Jurgenson, N. (2012) "The IRL fetish", *The New Inquiry*, June 28. Available at: https://thenewinquiry.com/the-irl-fetish/ (accessed October 12, 2020).
Keightley, E. and Reading, A. (2014) "Mediated mobilities", *Media, Culture and Society*, 36(3): 285–301.
Licoppe, C. (2004) "Connected presence", *Environment and Planning D: Society and Space*, 22(1): 135–156.
Ling, R. and Campbell, S. (eds.) (2011) *Mobile Communication*. London: Transaction.
Madianou, M. (2016) "Ambient co-presence: Transnational family practices in polymedia environments", *Global Networks*, 16(2): 183–201.
Massey, D. (1994) *Space, Place and Gender*. Cambridge: Polity Press.
Miller, D. (2012) "Social networking sites", in H. A. Horst and D. Miller (eds.) *Digital Anthropology*. London: Bloomsbury Publishing.
Miller, D. et al. (2016) *How the World Changed Social Media*. London: UCL Press.
Milne, E. (2010) *Letters, Postcards, Email: Technologies of Presence*. New York: Routledge.
Morito, B. (2012) *An Ethic of Mutual Respect*. Vancover:UBC Press.
Morley, D. (1991) "Where the global meets the local: Notes from the sitting room", *Screen*, 32(1): 1–15.
Morley, D. (2000) *Home Territories: Media, Mobility and Identity*. London: Routledge.
Morley, D. (2009) "For a materialist, non media-centric media studies", *Television & New Media*, 10(1): 114–116.
Morley, D. (2017) *Communications and Mobility*. Hoboken, NJ: Wiley-Blackwell.
Morley, D. and Robins, K. (1995) *Spaces of Identity: Global Media, Electronic Landscapes and Cultural Boundaries*. London: Routledge.
Nowicka, M. (2006) *Transnational Professionals and their Cosmopolitan Universes*. Frankfurt: Campus Verlag.
Papacharissi, Z. (ed.) (2010) *A Networked Self: Identity, Community, and Culture on Social Network Sites*. New York: Routledge.
Pink, S., Horst, H., Postill, J., Hjorth, L., Lewis, T. and Tacchi, J. (2016) *Digital Ethnography: Principles and Practice*. London: Sage.
Polson, E. (2016) *Privileged Mobilities: Professional Migration, Geo-social Media, and a New Global Middle Class*. New York: Peter Lang.
Rainie, L. and Wellman, B. (2012) *Networked*. Boston: MIT Press.
Sennett, R. (2002) *Respect: The Formation of Character in an Age of Inequality*. London: Penguin.
Sheller, M. (2018) *Mobility Justice: The Politics of Movement in an Age of Extremes*. London: Verso.
Sheller, M. and Urry, J. (2006) "The new mobilities paradigm", *Environment and Planning A*, 38(2): 207–226.
Sheller, M. and Urry, J. (2016) "Mobilizing the new mobilities paradigm", *Applied Mobilities*, 11: 10–25.
Strathern, M. (1996) "For the motion (The concept of society is theoretically obsolete)", in T. Ingold (ed.) *Key Debates in Anthropology*. London: Routledge.
Thielmann, T. (2010) "Locative media and mediated localities", *Aether: The Journal Media Geography*, 5A: 1–17.
Tomlinson, J. (1999) *Globalization and Culture*. Cambridge: Polity Press.
Urry, J. (2007) *Mobilities*. Cambridge: Polity Press.
Wilken, R. and Goggin, G. (2015) *Locative Media*. London: Routledge.
Wittel, A. (2001) "Toward a mobile sociality", *Theory, Culture and Society*, 18(6): 51–76.

PART I

Understanding mobile socialities

INTRODUCTION PART I

David Morley

In this opening Part I to the *Handbook*, authors address broad themes of mobilities and socialities within particular trajectories, including past and present directions for work at the intersections of media and communications, ethnology and cultural geography, sociology and phenomenology. Authors emphasise the significance of research attuned to historical and geographical context, sensitive to processes and practices at the micro level, and at the same time alert to the political economic and social contexts of infrastructures and modes of communication. In particular, in this section, authors argue for the significance of non-media centric perspective to mobile socialities which highlights the contingency, limitations and contradictions of mobility and sociality.

David Morley's Chapter 1 begins with a critique of claims that the new technologies of our age will ultimately create a borderless world of unprecedented rates of mobility and time-space compression. That simplistic and technocentric perspective, which attributes such transformative powers to the 'new media', is subjected here to a sceptical reappraisal. Morley aims to demonstrate the virtues of a non-media-centric perspective, which addresses the very real limitations of various contemporary forms of mobility, whether virtual or actual. He further addresses the question of differential access to mobility (or 'connexity') itself as one of the key dimensions of contemporary modes of inequality. The chapter moves on to consider questions of the changing forms of home, community and sociality, in a world where multi-site households are increasingly common within migrant communities.

The chapter attempts to integrate the analysis of the mobility of persons with that of technologies, and considers various modes of symbiosis between contemporary modes of communications and transport. It considers questions of de-territorialisation and the functionality of various tele-technologies with the capacity of transcending distance. Just as Orvar Löfgren argues in Chapter 2, Morley insists that these issues must be placed in historical perspective, for instance by the recognition of the importance of earlier modes of extensive mobility (e.g. the massive trans-oceanic migrations of the late 19th century) which, in some ways, dwarf the mobility of the contemporary world. The chapter addresses the politics and symbolism of ideas of "homelands" in relation to the contemporary rise of populist nationalism and its appeal to the relatively immobilised sections of the now de-industrialised areas of previously affluent Western societies. In conclusion, the chapter also addresses how the experience of the COVID-19 pandemic illuminates issues concerning various forms of population control and regimes of exclusion. The limitations on all forms of mobility which the pandemic has imposed, may thus portend a future in which our presumptions about the future of globalisation have to be fundamentally reconsidered.

Orvar Löfgren's chapter (like that of Morley and of Shaun Moores in Chapter 3) also has a fundamental commitment to a non-media-centric approach, which examines the ways in which media technologies and everyday practices have long been interwoven. While the field of technology studies is strongly associated with a focus on the present, the new and the future, Löfgren rightly insists on the continuing relevance of a historical perspective. The chapter takes a micro-perspective, which examines in detail the processes through which media have been central to the process of modernisation in Sweden, just as they have elsewhere. It traces in detail the historical process through which first radio, and then later forms of broadcasting, became so closely interwoven with everyday household activities, as the mediated home was more thoroughly articulated with the world of the public sphere. He shows how, to a large extent, these media created feelings of empowerment, giving dispersed listeners access to forms of culture and information previously unavailable to them and creating new forms of mobility and sociality. Perhaps most importantly, Löfgren examines not only what he calls the "pioneer" periods – in which what we now think of as old ('legacy') media first emerged, bringing with them all kinds of fantasies of what futures they might enable – but also then goes on to trace the process of their gradual institutionalisation and routinisation. He thus examines the ways in which various forms of analogue media offered a new toolbox of possibilities for identity creation and enabled new skills for navigating the landscape of accelerating modernity.

In its close examination of the articulation of these earlier forms of analogue media with the transformation of the life practices of modernity, the chapter insists on recognising the extent to which contemporary digital media practices are themselves based on earlier media forms which have now been re-mediated into the digital world. The chapter refuses the temptations of any naive 'presentism' which imagines that the digital media can only be understood through an entirely new perspective (sometimes referred to as 'Media Studies 2.0'). By contrast, the perspective developed here recognises that 'newness' (while always appearing in original forms) is itself a historical constant. Löfgren shows how an understanding of previous media eras is essential if we are to develop a firmer basis for the analysis of contemporary media than that which can be achieved by an approach focused myopically on today's visions of an imagined digital future, whether utopian or dystopian.

Like Löfgren and Morley, Moores also makes a strong commitment to a non-media-centric approach, preferring, like Löfgren, to set his analysis of the technologies at stake in the broader context of how they are interwoven in everyday life practices. His approach has similarities both with the ethnomethodological tradition of Garfinkel et al. and, as he observes, to the work of scholars like Georges Perec on the need to study carefully the mundane detail of the "infra ordinary" – that unconscious framework of everyday knowledge and life-skills which invisibly underlies and enables our social interactions.

Moores' approach is also informed by the recent development of studies of the cultural forms of embodiment. Thus, when he argues that new media should be seen as "doubly digital", his implication is that we must attend not only to questions of cognitive ability, but also of physical skills – so that we must attend to the embodied forms of the (literally) manual skills that these technologies require of us. He further argues for the importance of non-representational theory if we are to better understand embodied forms of knowledge. Thus, rather like McLuhan's idea of media as extensions of the body, Moores argues that the fingers feeling their way around the computer, typewriter or piano keyboard, thereby incorporate those technologies into bodily space. Thus, the relevant forms of knowledge are best seen as being incorporated into the hands, rather than just the brain. To that extent, he argues that we should see these technologies as prosthetic extensions of the body, rather than as external objects (just as many people now regard their mobile phone as a body part, without which they are unable to function effectively).

The argument here further emphasises the importance of analysing how people develop orienteering skills, learning to find their way around – to be "wayfully oriented" – as these skills allow them to negotiate both material and virtual environments at different geographical scales. Here Moores' argument can be seen to also offer connections with earlier work in human geography including what the pioneering Swedish geographer Torsten Hägerstrand called "activity spaces" (what are called here "manipulatory areas"). In developing his analysis, Moores draws both from the phenomenological tradition and also from the urban geography of David Seamon and Jane Jacobs on the choreographies of public life worlds – the "spatial dances of bodies and things" in everyday life and conversation. For many of us, the recent experience of the relative clumsiness of technologically mediated group conversations conducted during the COVID-19 pandemic via technologies such as Zoom or Microsoft Teams has vividly demonstrated just how much face-to-face interactions depend on the highly sensitive modalities of non-verbal communication – which are extremely hard to replicate in a virtual space. For Moores, these are issues which we must integrate into our analyses of technologically mediated communication practices.

In her Chapter 4, Leopoldina Fortunati offers a sophisticated re-theorisation of the relationships between spatial mobility and social mobility (or between physical movement in geographical space and symbolic movement across social space). In doing this, she stresses the need to address both the positive and negative dimensions of different modes of social interaction. Thus, beyond the analysis of processes of association (whether or not technologically mediated), she also insists on the importance of attending to processes of separation, distancing and exclusion (cf. also Morley, Chapter 3). Her objective, as she explains, is to develop an integrated perspective on the issues surrounding the changing place of mobility and communications in social life (a perspective in some ways related to Raymond Williams' concept of mobile privatisation and his stress on community as being itself dependent on and always constituted through communications practices).

One of the most distinctive features of Fortunati's contribution concerns the way in which she draws on the theoretical frameworks developed within Marxism and classical sociology (Simmel, Tönnies, Tarde, Bourdieu) to provide the conceptual grounding for her analysis. She attends closely to the dynamic – and at times contradictory – relationships between the different modalities of communication, labour (both productive and reproductive) and mobility. Thus, Fortunati goes back to Marx' stress in *Capital* on the constitutive role of transport and mobility in the construction of markets for both commodities and labour. Here she also explores the concept of mobility as itself a commodity and discusses how contemporary workers increasingly have to sell not simply their labour power but also their potential mobility via their own network of personal communications technologies. Among the chapter's other concerns are issues such as the differential modes of integration of communication with short-, medium- and long-range forms of mobility and sociality. Fortunati also shows how any recent growth of spatial mobility might best be understood in relation to the relative blockage of opportunities for upward social mobility in previously affluent industrialised countries now languishing in economic recession. In addressing these issues, the chapter's incisive analytical approach moves well beyond the realms of superficial description of the empirically observable phenomena of spatial mobility on which much contemporary research in this field remains too narrowly focused.

1
MOBILE SOCIALITIES
Communities, mobilities and boundaries

David Morley

Introduction

In this chapter, I aim to offer a commentary on the concept of mobile socialities, which begins by considering its origins in relation to a particular vision of the role of contemporary technologies of communications in transforming our relations to physical geography. It is often claimed that under the impact of the new technologies of our globalised age, we now live in a borderless world, characterised by unprecedented rates of mobility and time-space compression – where our burgeoning virtual links can free us from the narrow constraints of our embodied, local existences. While presented as the epitome of what it is to be modern, this is actually part of the rhetoric of a very old vision of progress – as being driven by improvements in the efficiency of transport and communication. This vision goes back to the physiocrats of 18th-century France, who believed that the removal of all barriers was the most efficient way of improving the free flow of trade and enhancing the productivity of the nation. Just a few centuries later, the nation itself was declared by some to be outmoded, with commentators arguing that the coming of air transport would, in itself, lead to the dissolution of national boundaries. Whatever the problem, increased mobility is often deemed to be the answer. Indeed, in Britain during the 1980s, in the early days of Thatcherism, the Conservatives blithely suggested that the solution to the problem of unemployment caused by the destruction of manufacturing industries in the North, was for the unemployed of those rustbelt regions to simply "get on their bikes" and travel more widely in search of employment. Today, it is the UK Government's initiative to build a high-speed train network (HS2), which currently plays the rhetorical part of being the magic bullet that will singlehandedly repair the economic damage to the industrial North done by which 40 years of austerity politics. In fact, the evidence mainly points the other way and HS2 may well do rather more harm than good (in facilitating the commuter brain drain away from the North, rather than taking business there). However, from the UK Government's point of view, it has the enormous political advantage of being a highly visible initiative, designed to garner kudos and positive publicity, exactly in the way that President François Mitterand's 'Grand Projets' did in France, in an earlier age. In all these visions, speed and mobility are unalloyed goods and stasis is the enemy of progress.

I am rather sceptical of these claims, and have a fundamental reservations about the centrality given to communications technologies as the supposed causes of all these transformations. One can perfectly well consider the matter the other way round – by looking at the extent to which such technologies were only invented as attempted 'solutions' to the problems caused by the socio-economic transformations of the age. Thus sociologists, like Elizabeth Shove talk about the invention of mobile communications devices as having been necessitated by the problems of the liquid, post-Fordist gig economy, in which our precarious lives, both at work and in leisure, can only be sustained by the constantly updated multitasking enabled by these "coping devices" (Shove, 2003). I would argue for a similarly 'non-media-centric' approach to these matters, as I think that the role of media and communications technologies can only be understood in the broader context of transformations in socio-economic circumstances (Krajina et al., 2014).

Furthermore, these issues need to be placed in a historical context if, in our anxiety to be maximally up to date, we do not want to risk failing to see the present itself in a historical perspective. Thus, one might say that the capacity of today's new media to enable one to be virtually present elsewhere is simply an extension of broadcasting's role in allowing television viewers to transcend the emotional and informational confines of their physical location. In any case, the transformations at stake are perhaps best seen in relative terms. If, as the old saying goes, trying to understand what is happening in the world by following the broadcast news is like trying to tell what time it is by watching the minute hand on the clock, then trying to tell what is happening by following Twitter would be rather like following the constantly flickering movement of the clock's second hand.

None of this is to suggest that matters of mobility are less than central to our contemporary predicaments. However, in attempting to understand them, we should resist any easy generalisations – such as presuming that 'we' are all becoming more mobile. To take a simple example, while the USA is often presumed to be inhabited by the very epitome of a mobile population, rates of mobility there have, in fact, almost halved among the working class over the last 30 years (from 6.1% to 3.6% of people moving between districts per annum – Tett, 2019). While *some* of us are becoming more mobile, others are being systematically de-mobilised – or even immobilised. To that extent, we might understand mobility as a differentially distributed privilege.

In what follows, I pursue a number of recurring themes in relation to the analysis of mobilities, the continuing significance of place-based sociality, the dialectical relationship between inclusion and exclusion in communities at different scales, and the co-dependence of enhanced social connexities (for some) with the exclusion (or ex-communication) of others. However, while it is important to scrutinise the actual patterns of differential mobility among any given population, it is also crucial to consider the question of *motility* – i.e. the capacity or *potential* for being mobile. Only if we address the issue from that perspective can we highlight the question of power – especially the question of who has power or control over their own mobility – and that of others.

Questions of home, community and sociality

If, in contemporary social theory, images of exile, diaspora, time-space compression and 'nomadology' abound, the concept of home – the obverse of all this hyper-mobility – often remains uninterrogated. Thus, I want to return here to the themes of my book *Home Territories* (Morley, 2000) which focussed on how ideas of home and community need to be adapted in the context of contemporary changes in patterns of communication and physical mobility. In such a world, the very words family and household have to be redefined, as some people now inhabit multi-sited households where at least one member lives thousands of miles away, sending back 'remittance' money to support their family back home. Moreover, despite their geographical distance from home, today's tele-technologies may allow them to conduct everyday forms of long-distance parenting, participating in family life from a distance. In this connection, the Black British photographer Vanley Burke recalls how, on being left behind in the Caribbean with his grandma when his parents emigrated to the UK in the 1950s, it was as if they had just disappeared into the ether, whereas, these days, geographically absent parents are often virtually present to their children on Skype every weekend (Burke, 2014; Madianou and Miller, 2012).

When I speak of 'home,' I mean both the physical place – the domestic household – and symbolic ideas of homeland or Heimat, i.e. spaces of belonging and identity, at whatever geographical scale. If we follow the classical anthropological analysis of home territory – whether domestic house, neighbourhood or nation – we discover that home is always a sacred space, which must be regularly cleansed via rituals of purification. These rituals rid the home of dirt, i.e. all forms of matter which are seen to be 'out of place,' whether grime in the kitchen, strangers in the neighbourhood or unwanted 'foreign' elements in the culture. Crucially, from the point of view of analysing sociality (whether fixed or mobile), its construction depends not simply on linking those who are included, but equally, on the exclusion of those people deemed to not belong (cf. Sibley, 1995).

Raymond Williams argued that a community is not a group of people who happen to communicate but rather, that a community is only constructed in and through patterns of communication (Williams, 1976). In that light, we can see that the construction of the European Community and its single market has followed exactly same logic advocated by the physiocrats of 18th century France, referred to earlier, by focussing on removing internal barriers to trade and communications, but this time at a transnational level (Mattelart, 1996). To that extent, Europe has to be understood as a zone which has been systematically created by means of the gradual standardisation of a whole series of technical, administrative and financial procedures. Moreover, the policy of taking down the internal barriers between the nations within the European Community, so as to increase the trans-border flow of trade and communication, has been predicated precisely on the simultaneous strengthening the external borders of Fortress Europe. Enhancing the mobility of certain types of largely white, Christian, property-owning Euro-citizens has been systematically accompanied by a tightening of the procedures governing the exclusion of others from its borders (Morley and Robins, 1996; Verstraete, 2010)

Communications and mobility

In considering these issues, I shall argue for a materialist perspective which focuses on questions of physical movement long annexed to the neglected realm of transport studies and asserts the continuing relevance of material geography. In earlier times, communication was understood as question of both physical movement and exchanges of information and meaning, and I think that there is much to be gained from re-linking the analyses of the symbolic and physical modes of communication. In all of this, mobility (whether of people, data or commodities) is clearly a key term. If mobility might be defined as movement ascribed with meaning, one might correspondingly define media as meaning ascribed with movement. But as noted earlier, we must also consider who has the power to control their own mobility (and/or that of others) and who has the *capacity* to move, regardless of whether they are in motion at a given moment, in relation to what Doreen Massey called the "power geometry" of mobility (Massey, 1994). The issue of whose mobility is given meaning by others clearly lies at the heart of debates in Europe about the refugee crisis. We certainly need to consider the concept of "connexity" (Mulgan, 2020) as a form of unevenly distributed cultural (and technological) capital, involving the capacity to communicate with others who are geographically elsewhere. There are parallels here with Torsten Haagerstrand's work in human geography on the differential scale of the "activity spaces" (both geographical and social) across which particular categories of persons feel they have the capacity to act meaningfully on the world (Haagerstrand, 1981).

The further question is who has access to which modes of transport and communications and who is allowed into (or seen as properly belonging to) particular material or virtual spaces. Thus, we need to examine the extent to which different categories of persons are visible, invisible or hyper-visible in particular categories of representation (e.g. news and documentary programmes, as opposed to comedy or sport) or particular geographical spaces (upmarket shopping malls, as opposed to public bus stations).

One further step involves returning to Arjun Appadurai's much quoted observations on the role of disjunctures between different modes of circulation in the contemporary world (such as those between mediascapes and ethnoscapes, in his terminology) (Appadurai, 1996). The trans-border mobility of information and commodities has certainly increased rapidly in recent years, as fibre-optic cables and container boxes have so much reduced the costs of transcending what geographers call the friction of distance. However, the mobility of persons – with the exception of affluent, well-qualified members of the business elites – has been systematically declining, and for the world's poor, that immobilisation also entails emiserisation. Thus the Morrocan writer Tahar Ben Jalloun describes a migrant on the docks in Tangier who wishes that he were a parcel, as he would then gain entry to Europe so much more easily than he can as a poor man (Ben Jalloun, 2009).

De-territorialisation and technologies of distance

It is sometimes argued that today's transnational media technologies mean that *national* perspectives are no longer relevant. The difficulty here is that an awful lot of media (and not only 'old' media like TV, but also a vast proportion of internet traffic) still, in fact, operate within a national framework – and are likely to continue to do so for the foreseeable future (Hafez, 2007). Similarly, if you go to the Facebook app which shows you where exactly people have friends, you see that despite the *technical* capacity which the medium offers you, to 'friend' people anywhere in the world, the vast majority of people *only* have friends in their own (or in a geographically adjacent) country – or in places with which their country has a well-established historical (and often colonial) relation, by virtue of which they have a shared cultural and linguistic background (available at www.facebookstoriesw.com/stories).

To take another example, the growth of 'offshore' telephone call centres, designed to handle the customer service functions of many Euro-American businesses in distant territories, is often presented as an example of *de*-territorialisation. However, while their locations may be transformed, they are by no means outside geography, i.e. they are better understood as *re*-terrorialised, rather than de-territorialised. Indeed, all this is perfectly explicable in terms of the history of imperialism – which is why British call centres tend to be in India, as a result of Britain's imperial 'bequest' of low labour costs and a high level of indigenous English language skills, while for similar reasons, French and Spanish ones tend to be in North Africa.

Beyond Year Zero – taking a historical perspective

Everything that we know about previous media technologies tells us that they rarely have straightforward effects on what people do. Nonetheless, much debate proceeds as if today's media are so much more powerful than those which preceded them, that we need to start again from the Year Zero of digitalisation, in order to recognise their special powers (cf. Gauntlett, 2011). The problem here, is that *all* ages have thought that the 'new' technologies of their day had overwhelming powers which would sweep everything else before them – and there is no reason to presume that will turn out to be any more true for us than it did for our predecessors (cf. Marvin, 1988).

Rather than proceeding abruptly from one era of communications to another, we find a number of continuities, overlaps and modes of symbiosis between old and new technologies of symbolic and material communications. It is this shifting landscape of differentiated forms of virtual and actual connectivities, in which old ghosts from previous historical eras still haunt the byways of cyberspace, and where the effectivity of even the latest technologies still depend, ultimately, on material infrastructures, that is the object of my concerns (Parks, 2005).

A historical perspective on these matters is crucial. The *proportional* increase in the speed of transport in the 19th century – with the transition from horse-drawn carriages in its early days, to the first modes of air travel at its close – was far greater than that experienced in recent years and ours can reasonably be considered, in historical terms, to be an age of technological stasis, relatively speaking (cf. Edgerton, 2006). As for the future, the notion that the world's population is becoming *more* mobile just does not hold up: the proportion of the world population involved in long-distance migration in the late 19th century was rather higher than anything seen today (with vast numbers of people migrating across both the Atlantic and Pacific oceans, en route to North America – see Kern, 2003). In any foreseeable future, the majority of the world's future population will live in the shanty towns and slums of the Third World megacities. Very few of them will have either the political rights or the money to travel – not least because it is even more expensive to travel as an illegal migrant than it is to travel first class, and the potential bills can leave your family in debt to the people-smugglers for a generation.

New mobilities – beyond nomadology

Contrary to the discourse of nomadology (cf. Deleuze and Guatari, 1986), we are *not* all nomads in the contemporary universe, as the social and cultural patterns within which we live offer each of us particular ranges of mobility options in both the virtual and material realms. Some of us, in privileged positions, with high levels of cultural capital, a good internet connection, legitimate passports issued by the governments of powerful countries, and plastic cards in our wallets backed up by a good credit rating, may well be increasingly mobile in both the material and virtual senses. However, others increasingly find themselves, *more* confined to their physical location than they ever were before – and for them, locality is still destiny. In many areas, a person's geographical postcode provides an excellent predictive index of both their lifespan and the 'quality' of that life (de Blij, 2009).

Many years ago, Doreen Massey questioned the extent to which the "space-time compression" supposedly made available by enhanced modes of communication was available to the poor (Massey, 1994). Her approach opens up the crucial question of what we might call the politics of waiting – which is, of course, often the fate of the poor, or those who lack the qualifications which give access to the relevant fast-track or priority lane. The amount of waiting which a person has to do (for a bus, for a doctor's appointment, for a visa, for a good broadband connection, etc.) has to be seen as functioning both as an effect of impoverishment and a form of relative exclusion from the realm of the privileged which further exacerbates the poverty of their lives.

Virtual and material geographies

In recent years, the 'death' of geography has been widely announced, following Meyrowitz's argument that, as the media now allow us to inhabit a range of "virtual elsewheres," we are no longer 'in' places in the way our forbearers were (Meyrowitz, 1985). However, we should not mistake the emergence of today's virtual/electronic landscapes for the death of material geography, and we should be alert to how material geographies retain significance even under changing technological conditions. Thus, we might better see the 'virtual' and the material as overlapping and intertwined, with physical geographies now overlayed and criss-crossed by information flows.

Most of the work that has been done on the role of the communication industries in time-space compression has focussed on the capacities of internet technologies to *transcend* physical distance (Wark, 1994; Khanna, 2016). However, the second largest category of e-mail traffic after 'spam' is that sent between people in the same organisation, and often in the same building. Rather than enabling the transcendence of distance, the primary role of these technologies is to intensify communications between those who are already physically – or organisationally – proximate. Naturally, all this intensification has a price: I noted earlier that the construction of community always involves the exclusion of those who are defined as being outside it. By the same token that these technologies serve to intensify communications between members of a network, they simultaneously tend to enhance the greater separation of different networks. Because the mobile phone enables people can 'take their network with them wherever they go,' they increasingly travel within the psychic carapace provided by that mobile sense of community – and are less open to communication with others outside their own public "sphericules" (Gitlin, 1998).

Contemporary digital forms of communication were initially lauded precisely for their capacity to transcend geographical distance. However, outside of the business elite, most people's lives are still conducted on a very localised basis, and studies of the actual usage of mobile phones tend to show that their primary use is for coordinating quotidian activities within a surprisingly small area (e.g. Who will pick up a child from school? What time to meet in the usual café? How late will the train be?).

Moreover, critical work on the mapping of the material geography of the internet industry shows that cyberspace itself has a perfectly identifiable geography, in which its routes and locations largely replicate the structure and patterns of earlier modes of communication. This work also alerts us the dramatic variations in the accessibility, speed and density of internet connectivity per square kilometre, in different geographical locations. Further, while the internet industry is commonly thought to be about the *de*-territorialisation of communications, it actually tends to cluster tightly in very particular places (such as the area now known locally as 'Silicon Roundabout' in Shoreditch, East London).

Rather than imagining that we stand on the cusp of an era in which the virtual geography of cyberspace will entirely replace the physical geography of the world, we would do better to accept that the virtual is now increasingly becoming a banal overlay to the world of the actual. Moreover, speaking of cyberspace in the singular may be less useful than recognising that cyberspaces are best characterised by reference to the particular ways in which the virtual and the actual are mixed together in different locations, according to localised cultural rules and priorities. Thus, if 'connexity' is considered an unevenly distributed asset, the questions are about who has access to which levels and modes of transport and communications. To this extent, one might argue that the amount of waiting which a person engages in is, increasingly, both a good index and simultaneously a reinforcing determinant of their social status.

The new enclosures

If hyper-mobility is one of the key figures of our contemporary condition, then its correlative is surely the gated community. There is a growing tendency towards residential segregation throughout the affluent societies of the West, as those who can afford to do so remove themselves from the fractious world of the decaying public sphere. Here we confront the politics of withdrawal and separation, both within the city, and in the flight of privileged groups to the suburbs, or the countryside, to escape from the burgeoning multiculturalism of city life.

We might see the rise of the gated community as part of a wider pattern, also involving a set of technologies like the private car, the mobile telephone and the internet. Leuven de Cauter describes all of these technologies as "capsular" in tendency, in allowing people to withdraw from the fractious realm of the public into more defensible spaces, containing only like-minded fellow citizens (de Cauter, 2006). They can thus heave a sigh of relief, as they escape what Kobena Mercer once called the "sheer difficulty of living with difference" (Mercer, 1994).

I referred earlier to how multi-sited households use tele-technologies to invent new forms of long-distance parenting. But this process is far from free of conflict: technologies like Facebook and Skype, which help absent mothers feel like they are still being good parents from a distance, are often resented by their children as unwanted forms of surveillance. And 'being there' via Skype can easily backfire, as my colleague Mirca Madianou reports when a Filipina woman working in the UK finds that being only virtually at her daughter's distant birthday party was so distressing that she broke down (Madianou and Miller, 2011).

The question is exactly how a variety of communications technologies are articulated across online and off-line territories. Natalie Dixon shows how the affluent (and of course, largely white) population of gated communities in the suburbs of South African cities have deployed WhatsApp technology to monitor the presence of any strangers (especially Black ones) who wander across their boundaries and more effectively police access to their own "home territories" (Dixon, 2019). Of course, WhatsApp is capable of other, more laudable, uses – but mobile technologies *can* readily enable one group of more powerful people to better restrict the mobility of others.

Virtual inequalities

In the face of all these complications, we need even more ethnographically informed, empirical research of a comparative kind, based on work both within and without the rich Northern part of the world (Larkin, 2008). That work needs to focus on how, in different polymedia environments, in varying cultural contexts, specific technologies are used by different categories of more – or less – mobile persons (Burrell, 2012).

One example here concerns a set of photographs which appeared in the British news media in 2017–2018, showing images of Syrian refugees using smart phones. These images provoked an outcry because they were seen to 'prove' that these were not real migrants – on the assumption that such people surely could not afford those types of phones. However, it is well-attested that all migrants tend to spend a high percentage of their disposable income on communications, as the maintenance of long-distance links is such a priority to them. If you are a refugee (or indeed, a homeless person), having a mobile phone is an absolute necessity. Without any stable physical 'home territory' around which to organise your life, your survival is, in fact, completely reliant on your access to up-to-date information about local sources of food and shelter – and without a phone, you may well not be able to get it.

The more fundamental point is that it is often very expensive to be poor. A recent collection on the relations of media and class shows how being poor makes it very difficult to access to good quality mobile communications (Poulson et al., 2020). Not being a homeowner, or being only in temporary housing, may disqualify you from broadband or cable contracts. On a 'pay-as-you-go' phone with a very expensive per-minute call charge, you may not be able to afford to hang on long enough to get through to the particular bureaucracy you need to contact about your welfare entitlement – in which case, you may just have to walk there to argue your case. As for the supposed de-territorialisation of nomad lifestyles, Arjun Appaduarai points out that the first thing a homeless family must do on arriving in Mumbai is to make a deal with whichever gang of protection racketeers controls the pavements of the area of the city where they want to sleep (Appadurai and Morley, 2011).

After globalisation? Brexit in the rust belt...

Let me now try to connect these concerns with mobility and sociality more directly to questions of politics. At the time of the 'Brexit' referendum in the UK in 2016, one of the regions with the largest percentage of voters in favour of leaving the EU was Birmingham. Its problems parallel with the dilemmas currently being experienced in many other parts of Europe. The key issue here concerns people who, until now, have largely been both immobile and invisible. Rather than exploring *any* kind of 'mobile sociality,' they are now retreating behind sedentarist, particularistic boundaries which are certainly based on xenophobia, if not always on racism.

Since the 1960s, one of the main effects of globalisation in the now post-industrial areas of the UK has been the destruction of manufacturing industry by cheap imports. The consequent high level of unemployment in these areas has paved the way for a variety of racist politicians to exploit the anxieties felt by the local working classes about their declining prospects. These are people who are now trapped by their local circumstances and consigned to immobility. Even the cheap annual package holidays to Spanish coastal resorts which they enjoyed in earlier, more affluent times are now a thing of the past – they are stuck in their fast-declining local communities, with little prospect of any escape and for them locality is increasingly destiny.

This Brexit backlash represents a kind of political revenge on the part of people who have come to regard themselves as 'left behind' by globalisation. Their demands to "take back control" over their own communities represent a yearning for some sense of security in this liquid world of flux. If these destabilising forces impact most sharply on those at the bottom of the social hierarchy, their response increasingly takes the form of a reactionary localism. For many years, these people have not voted at all, as they have felt alienated from all the UK's mainstream political parties – and so have been politically invisible. Their enormously disruptive potential consists in the fact that they feel they have nothing left to lose and their accumulated resentments are now articulated through sentimental modes of nationalism, "post-colonial melancholia" (Gilroy, 2004), moral indignation, anti-elitism and the constant scapegoating of all that is 'foreign.'

The moment when it became clear just how much trouble was looming over the horizon was the evening before the European Referendum in 2016. Our local Midlands television station was doing '*vox pop*' interviews that evening in the streets of a particularly depressed part of the Black Country called Rowley Regis, where my family still lives. The poor women being interviewed announced that they were planning to vote for Brexit because they had "had nothing" for years and so felt that things simply could not get any worse, so they might as well just make a "protest vote" – without even seeming to care where that might lead. This is the politics of disenfranchisement which, rather than being guided by rational calculations of the relative advantage to be gained by supporting this or that political party, functions more in the spirit of 19th century Russian nihilism, simply wanting to exact some form of painful revenge on everybody else for the wrongs which have blighted their lives.

We now perhaps enter, as Pankaj Mishra has put it, an "age of anger," where, in the "swamp of fear and insecurity" in these de-industrialised regions, we see what Hannah Arendt once called a "tremendous increase in mutual . . . irritability of everybody against everybody else"

(Mishra, 2017; Arendt, 1951). This is a kind of "negative solidarity" as she puts it, of the type of which Nietzsche spoke, when describing "the tremulous realm of the inexhaustible and insatiable *ressentiment* of the dispossessed" (Nietzsche, 1887/1998). This, in the political shorthand now commonly used in the UK to describe this situation, is the problem of those who feel 'left behind.' They have all the anxieties that come with the chilling recognition inscribed in the old Soviet phrase which announced – if in a different context – that 'not everyone will be taken into the future.' Their enormously disruptive potential consists precisely in the fact that, feeling they have nothing left to lose, they have no specific political objective other than to try to make sure that others will now "hurt" like they do, to quote one of Randy Newman's 'redneck' songs (Newman, 1974).

Mobilities in a post-Wuhan world

All of this chapter so far was written before the end of 2019, when the world as we then knew it was turned on its head by events originating in Wuhan, China. Nonethless, even before that date, there were some emerging elements of a fundamental critique of the overblown encomia to the wonders of global mobility which had dominated much of contemporary discourse in the first two decades of the century. Indeed, in an earlier book, I laid out the fundamentals of such a critique (Morley, 2017). In that book, I was concerned to re-integrate the study of the mediated forms of communications (on which our discipline has come to concentrate almost exclusively in recent years) with material questions of transport geography and the mobilities of people and commodities, as well as messages. As I explained in the book, in taking this approach, I was influenced by the development of the 'new mobilities' paradigm associated with the work of scholars such as John Urry and his colleagues at Lancaster University (Urry, 2007). Most notably, in the founding statements of the 'new mobilities' approach, reference was sometimes made – in what now seems like an extraordinarily perspicacious manner – to the necessity to pay attention not only to the seemingly desirable contemporary increases in the speed of circulation of 'Goods' (be they commodities, or tourists with passports visas and credit cards) but also to the circulation of 'Bads' (whether drugs or illegal migrants, often smuggled in the sealed container boxes which have become the standard medium of global trade). Beyond that, attention was also drawn by some scholars, to the potential of high-speed global forms of air transport to 'mobilise' undesirable organisms such as viruses across vast distances in a rapid and potentially problematic manner. Of course, at that point, these difficulties were merely hypothetical and for things to 'go viral' still had a positive connotation. Indeed, it was the ambition of celebrities, politicians and advertisers for their messages to achieve viral status – ideally, on a worldwide basis, as an index of hyper-success.

Looking back now, the optimistic perspective on globalisation articulated by its cheerleaders looks somewhat naïve. But at the time, it was still possible for people to mistake those forms of unpoliced hyper-global mobility as representing the beginning of a historical period set to continue onwards in that same style, far into the future. Evidently, we see now that it was no more than a short lived – and not only exceptional, but unsustainable – form of social organisation.

I will soon address the significance of the return to highly *immobilised* forms of sociality necessitated by the 'lockdown' of social life worldwide, at the height of the coronavirus crisis. However, in the first place, I want to go back to its beginnings in order to clarify the forces in play in the conjuncture in which the virus emerged. Doing so will involve an interdisciplinary perspective which pays attention to questions of sociology, geography, demography, transport, mediated communications and the forms of political authority.

Wuhan – a conjunctural analysis

For some considerable time now, we have been told that forms of sociality based on the physical co-presence of persons were merely an outmoded throwback to an earlier era which would gradually fade away, as a matter of generational change – especially once we got to a position where even those who were elderly would themselves have been born within the digital era and thus be more fully accommodated to its virtual modalities. However, there are deep foundations to the prescription of co-presence as a necessary element in social interactions at key moments (principally to do with the positive virtue of the 'redundancy' provided by the greater bandwidth of actual over virtual communications – which provides for a greater mutual sensitivity between participants). The conduct of family life in all societies has at its heart the performance of certain key rituals which, for their proper observation, require the physical co-presence of all household members. In predominantly Christian countries in Europe, it is what happens at Christmas; in the United States of America, that is what happens at Thanksgiving; in many Islamic cultures, a similar thing happens at Ramadan/Eid; and in China, at the Lunar New Year. At those moments, custom requires that all family members are required to gather together in the home – and at that point, a letter or a phone call or a video conference just will not do. As noted earlier with reference to Madianou's work on the creative forms of electronic long-distance parenting developed by some migrants, virtual interactions can often be usefully substituted for actual presence in routine interactions but are noticeably less successful at key symbolic moments (a wedding; a child's birthday party, etc.). At those moments, you simply have to *be* there (actually, not virtually) to physically demonstrate in a satisfactory manner, to all concerned, your continued commitment to full membership of the household.

So, the first point to make about the crisis that emerged in Wuhan is that it happened in a very particular point in the calendrical year when a substantial proportion of China's very large population was in movement to visit their families for the holiday. As the (as I write, still developing) narrative stands in May 2020, it seems that the first cases of coronavirus infection probably took place in Wuhan in December 2019. However, the city was only locked down to prevent further infection on 23 January 2020, just two days before the New Year, which fell on 25 January. By that time, many (unknowing, symptomless) infected people had already begun their journeys to and from Wuhan, so it was too late to prevent the disease from spreading. Moreover, given China's extensive geography, this process was taking place over very long distances. The speed of industrialisation – and consequently, urbanisation – in China in recent years has dictated a massive population movement from the countryside to the cities, often across very large distances. Indeed, one could make an argument that the scale of this internal migration within China's national boundaries effectively dwarfs the scale of global transnational migration in all other parts of the world. Thus, we can now begin to identify both the potential significance of the calendrical moment when the crisis occurred and how its timing interacted with the demographic and geographical conditions characterising the conjuncture in which it was situated.

Wuhan is the most populous city in Central China (11 million people) and is at the heart of the economic, industrial and geographical structure of the country. In terms of transport geography, it is one of the major interchange points in the whole Chinese transport system for both passengers and freight, travelling by road, train, boat and aeroplane. It is the largest transport hub in inland China and sits at a strategic position at the intersection of the Yangtze and Hanjiang rivers, both major inland waterways. It has an extensive network of roads and expressways leading to other provinces and is one of the major hubs in the Chinese railway system, giving direct access to many far-flung destinations via high-speed trains. So far as air travel is concerned, the city's airports are among the busiest in China. In effect, Wuhan holds an analogous position in China's transport system as does Chicago in North America – in either case, if you are going from one part of the country to another, it is very likely that is where you will have to change planes.

The flow of people travelling back and forth to visit their family in Wuhan (or travelling from the city to their homes elsewhere) for the holidays will have readily mingled with the flow of those changing planes there en route to and from quite different parts of China. Overall, five million people are estimated to have passed through the city for the New Year. These flows will themselves have merged with the transnational flows of travellers to and from the rest of the world going through other Chinese airports. Wuhan itself has an excellent range of direct flights to most other 'global cities' worldwide. These include Rome, where the authorities only began examining flights arriving from China on 23 January – a month after the disease had first been suspected in Wuhan. It may or may not be coincidental that Italy subsequently had one of the worst outbreaks of the infection in Europe.

To move to questions of political authority and mediated forms of communication, the further determining factor concerns the way in which the Chinese authorities attempted to suppress recognition of the crisis in Wuhan in the early period. Thus, they unfortunately emulated the way that the authorities at the Chernobyl nuclear power station in the USSR during the 1986 crisis refused to recognise and report the difficulties until too late, when things had already got completely out of hand (Alexievich, 1997). Here we readily see what, in another context, one might perhaps call the conditions for a 'perfect storm' as a result of the Chinese government's refusal to recognise the seriousness of the problem when it first emerged. When the authorities later moved to close down all social life in the city, and thus to control the outbreak in a highly efficient (if authoritarian) manner, that was certainly very effective within Wuhan itself. However, it was by then far too late to help the rest of the world as, by that time, the virus was well on its way, courtesy of long-distance airline travel, to many other countries.

Post-Wuhan / post-mobilities?

Writing as I do (May 2020) while the lockdown in the UK still persists, it is hard at this stage to make predictions as to what post-lockdown life will be without giving dangerous hostages to fortune. However, the one thing that seems clear already is that there will be no 'return to normal' in respect of mobility as much (or indeed more than) anything else. That is for the very good reason that, as one contemporary piece of graffiti has it, that 'the old normal was part of the problem' – not least in its consequences for our ecological prospects, given the extent of the poisonous emissions generated by the patterns of mobility that have been taken for granted in recent years. What we now see is that – in a world which will continue to be vulnerable to pandemics such as the current one – the forms of hyper-mobility to which we have recently grown accustomed may also exact a higher price than can be afforded in terms of the dangers to public health. To that extent, we now face a double danger – on the one hand, getting people out of private cars and onto public transport is a vital aim in terms of cutting emmissions, but on the other hand, the old forms of overcrowded public transport are no longer going to be a viable alternative.

One of the most striking things during the lockdown in the UK has been the statistics and visual maps showing the dramatic falls in levels of pollution consequent upon the curtailment of aeroplane flights and motorised transport in our cities, all of which will strengthen the environmentalist case in the coming period. It does also seem that the health considerations raised by this pandemic dictate that it will be necessary to re-design forms of public transport to a much lower density of occupation before they are safe for the future, if we are to reduce patterns of cross-infection in overcrowded aeroplanes, buses and trains. That consideration has severe implications if our cities are not to grind to halt in even worse traffic jams, as people choose the private car as a safer/more hygienic mode of transport. The combination of those difficulties may simply preclude the possibility of the resumption of the patterns of commuter travel necessitated by contemporary modes of work.

All this also has serious implications for the future of the tourist industry on which so much of the global economy has rested, for some years now (being the largest single industry in the world and contributing 10–15% of world GDP – cf. Urry, 2007). In this situation, it seems that what is at stake is rather more than a question of when the global air transport system might return to what it has come to understand as normality. At some point, we shall have to recognise that the viability of the kind of cheap air travel which has underpinned a great deal of tourist travel in Europe over the last 20 years was, in fact, circumstantially dependent on quite particular historical conditions which no longer pertain. These included the almost cost-free availability of unused old military airports across the continent after the end of the Cold War – which effectively subsidised the birth of many budget airlines. More broadly, it also presumed the continued willingness of governments to subsidise air travel fuel costs to a massive extent, so as to not to fall behind in the global competition to keep their countries' airports in key positions within the international transport system – in order to attract the maximum number of both tourists and business travellers to their territories. All those assumptions would now seem to have quite insecure foundations.

Substitutability – the actual and the virtual

In the UK, the lockdown has certainly functioned as a crash course in the acquisition of digital skills for many people. This has certainly been the case for the many businesses and organisations which have discovered that for many routine purposes, they can in fact operate via virtual rather than actual meetings. Many ordinary people, otherwise starved of social interaction during lockdown, have also discovered how much of their social lives can, at a pinch, be conducted through virtual interaction technologies such as Skype or Zoom. To this extent, one might expect that the population at large will rapidly now discover for themselves the possibilities already explored by many migrants concerning the viability (and problems) of various forms of virtual communication.

All this will doubtless encourage some greater sense of the substitutability of virtual forms of interaction for physically co-present ones, in both the domestic and business spheres. However, what is not predictable at this point is what mix of these activities will ultimately be deployed by which people in which situations. The further issue here concerns the extent to which the lockdown has vividly highlighted the degrees of inequalities in access to the virtual world available to different categories of persons. Thus in the UK, there are now serious concerns about the lockdown exacerbating inequalities of achievement among UK schoolchildren, given the vastly greater economic and technological forms of 'cultural capital' available in the homes of the children of more affluent parents during any period of home schooling. As I noted earlier, in the digital realm – as in most others – it is very expensive to be poor. Paying for the data your child needs to keep up with their home schooling on a 'pay-as-you-go' phone (which is all anyone without a good credit rating can hope to get) may mean that you cannot also afford to eat adequately

More broadly, the general issue will be the emergence of radically new and enhanced forms of inequality in many spheres. Thus, some people will have very wide menus of choice available to them of both virtual and actual forms of communications and mobility, while many others will face a much more restricted set of choices. For the poor, life may well become more locally circumscribed as they are increasingly confined to their localities – and that is certainly going to be true of the majority of the world's future population for whom it will be increasingly difficult to find exit routes from the mega-slums of the Third World cities where they live. In that context, mobility may, in fact become a luxury good, available only to the affluent. Certainly, to go back to the key case of air travel, the most likely outcome seems to be its re-invention as the expensive and therefore exclusive option for the rich that it was in the earlier 20th century.

Likewise, it has now been demonstrated throughout the Global North that basic healthcare of a routine kind can be administered much more cheaply in virtual form, by means of the use of call centres and automated patient-advice systems. There will thus be a continuing temptation for national governments to reduce the healthcare costs in servicing their (increasingly elderly, and thus needy) populations by providing them in future with mainly virtual, rather than actual, modes of interaction with healthcare, whenever possible. In this context, the forms of private medicine now available to the more affluent may perhaps be the only ones which will continue to involve actual consultations with physically co-present doctors. Non-virtual forms of healthcare thus may also become a luxury good, available only to the affluent, except perhaps in cases of emergency

The re-design of social life so as to reduce overcrowding on transport (and to reduce carbon emissions, which are now coming under suspicion as playing a potentially dangerous role in enhancing the effects of viruses like the one at the heart of the present crisis), along with a range of other factors, may function to return many of us to a more localised forms of life in many respects. This is not simply because the experience (even if compulsory) of having to live one's life within the confines of the physical communities around one's domestic home may have served to highlight, for some people, the social attractions of the 'neighbourliness', but more fundamentally, because of the increased attractions, in any crisis situation, of more predictable and reliable localised supply chains in respect to food and medicines.

The rise and fall of globalisation

This should not come as too much of a surprise. The key role of container box shipment in the transformation of global trade was identified as long ago as 2006 by Mark Levinson in his book on how the "Box" made the "world smaller and the world economy bigger" (Levinson, 2006). However, in more recent years. it has become clear that these boxes' distinctive capacities – as sealed containers which enable freight traffic to pass through ports much more rapidly (and therefore more cheaply) than was ever possible before – also has serious downside. This very feature is also the source of its capacity to act as a form of 'customs-control-avoidance' transport. If any attempt is made to scrutinise more than a tiny proportion of the boxes going through any port, then the whole system simply grinds to a very expensive halt. It is precisely the fact that smugglers can depend on it being so unlikely that any one box will be scrutinised that makes them a smuggler's dream – perfect for the distribution of Bads every bit as much as for Goods.

The problem with the long-distance supply chains on which the whole just-in-time logistics of the distribution industry is now based is that, in order to make the industry so much more profitable than it was before, what had to be sacrificed was any spare capacities in the process. Thus, it has become a matter of pride for logistics operations to always eliminate any surplus of time or space, cutting things down to the bone, so that the whole system becomes maximally profitable by virtue of the fact that it is, in one of the catchphrases of the industry, so very "lean and mean." In many countries, the same principle has in recent years also been applied to hospitals. These institutions were originally designed specifically to run at no more than 85% of their total capacity – precisely in order to have the room to spare that might be needed in a crisis. The tragedies we have seen in this recent period have often been where hospitals whose surplus capacity has been deliberately stripped out in recent years by the 'austerity' politicians determined to cut costs everywhere. In the case of the health crises we are currently seeing, many have been overwhelmed by demand – precisely because when most needed, they had insufficient staff, space or equipment to cope with the emergencies that have ensued.

We have also seen a return to forms of nationalist protectionism in many fields, with particular governments competing with each other to monopolise the supply of vaccines or protective equipment for the exclusive use of their own citizens. To this extent, protectionist boundaries are clearly on the increase and are likely to continue to be hardened. Moreover, the wartime principles which many governments have had to adopt in recent times hark back to an older set of geopolitical imperatives. In that context, it was always regarded as axiomatic that national governments should ensure that their citizens could, in times of crisis, be provided for in a self-sufficient manner by the capacities of their own nation. Manifestly, that approach is the complete antithesis to the modes of globalisation which have dominated our lives until recently.

However, the further problem with the globalised/ long-distance supply chains – and the 'enhanced mobility systems' of the recent past – was also identified some years ago by Levinson in his later work on the shipping industry. Having published a book on the 'heroic' role of the container box as the centrepiece of the story of globalisation in 2006, within two years, he was having second thoughts about the matter. Thus in a 2008 article on "freight pain," addressing what he already identified then as a potential cause of the fall of globalisation, he argued that it was "in retreat" because long-distance supply chains were losing their appeal, given their diminishing reliability (Levinson, 2008). This he attributed to the increasing anxiety which companies and governments had begun to feel as a result of developments such as the increasing impact of modern high-tech piracy in key shipping lanes around the globe. It was factors such as these which drove up the insurance costs of shipping transport and, along with uncertainties about fuel costs, led to a crisis of profitability in the global transport industry. At this stage, increased anxieties about security issues were also forcing many manufacturers to wonder about the wisdom of a system in which production was outsourced to distant countries which, while offering the advantages of low wage economies, also had the considerable drawbacks of unstable political and legal institutions. This concern was in part responsible for the increasing tendency of retailers and manufacturers, worried about whether the goods will arrive on schedule, to try to reduce the uncertainty in their supply chains by producing their goods closer to home.

In his conclusion to his later, prescient article, Levinson claims that for future historians, globalisation may perhaps appear not as an inexorable trend, but as a merely temporary stage in economic development which occurred around the turn of the 21st century. Before we project any of our previous experiences and assumptions into the future, we also need to consider whether the same might be said of the forms of mobility – and their associated mobile socialities – to which we have become accustomed in recent years. The sedentarist (and protectionist) rhetoric of slogans like "Stay Safe/Stay Home," and the expected continuation of forms of 'social distancing' (however measured) into any forseeable future, are fundamentally at odds with the visions of fully networked hyper-mobility which have dominated public discourse in recent years. All this, in combination with the revival of forms of authoritarian populism, nationalism and protectionism, not only in the UK, as discussed earlier, but on a worldwide basis (Trump in the USA, Putin in Russia, Mohdi in India) means that the future of 'mobile socialities' may take a rather more constrained shape than would have been anticipated only a short time ago.

I can only now end by recognising that this has been an attempt to produce a synthetic account of a set of complicated questions towards which, in conclusion, I can now only gesture back, by rehearsing the itinerary of the journey we have undertaken. One key starting point was with a critique of simplistic projections of a boundary-less and mobile future, in which material geographies would be dissolved in the realms of the virtual and the nation would account for little, in a thoroughly globalised world. One of my principal objectives has been to try to place these issues in a broader historical perspective and to address the value-laden terminologies of both nomadology and sedentarism in a critical spirit – which recognises the politics of mobility and the extent to which the various forms of mobility itself now constitute a crucial dimension of inequality. In doing this, I have at various points attempted to ground meta-discussions of questions such as historical periodisation in micro-perspectives relating more closely to lived experience – such as in relation to contemporary forms of healthcare and transport. If, in doing all this, I have raised more questions than I have succeeded in answering along the way, I can only hope that these will prove to be fruitful questions for future research.

Bibliography

Alexievich, S. (1997) *Chernobyl Prayer*. Harmondsworth: Penguin.

Appadurai, A. (1996) *Modernity at Large*. Minneapolis: University of Minnesota Press.

Appadurai, A. and Morley, D. (2011) "Decoding, diaspora and disjuncture", *New Formations*, 73: 44–55.

Arendt, H. (1951) *The Origins of Totalitararianism*. New York: Schocken Books.

Ben Jalloun, T. (2009) *Leaving Tangier*. Harmondsworth: Penguin.

Burke, V. (2014) "My best shot", *The Guardian*, G2, June 26.

Burrell, J. (2012) *Invisible Users: Youth in the Internet Cafes of Ghana*. Boston: MIT Press.

de Blij, H. (2009) *The Power of Place: Geography, Destiny and Globalization's Rough Landscape*. Oxford: Oxford University Press.

de Cauter, L. (2006) *Capsular Civilisation*. Amsterdam: NA Publishers.

Deleuze, G. and Guatari, F. (1986) *Nomadology*. New York: Semiotexte.

Dixon, N. (2019) "Attached to my phone; a study of affective mooring in mobile practice". PhD Thesis, Goldsmiths University, London.

Edgerton, D. (2006) *The Shock of the Old*. London: Profile Books.

Gauntlett, D. (2011) *Media Studies 2.0*. London: Routledge.

Gilroy, P. (2004) *Postcolonial Melancholia*. New York: Columbia University Press.

Gitlin, T. (1998) "Public sphere or public sphericules?", in T. Liebes and J. Curran (eds.) *Media, Ritual and Identity*. London: Routledge.

Haagerstrand, T. (1981) *Space and Time in Geography*. Sweden: Lund Studies in Geography, University of Lund.

Hafez, K. (2007) *The Myth of Media Globalisation*. Cambridge: Polity Press.

Kern, S. (2003) *The Culture of Time and Space 1880–1918*. Harvard: Harvard University Press.

Khanna, P. (2016) *Connectography: Mapping the Global Network Revolution*. London: W&N/ Orion Books.

Krajina, Z., Moores, S. and Morley, D. (2014) "Non media-centric media studies", *European Journal of Cultural Studies*, 17(6): 682–700.

Larkin, B. (2008) *Signal and Noise: Media, Infrastructure and Urban Culture in Nigeria, Raleigh*. Durham, NC: Duke University Press.

Levinson, M. (2006) *The Box: How the Shipping Container Made the World Smaller and the World Economy Bigger*. Princeton: Princeton University Press.

Levinson, M. (2008) "Freight pain: The rise and fall of globalisation", *Foreign Affairs*, 87(6): 133–140.

Madianou, M. and Miller, D. (2012) *Migration and the New Media: Transnational Families and Polymedia*. London: Routledge.

Marvin, C. (1988) *When Old Technologies Were New*. Oxford: Oxford University Press.

Massey, D. (1994) *Space, Place and Gender*. Cambridge: Polity Press.

Mattelart, A. (1996) *The Invention of Communications*. Minneapolis: University of Minnesota Press.

Mercer, K. (1994) *Welcome to the Jungle*. London: Routledge.

Meyrowitz, J. (1985) *No Sense of Place*. Oxford: Oxford University Press.

Mishra, P. (2017) *The Age of Anger*. Harmondsworth: Penguin.
Morley, D. (2000) *Home Territories*. London: Routledge.
Morley, D. (2017) *Communications and Mobility*. Hoboken, NJ: Wiley-Blackwell.
Morley, D. and Robins, K. (1996) *Spaces of Identity*. London: Routledge.
Mulgan, G. (2020; orig, 1998) *Connexity: How to Live in a Connected World*. Harvard: Harvard Business Review Press.
Newman, R. (1974) *Good Old Boys*. Burbank, CA: Reprise Records.
Nietzsche, F. (1887/1998) *On The Genealogy of Morals*. Oxford: Oxford University Press.
Parks, L. (2005) *Cultures in Orbit*. Durham, NC: Duke University Press.
Poulson, E., Schofield Cark, L. and Gajjala, R. (eds.) (2020) *The Routledge Companion to Media and Class*. London: Routledge.
Shove, E. (2003) *Comfort, Cleanliness and Convenience – the Social Organisation of Normality*. London: Berg.
Sibley, D. (1995) *Geographies of Exclusion*. London: Routledge.
Tett, G. (2019) "Fear, loathing and automation", *Financial Times*. Available at: www.ft.com/content/ab94fb30-cea2-11e9-b018-ca4456540ea6 (accessed January 10, 2020).
Urry, J. (2007) *Mobilities*. Cambridge: Polity Press.
Verstraete, G. (2010) *Tracking Europe*. Durham, NC: Duke University Press.
Wark, M. (1994) *Virtual Geography*. Bloomington: Indiana University Press.
Williams, R. (1976) *Communications*. Harmondsworth: Penguin.

2

MEDIA AND MOOD WORK

Routines, daydreams and micro-moves

Orvar Löfgren

On the move

What kind of skills and competences are needed to produce a snappy Instagram post, taking photos with a smartphone or zapping between film clips on YouTube? How does one create a Facebook profile or organize a blog? What about the techniques of surfing and scrolling, or the skills of multitasking and combining media uses while on the move?

There are so many digital media skills that have been turned into everyday micro-moves, habits and reflexes, resting in the body rather than in the mind, but they all have a history of routinization. The intensive discussions of the digitalization of everyday life have often focused on how media have invaded the home, reorganizing many mundane activities and sociabilities and creating new dependencies and possibilities. In such narratives, the historical perspective may become too short. Many contemporary digital habits, such as skills of navigation and communication, are based upon competences developed earlier. (In a sense, this historical legacy is symbolized by the use of analogue icons on the computer screen: the scissors icon for 'cut and paste', and the wastepaper basket for deleting, echo the ways in which analogue techniques have been remediated into the digital world.)

This chapter moves back to a pre-digital era and looks at some of the aspects of the mediatization of everyday life in the three decades of the 1930s, 1940s and 1950s, when people learned to handle new media such as the radio, the cinema, the box camera and the weekly magazine. Texts, images and sounds flowed in and out of the home in ways that did not make the domestic scene more privatized, but turned it, rather, into a laboratory for handling the outside world and organizing social relations. Novel skills and routines of media use and multitasking were created, as well as new potentials for intensive daydreaming that could be worrisome as well as pleasurable. Discussions of mobile sociabilities should not overlook this changing world of micro-moves in everyday life.

My chapter will focus on such transformations, mainly in terms of acquired competences of combining moving bodies and roving minds. My background is not media studies but the anthropology of everyday life. During recent years, there has been a growing interdisciplinary dialogue in an emerging field of mobile socialities in media and everyday practices (for a few examples, see Borgerson and Miller, 2016; Highmore, 2017; Koch, 2017; Moores, 2018; Morley, 2017; Pink, 2013), a dialogue I would like to further.

My analytical starting point is a concept developed by the anthropologist Kathleen Stewart (see Carlson and Stewart, 2014). Stewart uses *mood work* as an explorative concept to catch the new mood of the good life among a rising American middle class in the decades following the Second World War. 'The good life' was the elaborate design of an everyday life that slowly turned from dreams into something real as a new economic affluence was felt in both working- and middle-class settings. Mood work was shaped by new patterns of consumption and popular culture and media use, and included tips for living, daydreams with scenes of pleasure floating by, and smells of newness, but also feelings of uncertainty and uneasiness. People felt they were part of a grander and challenging project, often in vague and non-verbalized ways.

The concept focuses on the ways in which senses and the world co-exist. "Mood work marshals bodies, objects, technologies, sensations and flights of fancy into forms of partial coherence. . . . They press, promise, enfold, and power worlds being thrown together" (Carlson and Stewart, 2014: 114). These aspects of intertwining can be developed further by using the analytical perspective the cultural geographer Doreen Massey (2005) has termed *throwntogetherness*: the ways in which materialities, bodies, sensibilities, media flows and activities co-exist. Her concept is useful for understanding mood work as it explores the ways in which diverse elements come to cohabit in a setting or a situation, often as unexpected neighbours, which calls for ongoing processes of confrontation, negotiation and accommodation. Co-habitation may hide ways of non-communication and disintegration and of being out of sync.

Mood work is about modes of being in the world; it is about feelings, rhythms and atmosphere, as well as performativity and affective energies. In this sense, there is a strong focus on the affective dimensions of everyday life – the in-betweenness not only between human actors, but also between humans and objects. Affect is about reactions and communications, which are often unconsciously driving us toward movement or thought (see the discussion in Wetherell, 2015; Frykman and Povrzanović Frykman, 2016). In mood work it can, for example, be a sudden sensibility, a creeping irritation, a shadow of guilt or a vague anxiety.

My discussion will be organized around these two perspectives of mood work and throwntogetherness, because they bring out the complex interaction of media stuff and everyday life. I begin by exploring how the radio developed new domestic soundscapes, social relations and management of affect. Second, I explore the ways in which media became a tool box for identity work among young people during the period, mixing and combining experiences and materialities.

My examples are mainly fetched from the Swedish experience and an ongoing research project on domestic media use in the period 1930–1960, drawing on a wide range of materials, with some international comparisons.[1] These three decades saw a radical transformation of everyday life and living standards. During the interwar years, and following the austerity of the Second World War (which Sweden stayed out of, trying to maintain an often problematic position as a neutral nation), life for many was characterized by economic hardships and poor housing standards. Together with Finland, Sweden had Europe's most crowded living conditions. In the 1930s, most people still only had access to an apartment with one room and a kitchen. Even in the 1940s, many Swedes still struggled with an everyday of wood-fired stoves, cold water, an outside privy and makeshift sleeping arrangements.

It was also the brief era of the housewife. In the interwar years, most women started staying at home, with the husband as the breadwinner. In the 1960s, they began to return to the labour market, and the era of the housewife changed drastically with the regendering of domestic life.

There were also great class differences in living standards, which did not begin to diminish until the growing affluence of the working class in the 1950s. Cramped living and tight budgets thus came to shape media use for many, but media were important in framing the longing and the impatient waiting for a modern life that was presented in political blueprints, and in the imaginaries of popular culture. For most people, this throwntogetherness of difficult living conditions and the constant media barrage of the dreamscapes of modernity and affluence created an uneasy co-habitation, a strong out-of-sync-ness that produced not only micro-moves between longing, waiting and frustration, but also feelings of guilt and shame. In many respects, what characterized mood work of the austere 1930s and 1940s was the great divide between messages of modernity and actual everyday worries. 'The good life' was still out of reach for most Swedes – as, for example, the author Immi Lundin (1987: 106) remembers from her childhood in a working class setting in the early 1950s:

> We lived in a town house, and that sounded grand. But there was another word that began to hurt: unmodern. Our home was not modern. It meant firewood that had to be carried up the steps to the tiled stoves and the cooker. It meant the toilet in the backyard. . . . It meant the scullery. I understood this when I went to the housewives' film show with my mother and saw her crying over the modern kitchen and the fancy washing machines they demonstrated. Things like that existed. They could be wished for. The difference between modern and non-modern had to be greater than what could be understood from the words alone. It was worse than what happened that baking hot afternoon when we fancied some Danish pastry while we were out shopping, and decided to try the modern frozen kind. When we opened the packet they were pale and unbaked, and we were so surprised that we burst into laughter. It was crazy, for the picture on the outside showed beautifully baked brown pastry, and it was the picture that had given us the craving in the first place. We lit a fire in our stove so that we could bake them, and as the sweat poured from us, we laughed at ourselves and at the ridiculous pale-faced pastry of modern times.

The mood work of longing and disappointment Lundin captures fed on the many images of modern life, in films as well as in the advertisements for novelties such as frozen food. The good life was out there, depicted in popular culture, advertisements and mass media, but the actual everyday looked very different.

There was an important mood shift as the 1940s turned into the optimistic 1950s, and longings slowly materialized. People moved into new modern housing estates and the children, with their future-detecting radar, scrutinized the kitchens and sitting rooms of the neighbours and returned home breathless to report that they have a fridge and a radio in the kitchen, but no radio gramophone in the parlour. Children learned to read old and new class markers. (In school catalogues of the early 1950s, a grey streak after the address meant "no telephone", a shameful sign of lagging behind.)

So how did media become a part of mood work in the 1930s and 1940s? Let me start with one example: the radio and the ways in which it became an important co-actor in domestic life, creating radio families as well as new forms of national belonging.

Domesticated soundscapes

A farmer remembers his first contact with a radio receiver as an 8-year-old boy, back in 1926: "I was lent one of the headphones. Oh, Jesus, damn it! You entered directly into heaven as you listened!" By then the Swedish government had taken a firm grip of the medium, copying the BBC system of a national broadcasting company. (For a discussion of the development in Sweden in relation to the radio politics of other nations during this early era, see Löfgren, 2000.)

Many have described the magic of the first confrontation with this new media: "We felt that something new and grand and mysterious had come into our lives . . . to be able to sit at home and listen to speech and music from a long distance, it was something of miracle". New media are (too) often described as revolutionary, but the term might still be apt for radio. Perhaps like no other medium before (or after), it rapidly colonized domestic life. Today, it might be hard to understand how much the medium affected everyday life. In comparison with television, and later digitalization, the radio was something drastically new in a time when media still had a limited impact. (Television was a medium for which the radio, in many respects, paved the way.)

The magic of the radio also had to do with what Rachel Plotnick (2018) has called "the power of the button". Just a turn of the knob or a push on the button and the world flooded into the home, shaping activities, daydreams and feelings. Radio listening started with cumbersome crystal receivers, a technology best administered by the young who had to patiently guide the older people. During the first years of radio in the 1920s, the need for earphones made listening a personal experience and one that demanded intense concentration. An early listener remembers: "At the start, it was hard to fully understand what was being said, but you soon relaxed and got used to it and could sit and listen for hours, although the headphones pressed on your ears and made them sore". Another man recalls the hostile looks from other listeners if you did not keep dead quiet. He worked as a farmhand, and "the farmer and his wife sat evening after evening sharing earphones; they looked a bit silly and gazed absentmindedly into space". In one advertisement, it was pointed out that bananas were the perfect food for listening, providing a soundless meal.

As radio sets with loudspeakers began to replace the crystal receivers in the late 1920s, the sound colonized the whole room. At home, everybody had to listen or learn the art of just listening with half an ear. Slowly the radio could also develop into a background sound, ready to be combined with other activities.

Synchronizing the home and the nation

During the pioneer era of radio, much of the fascination focused on the strange international world discovered through the new skill of dial-twisting. Magic places like Hilversum, Bordeaux, Königsberg, Murmansk and Kalundborg were visited in a novel soundscape. One could listen to music from the Savoy Ballroom in London or an unintelligible lecture from Brno. During these early years, we encounter a metaphor that has been recycled with the advent of all new mass media: 'at last, the world has entered the living room'. This feeling of having an international world at your fingertips soon faded away, as national broadcasting established a firm pattern of daily listening. Radio became not international but firmly nationwide, linking the peripheries to the centre. A woman remembers the feeling of magic enchantment when as a child she heard the radio announcer in Stockholm say: "now children, I will put my microphone outside the window to give you a chance to experience the traffic here in the capital". In that magical movement, the national centre was materialized in living rooms all over the country. It was the same technique Franklin D. Roosevelt used when he paused in the middle of one of his famous national "Fireside chats" and audibly took a sip of water, saying "My friends, it is very hot here in Washington" (after Levine, 1992: 1394).

Radio became a powerful mode of nationalizing everyday life, very effectively creating what Benedict Anderson (1983) has called "imaginary community". For the broadcasters, it was quite clear that radio was a tool for a nation building, which should transcend class divides, and called for the education of an audience, whose members were often perceived as not knowing what was best for them. The Social Democrats who led the government in these decades had a vision of a new, classless Sweden, but the framing and content of the messages and the admonitions from the state were strikingly middle class. The new and growing middle class was seen as the vanguard of mobility, flexible and open, not like the working class, which was often talked of as clinging to traditions and suspicious of the new.

Sweden achieved the highest percentage of licence holders in Europe in 1937, and until 1955, there was only one channel available. This turned the radio into a strong medium of social and cultural integration, as well as persuasion, territorializing both attention and feelings of belonging.

In Sweden, as with most other European public service institutions elsewhere, a battle ensued between the rather highbrow and educational tone of the early broadcasting and the popular appetites of the listeners. People talked sarcastically of the "radio gentlemen in Stockholm" with their high cultural ambitions and elitist aura, as well as the patronizing and invasive tone of broadcasting. One listener made the following point in a letter: "We have democracy in this country, and so there should be more accordion music on the radio". Accordion music did, in fact, became a symbol in the struggle between popular and highbrow programmes. Following the first years of complaints, the board of directors finally took a decision to broadcast 30 minutes of extra accordion music each week.

People might have resented much of what was offered and kept complaining of too little popular music, but they still flocked to the receivers. A novel from 1931 describes the radio day of a police inspector:

> The inspector stayed at home in the evenings. The reason was that he had acquired a radio year. He woke up to it on winter mornings at 7.15, when a captain in Stockholm ordered him to do knee bends, and as he dressed he switched to the morning service, sometimes listening to it, sometimes turning it off again. Before he sat down to dinner at six in the evening, he would turn the knobs: gramophone music, recorded by the world's leading symphony orchestras . . . kept him company.
>
> (Hellström, 1931: 269)

Similar patterns emerge in many recollections of the new national rhythms of listening. Programming was extremely stable. Some daily favourites ran for many decades at the same hour, creating a very predictable rhythm. Captain Uggla, for example, led the national morning gymnastics from 1929 until his death in 1945, with a firm military tone, framed in ideas of the healthy nation. Next to the 7 o'clock news, the most popular programme was the gramophone hour, from when it started in 1929 until into the 1950s.

The introduction of the weather forecast created another national institution: the Swedish weather. It was taken for granted that people living in southernmost Sweden needed to learn about the weather in Lapland, 1,500 kilometres to the north, but not about the conditions in Denmark just across The Sound. In the daily shipping forecast, the names of the coastal observation posts of Sweden were read out like a magic chant as outposts encircling the nation. Some remember this chant from their childhood as something of a protective shield; one felt safe inside those national border posts.

Quite early on, the radio established traditions for seasonal celebrations. In the late 1920s, music began to be provided for the obligatory dance around the Christmas tree, and in the 1930s, a New Year's ritual started broadcasting that carried the whole nation into the new year. As Swedes gathered around their radio sets at midnight, there was an electric sense of Gemeinschaft in the air. Radio thus came to provide a stable national structure for many domestic activities and discussions, creating shared prime times with the news, the weather forecasts, the gramophone and the comedy hours, synchronizing bodies as well as minds in this process. The nation would dance together around the Christmas tree, raise their glasses and toast fellow Swedes at midnight on New Year's Eve, or join in programmes of community singing. It was not just passive listening, but active participation.

One man who grew up in the 1930s remembers how radio created a rhythm of morning routines. His father woke him up by putting on the morning news at 7, loudly. If he had not got up by the time the weather report had started, his father yelled up to him, "Time for school!" Out at the gate, his friend was waiting and sometimes shouted: "Hurry up, they just played the last hymn of the morning service when I ran out of the house!" (Höijer, 1998: 115)

In a large survey of domestic life from 1937, it is clear that radio listening had established a marked daily and weekly rhythm. Housewives listened the most, and as one remarked: "It is just like a sickness, that radio, it is on all the time, one could not live without it" (Åkerman, 1941: 194). One of the reasons for the great impact of the medium was that it could easily be combined with other activities, once people had learnt to listen with half an ear. Listening became woven in with the washing up, the sewing, homework, reading and small talk. A new world of multi-tasking emerged.

The radio became domesticated, and people increasingly felt at home with both the medium and the nation. Daytime broadcasting focused to a great extent on housewives. In the ideology of welfare nationalism, they were seen as the key makers of modernity, responsible for introducing good habits and ideas into the home. A new gendering of the nation emerged. For women at home, the radio became an everyday companion and special listening slots were created for them during the day, with everything from lighter listening to talks on 'rational dish washing'. Again, the message was often about the need to improve everyday life; it was not a question of 'mother knows best' but of her getting rid of what were seen as traditional values and wrong priorities. The flood of advice and admonitions could be anxiety producing. Am I doing enough for my family? Do I live up to the standards of a proper housewife?

Later in the evening and at weekends, there was a focus on family listening, when the whole household would gather around the receiver. This was a new experience for many, since family togetherness was not a traditional working-class project. In overcrowded lodgings, people socialized with peers outside the home rather than with family, men with other men, women with women, and youngsters with other youngsters. On Saturday nights, the radio pulled the family together, often in a contemplative or meditative mood.

There are many nostalgic memories of such idyllic listening, but in the background lurked more troubling thoughts. People could not help comparing their own domestic life to the ways in which a perfect and modern family life was presented in the soap operas and lectures. Listening in an overcrowded kitchen, which later at night would be turned into an improvised landscape of bedding and sleeping places, it was easy to start thinking about how far removed the broadcast images of domestic life were from one's own everyday. There was happy talk of the modern married couple based on companionship and sharing, while most housewives had a husband who at most took care of radio maintenance, but rarely considered helping with the daily chores.

In colonizing the home, radio could be experienced as an invasive and preaching media. A man who grew up in a working-class household of the 1940s remembers:

> The radio turned the sitting room into a public room. The voices from the ether spoke from the capital and united us with our rulers, but also with all the other radio listeners around the country. When the radio was on, the room was not really ours, the sonorous voices with their Stockholm sounds – the *a* in particular made the fruit

> bowl vibrate – pushed our own thick voices into a corner, where we commented in whispers on the cocksure statements from the radio.
>
> (Ambjörnsson, 1996: 15)

Radio moods

As part of everyday mood work the radio was an active co-actor in the making and unmaking of moods and feelings. As Jo Tacchi (1998, 2012) has pointed out, it played an important role in the affective management of everyday life. The radio became emotionally charged media, which in a way denied its structure as a mass medium. Cinema goers formed a physically close collectivity, while newspaper reading could produce a distance between reader and the text, but the radio voice found its way directly to the individual and into the most private corners of domestic life. The sound hit without any filtering mechanisms, but the affective impact depended on where, when and how people listened. The radio became a mood shaper, or as one man remembers about his radio childhood in the 1950s:

> A radio was like an animal with its own blood system and flowing electricity. Everywhere in the room and outdoors there were signals. Wherever one walked or stood in the room, one breathed this electricity from space.

He remembers how he learned to search the stations and how the slightest turn of the knob changed the mood. The ether became a sea full of voices and endless amounts of music. "The parlour was filled by opera singing or was drained by a dry and factual weather report" (Brunner, 2014: 50).

During the early years, listening to the radio was often surrounded not only by a sense of intense concentration, but also by a strong solemnity. One Swedish woman remembers how neighbours gathered together to listen to a charity gala in the early 1930s: as one of the princesses started to talk, there was total silence and a few of the women even curtseyed discreetly.

Broadcasters gradually learnt to create a more intimate atmosphere. Some began talking about the radio as a new friend; one did not have to be alone now. Intimacy could be developed in different ways. Singers developed a crooning style of addressing the listeners. One of the Swedish pioneers of this style was a popular singer, who in the late 1920s impressed a radio colleague with his new technique:

> Sandberg moves close to the microphone, like a lover, cautiously. Low key, and of course ingratiating himself in order not to seem frightening. With devastating efficiency, he starts to sing: "Are you listening to me tonight, mother dear?" And among the tens of thousands, the song pinches like a pair of pliers in the heart and the first radio hit is a fact.
>
> (quoted in Strand, 2019: 122)

Sandberg was aware of something many other broadcasters had yet not captured. You were not addressing a large auditorium, but a single listener: I am talking directly to you.

The ritualization, and later routinization, of listening also created different modes of listening. The 7 o'clock news called for concentration and a general hushing of other activities, while evening entertainment created a special family atmosphere. The location of the set also influenced listening. The solemnity of 'the best room' made listening in there more ritualized, while a radio set in the kitchen made for more a relaxed attitude and multi-tasking.

Radio created new forms of mobility and socialities. Movement was, of course, very dominant in the way the new medium worked. Music and sounds flowed into the home, drifting through rooms, and could change the mood, but there was also the constant oscillation between the outside world and domestic life. Listening to the news meant taking in big events, but it was in the domestic context that they were evaluated and digested – a constant training in handling the outside world. There was also a strange feeling of control, just by pushing a button or turning the dial the outside world was either let in or locked out. There was something empowering about silencing a boring speaker with a click.

The radio set was an affordance, a potential, which could be used for very different aims. The multi-tasking listener's attention wandered between the radio and the daily chores like washing, mending and schoolwork, transforming such mundane tasks into something different.

Listening created movements in time and space, back into memories or into fantasies of the future. Consciousness moved out into the world or into a very private daydream. The fact that the medium provided sound but not images gave new dimensions to daydreaming. One had to create one's own images to go with the sound, fantasies about the broadcaster or the setting described. There was a constant training in the skills of imagineering (see the discussion in Ehn and Löfgren, 2010).

Small and overcrowded domestic spaces turned the radio into a shared media, with few chances of any private listening as long as there was only one set in the home. And who decided when, how and to what the household should be listening? Who was allowed to turn the knob on or off? When Dad was at home, it was usually him who was in charge of listening, but it was also Dad who owned the new technology: "Don't fiddle with the mechanics!"

During the pioneer years in the late 1920s, there were many dreams of mobile radio. Advertisements showed heavy but portable radio sets, which could provide music outdoors, providing sound to different settings – a romantic situation, for example, in which a couple glide in a boat over a lake with a portable radio aboard, the antenna trailing in the water. Radio did, however, become mainly a stationary media due to the cumbersome existing technology, until the advent of the portable radio in the 1950s. For a long time, the positioning of the family set thus organized people's movements and activities. The radio had a central position in the room, sometimes placed on a new piece of modern furniture called a "radio table". But the radio became mobile in other ways. Manufacturers soon realized that the receiver should have an aura of modernity about it and designed radios with hypermodern knobs in plastic and shining panels with a warm and inviting light. And most importantly, the sets should have a streamlined design, as if they were made to be hurled against the wind; they showed that the proud owners were a family on the move.

Patterns of domestication

The role of radio in domestic life can be compared to that of the gramophone, with its creation of a parallel but very different listening world. The younger generation craved popular music and jazz, which was heavily rationed in broadcasting. Those who grew up in the 1930s and 1940s remember waiting impatiently for some popular music.

Radio was not the favourite medium of the young. The portable gramophone offered a different freedom and mobility; it was a liberating tool compared to the strict regime of the radio. Now one could choose one's own music, and also allow the gramophone to move into new territories. Teenagers could listen to their favourite jazz music up in the attic without the disapproval of the adults, or provide dance music for the outdoors during camping trips or at backyard gatherings. A new dimension was added to socializing.

Gramophone music also filled cafés, as well as the new world of music stores, where teenagers gathered to listen to the latest releases. A new teenage sociability was created (see Björnberg, 2019), but a new commodity world also. Swedish radio was an institution outside the market, but the gramophone depended upon acquiring records. An arena of investing, sharing and collecting emerged. One could become a collector and a connoisseur, and among the young, this often turned into a male project – particularly in the world of jazz, for example (see Fornäs, 2004). Family members bought and cherished their own records, reflecting personal tastes and frequently creating some jealously guarded private property.

When people are asked to remember their early radio days, there is always a risk of a nostalgic or ironic tone emerging ("How naïve we were as listeners!"). The transformation of broadcasting, as well as gramophone use, during the 1930s and 1940s shows, however, a similar pattern to the many other introductions of new media of the period, such as the telephone and later the television.

There is a pioneer period when the new media is seen as a fantastic potentiality, with all kinds of fantasies about what it can do or along what lines it will develop. The second phase sees the gradual institutionalization and routinization of media use (see the discussions on domestication in Haddon, 2017 and mundanization in Willim, 2017), from the early enchantment and often the feeling of a sacral atmosphere surrounding the new, into an acquired competence of listening as an everyday practice. Slowly, one learns to listen with half an ear and develop a rich array of multi-tasking, in which for example the radio has to co-exist with many other activities. As the media turns into something mundane, a critical and ironic distancing emerges. People begin to talk of the unfortunate effects of an exaggerated listening, complaining about "radio trash" and "radio gossip".

Identity work as cut and paste

In overcrowded homes, young people often choose to socialize in outside spaces, and media helped to create social arenas. A new world was opening up, and people from very different stations in life were confronting each other in urban life. Not only was there a new social thrown-togetherness in public places, but there was also a mix and co-habitation of media and popular culture – the gramophone, cinema, photography and the press. How did they work together as a toolbox for identity work and new socialities?

Take the case of cinema-going, for example. During the interwar years, Sweden became a nation of movie goers, with mixed audiences being created that transcended class and gender. Even in homes with tight economies, a visit to the movies – preferably once a week – became a spending priority. Children learnt to handle the new medium at the Sunday matinees. There are many recollections of the intense longing for the magic moment when the film started. Expectations had been building up over the week, and they exploded when the curtains were drawn back. Caps were thrown into the air and intense whistling was mixed with happy shouts. When the film started, "the dream of the week was there". Unlike the radio, the cinema created a new public arena, where strangers could share intense experiences:

> To sit so close together with so many, to be sucked into this collective darkness, which was caught up and woven together in the flickering grey rectangle on the wall; to look out into another world through this window of imaginaries and at the same time be contained, sitting in there with all the others – I only knew a fraction of them – taking in the presence of the bodies, hearing the breathing, seeing the shining eyes.
>
> (quoted in Sjöholm, 2003: 199)

A cinema ticket was thus an invitation to a new community, especially for the young. As Carina Sjöholm has pointed out, memories of cinema-going often show that the actual viewing did not have to be the central experience. "I don't remember the movie but rather the bike ride home", says one woman. For teenagers, the movie evening or matinee was in many ways a social experience. One biked together with friends and saved enough money for a cup of coffee or a hot dog. To stand around holding on to the handlebars and discuss the movie, and then ride home together in the dark, was very important. Youths gathered outside the cinema well ahead of time and there was an intense mingling in the street, checking out who was there and which of the local gangs appeared. The air was full of expectations, and occasionally friendly scuffles started, but they ended the moment the cinema doors opened.

Cinema-going and gramophone listening organized attention and sociability in new ways, but they were also linked to the rapidly expanding world of print media that was aimed at the younger generation: film magazines, comic books, sport journals, etc. A new rhythm of waiting appeared. The walk to the newsagent turned into something of a ritual, or as one woman remembers about how the early 1950s were for her: "I always longed for Saturday, because then I was allowed to go to the kiosk and buy my Western comic book and a piece of mint chocolate". It was wonderful to combine leafing through the magazine with the taste of the mint.

The gramophone, the cinema and the weeklies created rhythms of longing and anticipation, which became an important dimension of mood work among the young. There was an impatient waiting for the *next* – the next instalment of a romantic novella, a cowboy adventure, the latest soccer results or piece of Hollywood gossip, or eager saving up for the next movie night or for a set of collector's cards or a new record.

Media provided rich material for daydreaming. It could be stories of village boys who made it into the national soccer team and achieved star status in spite of many obstacles, or the shop assistant who became a film star. At the same time, there were new possibilities for shaping personal and individual mediascapes. The tone between writer and reader turned more intimate; the content in the weeklies talked directly to you, the individual reader. This feeling of dialogue was strengthened by the growth of Reader's Queries and Lonely Hearts columns. How should I handle my boyfriend? What might help my skin problems? How does one get a chance to work in the movie industry? Many anxieties of the period surface in these magazine exchanges. There are worries about the still heavily taboo theme of sexuality, and also questions of class, often hidden in issues about etiquette rules and behaviour in public places.

The new media habits of the period are often described in terms of taking in, in the form of listening, reading, viewing – but looking back at these decades, it is possible to see how a range of new activities developed, in which media offered a toolbox of possibilities or affordances, a range of working materials. Media gave a chance for experimenting with identities, mixing materials and genres.

The camera was one important tool. During this period, 'Kodakisation' offered a much more affordable technology with the inexpensive box cameras. "Just press the button and we do the rest!" (West, 2000: 41 ff). The camera was not only a technology for producing memories; the actual photo activities themselves become part of social life and self-representations. Certain situations and events called for a photo, helping to choreograph moods of intimacy or togetherness, fun or solemnity. Being photographed or photographing others heightened awareness of self-presentation, or as Roland Barthes (2000: 10) once stated:

> I have been photographed and I knew it. Now, once I feel myself observed by the lens, everything changes: I constitute myself in the process of 'posing', I instantly make another body for myself, I transform myself in advance into an image.

Daily posing in front of the mirror at home helped to create this choreography of body awareness and self-presentation. The mirror was a monitoring tool working together with other technologies, from photography to diaries, as Ulf Hannerz (2019: 116) has explored in his study of Nigerian youths in the 1970s.

There was an upsurge in the production of diaries, but above all in the making of albums and scrapbooks. All kinds of stuff went into such cut-and-paste projects at the kitchen table. Cinema visits were documented by pasting in ticket stubs, photos of film stars and cut-outs of reviews, complemented with one's own views and comments. Boys made albums with their favourite sport teams and players, while girls produced albums of film stars. The new low-price department stores had sections for paper products and stationery, which the youths loved. Here you could find all kinds of wonderful stuff for album products: coloured paper, with or without smells; funny erasers, envelopes, writing pads and notebooks.

Favourite photos went into the albums, with written commentaries: "Me and my brother clowning at a Sunday picnic, my best friend in her new outfit, the family in front of the new car, or my classmates at an outing". A new kind of album took shape, very different from the rather rigid and solemn genre of traditional family photo albums. Funny captions were produced, people clowned in front of the camera, and photos were mixed with other materials. This kind of documentation was not only about collecting a history of frozen moments; these albums became performative media with strong affective dimensions, constantly reinterpreted and on the move (Urry and Larsen, 2011: 155).

All kinds of media stuff supplied rich material for comparison and imitation: walking like Alan Ladd, smoking like Humphrey Bogart, trying out a Greta Garbo look or imitating Veronica Lake's hairstyle. Girls began to spend more time on make-up, using tips from magazines, and there was a fast-growing genre of beauty ideals in advertisements. Body movements and postures turned into more self-conscious choreography. Who am I? Who could I become? What do others think of me? New modes and moods of reflexivity emerged, as well as skills of enactment and staging. There was a lot of daydreaming, drawing on media material. Anything could get the imagination going – a line from a popular tune, an advertisement, a sequence from a movie or a fragment from a radio programme. Underlying all this experimentation were the anxieties and challenges of being modern and of having to move with the times. As a teenager, to be counted, you had to build up some social and cultural capital; it was important to be up to date on the latest film, the new hit record or a rising star, to make sure you did not miss anything. You had to show you could navigate in this landscape of accelerating newness.

Challenges of class

Class entered in many ways. The dream world of Hollywood represented a kind of classless universe; America was a world in which social boundaries seemed much easier to transcend. At the same time, there were the everyday confrontations with youths of other backgrounds. In urban settings of the 1930s and 1940s, class was often strongly territorialized. One kept to one's own neighbourhood, but in the new public spaces, you were confronted with people from other settings. Working-class memoirs of these decades often stress the feeling of being under surveillance. Public streets, parks, department stores, dance halls and restaurants are described as being policed by the middle class, represented by the snobbish waiter, the haughty shop assistant, the stern lady in the public library or the schoolteacher out for an evening walk. Under their watchful eye, teenagers felt out of place, lacking the middle-class codes of public behaviour. Memories of class in this urban setting are often memories of awkwardness, or even shame, of being judged and ridiculed. Although one dressed up to go out, there was the feeling that middle class kids could always spot what was wrong in your style or behaviour. This is what the painter who played in an amateur dance band, mainly to working-class audiences, had to say:

> We called ourselves *The Light of the Stars*, but we didn't know what it meant. And I sung in English and German without knowing a single word, but I sung it the way it was written on the music sheets and nobody understood. . . . Once during the summer we noted a white cap (worn by high school graduates) in the audience and then I stopped singing. But otherwise I sung "vorgon vells" just the way it was said and everybody thought we were a damned good band who could play in English.

Another worker remembers teenage attempts to visit a music café:

> It was a long time until we dared enter. Because of the doorman in uniform. We had a great deal of respect for him. Many times we went up to look, but when we saw him we didn't know how to handle ourselves. After all, he was supposed to take your coat, and then when you were leaving he helped you on with it again. Yes, and then at the same time he was the doorman. This feeling of not knowing how to behave was really tough.

The tables could be turned, however. Sometimes working-class kids were more at home with the world of popular culture. In an ongoing study of the 1940s and 1950s, Jonas Frykman is studying the confrontations at dance venues of the period, where smart working-class couples easily outdanced middle class kids, who were seen as clumsy and awkward: "They just can't let go".

New public places thus became training grounds, and sometimes battle grounds, for adolescents who were learning to read class and to transcend it, but popular culture was also used to carve out some kind of private media sphere, to find a time or place to listen alone to the gramophone or the radio or to read a magazine. One's cherished media materials, from photo albums to diaries, could be stuffed into a cardboard box or fill a drawer. Many thoughts and daydreams were crammed into these small spaces.

Moving bodies, roaming minds

My focus has been on the skills and competences people developed when new media were integrated into the everyday, co-habitating with other elements. This throwntogetherness and mood work created new combinations of moving bodies and roving minds, in even the smallest moves or barely noticeable reflexes: picking up a camera and suddenly seeing the world through a lens; a twist of the dial and the room is filled with new sound; flicking through a magazine with the scissors, cutting things out for the album; or lighting a cigarette, Hollywood style. Feelings were constantly on the move, often in unreflected ways. When discussing patterns of mobile socialities, such everyday micro-moves may be overlooked or seen as unimportant.

During the period from 1930–1960, new ingredients entered everyday life. New materials and technologies were mixed and created different experiences of speed and energy: neon light, asphalt roads, concrete structures, cellophane, Bakelite, nylon and many other synthetics shaped a new everyday of sights, sounds, smells, tastes and tactilities. Hands reached out for new forms and surfaces, bodies moved on novel foundations. The index finger learnt to open the world – a simple press on the electric bell in the new housing estate, finding the right hole in the telephone dial, pressing the camera button or the keyboard. Hands were trained to calibrate the radio dial or the gramophone needle. The senses were sharpened in new ways. Eyes got used to moving through an abundance of pictures, in albums and magazines, or to follow the quickening pace of a movie. There were all the smells and tastes of newness. Gasoline gave an adventurous and somewhat sweet odour to urban cityscapes, while electricity carried a new, slightly burnt smell, as well as bringing a muffled and hardly noticeable soundscape of humming into the homes.

Old everyday competences became obsolete, and new ones turned slowly into reflexes as the autopilot learnt to handle the new body moves. New kinds of throwntogetherness and mobilities emerged – not just the movie, but the bike ride, the bag of candy, cutting out the reviews. The kitchen sofa choreographed radio listening and newspaper reading. The cut-and-pasted album became a Gesamtkunstwerk, combining colours, images, texts, photos. At the café, music tasted of coffee and cinnamon rolls while you leafed through the piles of magazines.

As new worlds opened up, the gap between the possible and the desired widened, maybe to an extent that had not been experienced before or after the 1930s and 1940s. A yearning for modernity coloured the mood work of everyday life. The importance of being modern was stressed everywhere, in radio talks, popular tunes, in advertisements, magazines and movies. (The adjective "modern" was the most common sales argument in Swedish advertisements during the 1950s.) One was constantly reminded of the need to be on the move and to improve oneself.

There was a new restlessness built into this yearning for a better life. The sociologist Colin Campbell has discussed this accelerated production of newness. He points out that the project of modernization is not only about ideas of rationality, blueprints for change and improvement, but also an often underestimated emotional dimension: "The cultural logic of modernity is not merely that of rationality as expressed in the activities of calculation and experiment; it is also that of passion, and the creative dreaming born of longing" (Campbell, 2018: 328). Media use was crucial in this yearning for being part of the modern project. Keep on moving!

Leafing through magazines, watching the movies, visiting the new department stores or listening to popular tunes, this mood of longing searched for anchorage. It could be small things, like a shiny red plastic hair clip, a silvery ball point pen, a colourful holiday postcard – all kinds of seemingly trivial objects which could materialize the future, but the grander longings – like a modern apartment, a holiday trip or a radio gramophone – were still out of reach. The fellow travellers of longing, waiting and hoping were disappointment and frustration.

In the media, a focus on the pursuit of happiness and new freedoms was increasingly common. Are we a happy family? Is this a good home? Is my life modern enough? Modernity was also about taking individual responsibility for improvement, a message hammered home everywhere. Did one's life measure up to the life of others? Class was hiding behind such ideas of modernity and normality. Middle class reformers might complain about working-class opposition to aspects of modern life, but they did not realize that dreams of a modern future were constantly stalled by worries about tomorrow. What was called "working-class respectability" was actually a defence system of everyday routines and rituals – a battle on two fronts. On the one hand, there was the fear of failing and falling and becoming one of those families who had given up, where drinking, unemployment and illness meant drifting into an underclass; on the other hand was the wish to distance oneself from what was seen as the pretentiousness of the expanding middle class and their obsession about being modern. This battle created a mood work based upon a constant tension, but also a feeling of lagging behind, of being out of sync. Media supplied materials for dreams of a better future, but also a feeling of being judged and measured – you were never good enough. Examples of modern interior decoration, fashion tips and an endless stream of advice on better housekeeping could create feelings of guilt and shame.

At the same time, learning to live with new media also fostered skills of distancing oneself from messages and campaigns of persuasion – either from the market or the state. One learned that frozen Danish pastry was not as appetizing as the advertisement, and that Captain Uggla's military-style gymnastic instructions could be seen as comical. There was a growing skill of maintaining a critical distance to the authorities as much as to the consumer promises.

Media helped to develop new platforms for sociality. Broadcasting created radio families as well as feelings of national belonging and identity. For adolescents, mediated popular culture turned into a resource for organizing both social relations and identity management in novel ways, sometimes transgressing class boundaries, sometimes reinforcing them.

What I have tried to illustrate is that a given period or setting has its special forms of thrown-togetherness and mood work, with media as important co-actors. The 1950s differ from the two previous decades as one in which longings began to come within reach. Moving to another era – say, the 1970s and 1980s – we would find quite different forms of throwntogetherness and mood work (see, for example, the discussion in Löfgren, 2013, or David Morley, 2000, for the 1990s). The point I aim to make here is that everyday life can be seen as a machinery that drastically changes the forms, functions and futures of, for example, new media. It chews up and devours new technologies, with some spat out rapidly because they cannot be integrated into everyday practices and needs. Others are digested, tested, adapted and changed, often in the process of developing new forms of multi-tasking. New media have to fight for space in the crowded mess of the quotidian throwntogetherness in order to survive, often by being able to blend in or strike up alliances with other activities and technologies. Many of these processes are hard to notice, and are difficult to verbalize and to operate, like slow accumulations of change. That is why, analytically, a historical or contrastive perspective is helpful.

I began by talking about analogue skills and micro-moves that have made a digital everyday possible. What people learned during these analogue decades was, for example, the skill of grasping film narratives, with the rapid cuts between scenes and situations; developing the art of chatting over the phone; writing short messages on postcards; creating media albums and diaries with narrative skills of self-presentation; and not forgetting the competences to mix and combine media and other activities. These were all the kinds of basic media skills necessary to survive in a coming digital era and develop new forms of mobile socialities.

Note

1 Together with Jonas Frykman, I am currently working on a study of everyday life of this period. For the media examples, I have drawn heavily on the collections of materials in the Nordic Museum and the Folk Life Archives of Lund University, such as interviews, answers to questionnaires and oral histories. (For an earlier presentation of some of the materials, see Löfgren 2014)

References

Åkerman, B. (1941) *Familjen som växte ur sitt hem*. Stockholm: Nordstedt.

Ambjörnsson, R. (1996) *Mitt förnamn är Ronny*. Stockholm: Bonniers.

Anderson, B. (1983) *Imagined Communities: Reflections on the Origins and Spread of Nationalism*. London: Verso.

Barthes, R. (2000) *Camera Lucida*. London: Vintage.

Björnberg, A. (2019) "Det mobila musiklyssnandet före transistoriseringen", in U. Volgsten (ed.) *Musikens medialisering och musikaliseringen av medier och vardagsliv i Sverige*. Lund: Mediehistoriskt arkiv, 44: 57–71.

Borgerson, J. and Miller, D. (2016) "Scalable sociality and 'How the world changed social media': Conversation with Daniel Miller", *Consumption, Markets & Culture*, 19(6): 520–533.

Brunner, E. (2014) *Där går han.(1: 1950–1970)*. Sthm: Bonnier.

Campbell, C. (2018) *The Romantic Ethic and the Spirit of Modern Consumerism: New Extended Edition*. London: Palgrave Macmillan.
Carlson, J. D. and Stewart, K. C. (2014) "The legibilities of mood work", *New Formations*, 82: 114–133.
Ehn, B. and Löfgren, O. (2010) *The Secret World of Doing Nothing*. Berkeley, CA: University of California Press.
Fornäs, J. (2004) *Moderna människor. Folkhemmet och jazzen*. Stockholm: Nordstedt.
Frykman, J. and Povrzanovic´ Frykman, M. (eds.) (2016) *Sensitive Objects: Affect and Material Culture*. Lund: Nordic Academic Press.
Haddon, L. (2017) "Domestication and the media", in *The International Encyclopedia of Media Effects*. New York: John Wiley & Sons.
Hannerz, U. (2019) *World Watching: Street Corners and Newsbeats on a Journey through Anthropology*. London: Sage.
Hellström, G. (1931) *Malmros*. Stockholm: Bonniers.
Highmore, B. (2017) *Cultural Feelings: Mood, Mediation and Cultural Politics*. London: Routledge.
Höijer, B. (1998) *Det hörde vi allihop! Etermedierna och publiken under 1900-talet*. Stockholm: Skriftserien Etermedierna och publiken under 1900-talet.
Koch, G. (ed.) (2017) *Digitisation: Theories and Concepts for Empirical Cultural Research*. London: Routledge.
Levine, L. (1992) "The folklore of industrial society", *American Historical Review*, 1369–1399.
Löfgren, O. (2000) "Technologies of togetherness: Transnational flows, mobility and the nation state", in D. Kalb, M. van der Land, R. Staring, B. van Steenbergen and N. Wilterdink (eds.) *The Ends of Globalization: Bringing Society Back In*. Boulder: Rowman & Littlefield, 317–331.
Löfgren, O. (2013) "Changing emotional economies: The case of Sweden 1970–2010", *Culture and Organization*, 19(4): 283–296.
Lundin, I. (1987) "Att längta ut genom orden", in L. Persson (ed.) *Folkhemsprinsessorna: Om att växa upp på 50-talet*. Stockholm: Litteraturfrämjandet, 94–107.
Massey, D. (2005) *For Space*. London: Sage.
Moores, S. (2018) *Digital Orientations. Non-Media-Centric Media Studies and Non-Representational Theories of Practice*. New York: Peter Lang.
Morley, D. (2000) *Home Territories. Media, Mobility and Identity*. London: Routledge.
Morley, D. (2017) *Communications and Mobility. The Migrant, the Mobile Phone and the Container Box*. London: Wiley-Blackwell.
Pink, S. (2013) "Saturated and situated: Expanding the meaning of media in the routines of everyday life", *Media, Culture & Society*, 35(6): 677–691.
Plotnick, R. (2018) *Power Button. A History of Pleasure, Panic and the Politics of Pushing*. Cambridge, MA: MIT Press.
Sjöholm, C. (2003) *Gå på bio. Rum för drömmar i folkhemmets Sverige*. Stockholm: Symposion.
Strand, K. (2019) "Mikrofonsångaren: Sven-Olof Sandberg och gestaltningen av närhet och distans i de tidiga elektriska ljudmedierna", in U. Volgsten (ed.) *Musikens medialisering och musikaliseringen av medier och vardagsliv i Sverige*. Lund: Mediehistoriskt arkiv, 107–128.
Tacchi, J. (1998) "Radio texture: Between self and other", in D. Miller (ed.) *Material Culture: Why Some Things Matter*. London: UCL Press, 25–46.
Tacchi, J. (2012) "Radio in the home: Changing experiences of domestic audio technologies in Britain", in L. Bessier and D. Fisher (eds.) *Radio Fields: Anthropology and Wireless Sound in the 21st Century*. New York: New York University Press.
Urry, J. and Larsen, J. (2011) *The Tourist Gaze 3.0*. London: Sage.
West, N. M. (2000) *Kodak and the Lens of Nostalgia*. Charlottesville, VA: University of Virginia Press.
Wetherell, M. (2015) *Affect and Emotion. A New Social Science Understanding*. London: Sage.
Willim, R. (2017) "Imperfect imaginaries. Digitisation, mundanisation and the ungraspable", in G. Koch (ed.) *Digitisation: Theories and Concepts for Empirical Cultural Research*. London: Routledge, 53–77.

3

INVESTIGATING "COMMUNITIES OF CO-MOVERS"

Motricity, spatiality and sequentiality in social life

Shaun Moores

Contributors to this book have been invited to explore its overall theme of "mobile socialities" in light of their specific research concerns and perspectives. In the present chapter, I will therefore be offering a particular take on that theme – and I want to start by providing a statement of where I am coming from in academic terms, explaining the basics of my approach (a quotidian cultural studies perspective that has a non-representational theoretical emphasis). I also want to indicate, in an overlapping manner, where it is that I am heading for, pointing in fruitful directions for future conceptual and empirical investigations of mobility and sociality. Later, though, as a way of focusing my forward-looking considerations, I will spend much of this chapter looking back, recovering and re-evaluating some important research that was done long ago by David Sudnow, who is one of the lesser-known characters in the history of the social sciences.

Quotidian cultural studies, non-representational theories and keywords

The first thing for me to explain is that, although my academic background is in media studies, the present chapter is not mainly concerned with media or even with uses of media (at least as media are typically defined today in the field of media studies, which is still rather narrowly). So, for example, I will have very little to say here about mobile media and communications. Some limited attention gets paid to new media technologies, acknowledging that they have certain distinctive features, but this is to be in the wider context of considering, among other relevant matters, the skilful manual activity of piano playing and the ordered or ordering practices of talk-in-interaction. Indeed, in the past, my own research has been characterised as a form of "non-media-centric" media studies (Morley, 2007: 228; see Krajina et al., 2014; Moores, 2005: 3, 2012: 103–110, 2017a, 2018). While I realise that this may seem to be a contradiction in terms, I agree completely with David Morley (2007: 200), who initially employed the expression, when he recommends that "we need to 'de-centre' . . . media in our analytical framework, so as to understand better the ways in which media . . . and everyday life are interwoven". In my case, the extent to which I have come to de-centre media now leads me to prefer a label such as "everyday-life" studies (Moores, 2017b, 2018: 175–178) or "quotidian" cultural studies (Moores, in press), in an attempt to put a suitable name to the loosely constituted field that my work is contributing to. Of course, I would accept that the everyday is a far-from-new site of exploration within the social sciences, and also that cultural studies have long had a strong quotidian dimension, particularly evident within what became known as a culturalist or cultural-materialist tradition, which can be traced back to early writings on culture as a way of life and as "ordinary" (Hoggart, 1990 [1957]; Williams, 1989 [1958]).

A second thing to explain here is that my approach to quotidian cultures is informed by "non-representational theories", broadly conceived (Moores, 2017a, 2018, in press; and see Thrift, 1996, 2008), which now serves to distinguish it from, for example, Morley's insightful non-media-centric perspective on various contemporary "communications" (Morley, 2017). Among the defining features of the non-representational theories that I favour these days, as I discuss in much greater detail elsewhere (Moores, in press), are what I call their anti-rationalist and anti-structuralist tendencies. Non-representational theories pose a firm critical challenge to any intellectualist and constructionist claims about the centrality of representations to life, whether those representations are conceptualised as mental or symbolic. This sort of critique is itself far from new, having its roots in long-established philosophical traditions such as phenomenology and pragmatism, which I see retrospectively as non-representational theories. It also has a close connection with developments over the past few decades in practice theory, and with culturalism's stress on "the sensuous side of lived culture" (Willis, 2000: 20). For me, however, the critical challenge remains a necessary one, since the field of media studies, out of which my own work came, has never broken entirely with implicit rationalist assumptions and early structuralist influences (still evident in taken-for-granted notions that it is only through thought and language that the world can be made to mean, at least as cognition and signification have conventionally been understood). Importantly, I must add that non-representational theories are not to be defined solely in negative terms, as a critique of other positions in philosophy and social theory that assert "the primacy of representation" (Thrift, 1999: 300). They are marked, too, by a positive assertion of "the primacy of practices" (Thrift, 1999: 308) or "the primacy of movement" (Ingold, 2011: 12; Sheets-Johnstone, 2011).

In what I describe as the cross-disciplinary project of non-representational theories (Moores, in press), and specifically in the present chapter, the kind of movement that is given primacy is a pre-reflective, yet knowledgeable and responsive, bodily movement. It is for this reason that the first of the three keywords in my chapter's subtitle is "motricity", which I have drawn from the latest English-language translation of Maurice Merleau-Ponty's *Phenomenology of Perception* (Merleau-Ponty, 2012). As Donald Landes (2012: l) notes in his translator's introduction to that classic text, it is the nearest equivalent to the French word "*motricité*" originally used by Merleau-Ponty, which "indicates . . . motor activity . . . and the . . . faculty of movement" of the human body, previously translated as "motility". One of Merleau-Ponty's principal claims is to have discovered, precisely through the study of such bodily movement or motricity, a new meaning of meaning, or, as Landes chooses to put this, "a new sense of the word 'sense'" (Merleau-Ponty, 2012: 148). Contemporary philosopher David Morris (2004: 24; and see Landes, 2012: xlviii) explains that, in French, "'*sens*' not only connotes meaning and the senses, but direction", before adding that: "Merleau-Ponty plays on this link . . . there is *sens* within the body's . . . directedness toward the world . . . meaning within a movement that crosses body and world". The living, sensing, moving body and the inhabited, known-in-action, familiar world in which it is immersed and oriented can best be considered in terms of this "crossing" (Morris, 2004: 4). There is meaning in that basic contact and practical orientation, which does not somehow depend upon the intervention of mental or symbolic representations. "My body has its world", insists Merleau-Ponty (2012: 141), "without having to go through 'representations'".

For me, in the context of my contribution to this edited book, motricity is a valuable distinguishing concept, then, because it points to a specific type of mobility, which I want to focus on. Sociologist John Urry (2007: 48), when he outlines "the new mobilities paradigm" (and see Sheller and Urry, 2006), does help to open up issues of motricity, with his references to the "corporeal", to the body that "senses as it moves", to technologies that facilitate a sensuous extension of "human capacities" and, with particular reference to practices of walking and driving, to the "touch . . . of . . . feet on the pavement or . . . hands on . . . the steering wheel" (and see Moores, 2020). However, the Merleau-Pontyan (non-representational) case pursued in the present chapter, for the significance of the body in movement and of the continual crossing of body and world, is designed to push discussion of these issues somewhat further than Urry's. Indeed, I have previously sought to stress how what he calls the "imaginative", "virtual" and "communicative" travellings facilitated by media (Urry, 2007: 47), as well as mechanised "physical" mobilities, still rely upon the skilful activity and directedness, or the "motor intentionality" (Merleau-Ponty, 2012: 112), of the body. This is why, in what might seem to be an unusual fashion, I have argued for a broadened understanding of "digital media" also as "manual media", as "doubly digital" (Moores, 2018: 2, 176), since media technologies today tend to be operated by the deft tapping and sliding of fingers or "digits" on keyboards and touch-screens. In addition, it is why I have been involved, over recent years, in descriptive, qualitative studies of "skilfully moving around and negotiating environments, at different geographical scales ranging from that of a computer keyboard or smart-phone screen to that of, for instance, a complex contemporary transport system" (Moores, 2018: 4; and see Moores, 2014: 197–198, 204–205; Moores and Metykova, 2010).

That notion of negotiating environments, cited at the end of the preceding paragraph, brings me to the second keyword in my chapter's subtitle, "spatiality". So, for Merleau-Ponty (2012: 100–148) and, implicitly, for Martin Heidegger (2010) before him in the phenomenological tradition, concerns with spatiality are intimately tied to matters of motricity (I say implicitly, though, since Heidegger, in his classic text *Being and Time* – despite its highly relevant discussions of directedness, orientation and being-in-the-world, and of handiness and practical dealings with tools – makes just passing mentions of corporeality, while deeming it to be beyond the scope of his philosophical inquiry). The sort of space that is of interest to those phenomenologists is what Merleau-Ponty (2012: lxx) theorises as " 'lived' space" or as existential space, as opposed to geometrical, objective space. This is the spatiality of what is practically accessible, of the ready-to-hand, which is emergent in engagements with equipment that solicits handling but withdraws in use, and, more generally, in an everyday, absorbed "skillful coping" (Dreyfus, 2014) with environmental conditions. It is comparable with what has been named, in the pragmatist tradition, a "manipulatory area", which is similarly related to the "readiness to grasp" (Mead, 1972 [1938]: 103; and see Gallagher, 2017: 180). One of Merleau-Ponty's examples of the emergence of that meaningful spatiality, serving to illustrate an important point he makes about the body's incorporating of the technologies that it deals with on a habitual basis, is his short account of the skilled use of a typewriter keyboard (Merleau-Ponty, 2012: 145–146). He contends that the accomplished typist is not required to "posit the keys as objective locations" (Merleau-Ponty, 2012: 146), because, as he explains: "The subject who learns to type . . . incorporates the . . . keyboard into . . . bodily space". As a consequence of this incorporation, the typist's in-practice, engaged knowing of "where the letters are on the keyboard" (Merleau-Ponty, 2012: 145) is of much the same kind as knowing, in a pre-reflective way, "where one of our limbs is . . . a knowledge of familiarity" (and see Moores, 2009). Such knowledge, a tacit know-how, may be conceptualised as an element of "practical consciousness" (Giddens, 1984: xxiii, 7), which is distinguishable (but not entirely and impermeably separated) from "discursive consciousness . . . what can be said".

Next, I want to move on to the third keyword in my chapter's subtitle, "sequentiality", which serves to emphasise the temporal or processual aspects of action and interaction in everyday socialities. This is a term that I have borrowed from the recent work of sociologists Kai-Olaf Maiwald and Inken Suerig (2020: 16–31), although the issues they discuss there had been raised initially by Harvey Sacks and colleagues (1978; and see Sacks, 1995; Schegloff, 2007), the founding figures of conversation analysis. These figures could be regarded retrospectively, in my view, as non-representational theorists. Their primary research interests included detailing the sequential organisation of ordinary talk, often by way of analysing conversations held over the telephone (Moores, 2005: 89–90). I have written previously of an anti-structuralist tendency in non-representational theorising, and I have alluded, too, to a sensuous meaningfulness of bodily movement that is potentially extra-linguistic, and yet the type of approach that I am advocating retains a strong concern with the structuring of routine activity and with talk-in-interaction as an instance of meaningful co-movement, among many other such co-movements. In contrast to a structuralist perspective on language, however, conversation analysts consider the systematic, ordered features of talk, and most notably the organisation of turn-taking and turn-allocation practices, to be a collaborative achievement or accomplishment of skilled speakers in the ongoing interaction (as it unfolds in lived time), rather than being determined by an abstract, pre-given linguistic structure. Speakers' acquired habits of talk are one example of what Sacks (1984: 21), citing ethnomethodology, refers to as "methods persons use in doing social life" (and see Garfinkel, 1984 [1967]). Indeed, I would add that issues of motricity and spatiality, no less than those of sequentiality, are closely linked with this doing or performing of "social life", and as Lucy Suchman (2007) has demonstrated, through an appropriation of selected principles of conversation analysis in her innovative ethnography of the uses of a photocopier, the notion of sequentiality has relevance not only for studies of interpersonal interaction but also for empirical research projects on the sociality of "human-machine" or body-technology relations.

Sudnow: a significant figure in historical context

With a view to concentrating my exploration of motricity, spatiality and sequentiality, I want to turn now, and in the following sections of this chapter, to recovering the work of someone who is probably regarded today, by most academics who recognise the name at all, as having been a marginal, and perhaps rather an odd, character in the social sciences. Sudnow only had a brief career as a young university professor before proceeding to focus largely on matters of music pedagogy, and he went on, outside the academy, to develop a reputation for his novel piano-teaching technique. Previously in my writing (Moores, 2014: 199–201, 2020: 112–113), I have touched on Sudnow's *Ways of the Hand: A Rewritten Account* (Sudnow, 2001), in which he tells the tale of his learning to improvise jazz with the piano, and I return to that remarkable, yet far-from-well-known, book here. In addition, though, I will be considering two other authored books of Sudnow's. One of them, *Talk's Body: A Meditation between Two Keyboards* (Sudnow, 1979), which was published just a year after the original version of *Ways of the Hand,* provides a continued account of his piano playing, combined with extended reflections on practices of speaking and typing (the two keyboards there are those of his piano and typewriter, between which he was shifting on a swivelling chair). It is from near to the beginning of Sudnow's *Talk's Body* that I have extracted the three words quoted in my chapter's main title. After offering the bold assertion "that improvised music-making and ordinary talking are deeply alike", he states, with a hint of eccentricity, his take on the similarities:

> You go places, getting from place to place. You have to get places on time. You find places in the course of moving . . . learn how . . . to reach the whereabouts of places that form up the setting for such movements . . . there is a social world, an organization of ways of doing such movements. . . . There are *communities of co-movers*.
>
> (Sudnow, 1979: 3–4, my emphasis)

Sudnow's research concerns, as suggested by this cited passage, are with exactly the things that interest me, too, as I argue for a non-representational theoretical approach, since he raises issues of bodily movement, existential space and sequential organisation (of going and getting places on time, of finding places in the course of moving), while making it plain that these are profoundly social issues, bound up with acquired, performative, collaborative skills, or with the learnt and habitual ways of doing in an intersubjective life-world. His work is noteworthy, at least in part, as a mixing of this whole range of conceptual concerns, and, as I will explain, the mix was made possible by Sudnow's borrowings from phenomenology and his positionings in relation to conversation analysis and ethnomethodology.[1] Furthermore, completing a trilogy of monographs, which were initially published in a highly productive five-year period for him, is *Breakout: Pilgrim in the Microworld* (Sudnow, 2019), recently re-released in a series aimed at a readership of videogame enthusiasts, over a decade after his death. Long before the rise of game studies as a fashionable academic sub-field, *Breakout* provided extended reflections on his practical, embodied and sensuous engagements with what was, in that period, still a new media technology and cultural form. I will also draw in the pages ahead on that pioneering book, but less fully, more in passing, than on *Ways of the Hand* and *Talk's Body*.

Readers may be asking, even so, why I am choosing to revisit Sudnow's work in particular here, especially if, as I have already accepted, he is now likely to be recognised as a marginal academic figure, if at all, and if his writings are, as should become increasingly clear, somewhat eccentric in style. In reply, I would say that I find him to be a significant figure in historical context, because of the way in which his research on music, language, body and technology can be seen, in retrospect, as a sort of theoretical junction-point. What I am getting at with this term is that Sudnow unwittingly connects with later non-representational theorising, anticipating aspects of Tim Ingold's thought, about which I will say a bit more in due course, while also linking with several figures working before or alongside him, in philosophy and social theory (rather better known than Sudnow is there), most of whom I cited in my chapter's preceding section. For example, in *Ways of the Hand*, Sudnow (2001: 131–132) refers to Merleau-Ponty's *Phenomenology of Perception* as an important "source of . . . inspiration", revealing the use that he made of the "discussions of embodiment" in this philosopher's text, and there is the added revelation that Sacks "was a close friend and . . . colleague", with Sudnow admitting that "whatever acumen I have for appreciating . . . order in the tiniest details was substantially nourished by my . . . association with him". Sudnow and Sacks had been doctoral research students under Erving Goffman's guidance, when Goffman (1963, 1971, 1990 [1959]) was perhaps the predominant social theorist of everyday life, and both were associated, too, with Harold Garfinkel, the leading light in ethnomethodology, to whom Sudnow (2001: 132) gives credit for having shaped his thinking about "ways to study music". Subsequently, in the acknowledgements given at the back of *Breakout*, Sudnow (2019: 197) mentions Sacks's fellow conversation analyst "Manny Schegloff" and the philosopher "Bert Dreyfus", a major commentator on Heidegger's *Being and Time* (Dreyfus, 1991) who wrote a short foreword to *Ways of the Hand,* in which Sudnow's book on the seemingly extraordinary skills of jazz piano improvisation is characterised in helpfully general terms as "a phenomenology of how we come to find our way about in the world . . . a study of how our bodies gain their grasp of the world . . . finding order in . . . unfolding experience" (Dreyfus, 2001: ix). Interestingly, Hubert Dreyfus (2001: xii) goes on to claim that Sudnow's work was, at the time, with that of philosopher Samuel Todes (2001), "at the forefront in . . . Merleau-Ponty-inspired research". As an indication of Sudnow's range of connections in the social sciences, one final piece of evidence that could usefully be cited is a book he edited, *Studies in Social Interaction* (Sudnow, 1972), prior to the trilogy of authored monographs referenced previously. That collection includes individual contributions from Garfinkel, Sacks, Emanuel Schegloff and Gail Jefferson, along with an early, Goffman-influenced essay of Sudnow's on certain temporal features of relations in public.

There is another, crucial reason why I am choosing to revisit Sudnow's work in the present chapter. This has to do less directly with his theoretical anticipations and inspirations, or with his personal and professional associations. Instead, it is tied more to the methodological question of how empirical investigations of *motricity, spatiality and sequentiality in social life* might actually be conducted, so as to illuminate various communities of co-movement (attending to the co-ordination of activities and to the being-with of socialities, which involve practical dealings with technology, as well as interpersonal interactions). A key difficulty facing researchers in this area is the problem of how best to document and reflect upon sensuous, meaningful bodily movements and co-movements that tend to be, in the processual flow of everyday routines, pre-reflective. Conversation analysts have developed techniques for the transcription of tape-recorded speech, in their studies of sequenced talk-in-interaction, that go part of the way toward solving the problem – and I want to suggest that, in a related manner, Sudnow's writings demonstrate a "methodology of descriptive inquiry" (Sudnow, 1979: 23), which has many potential applications in quotidian cultural studies. He refers, then, to *Ways of the Hand* as a "descriptively close account" (Sudnow, 2001: xix), in which formerly "unanalysed aspects of the body's ways" are "closely depicted" (Sudnow, 2001: 3), and more specifically he sees his book on playing jazz piano as a kind of "production account" that is written from "the 'practitioner's' perspective" (Sudnow, 2001: 131). It is the detailed story of his gradual acquisition of manual, musical skills "in a terrain nexus of hands and keyboard" (Sudnow, 2001: 127), and also in wider situations of collective, public performance. I will be coming back to these matters of methodology in a while, in my concluding remarks, linking Sudnow's commitment to descriptive inquiry with other instances of close description, and trying to imagine future empirical research projects on finding ways about in the world or on getting a hold on things (with a view to bringing at least some elements of the pre-reflective to discursive consciousness).

Incorporation: a genuine extension of the body

Let me pick up now from the passage cited in the last section, from near to the beginning of *Talk's Body,* where Sudnow drew a rather unusual parallel between improvised music-making and ordinary talking. A little further on, he attempts to counter an obvious objection that could be raised, in response to his drawing of such a parallel:

> It might be proposed . . . that in the former there is an instrument lying outside the body, while in the latter all the parts being used are body parts. Yet a true incorporation . . . the instrument as a genuine extension of the body . . . does occur. . . . It is a

> recurrent experience of . . . jazz play . . . that the piano . . . is no longer . . . an external mass of ivory and wood and steel, but . . . seems to dissolve into a . . . space . . . to speak with.
>
> (Sudnow, 1979: 6–7)

The three words at the end of this extract, "to speak with", employed in that particular context, serve to confirm Sudnow's ambition to consider the practices of jazz play together with those of talk or speech. For him, both activities involve the production of "articulated gestures" (Sudnow, 2001: xix). Indeed, there are several occasions, in *Ways of the Hand* as well as in *Talk's Body*, on which he refers to his piano playing, metaphorically, as a sort of speaking or even of singing. He declares, for example, that "voice emerges in and through the fingertips" (Sudnow, 1979: 67) and that: "I sing with my fingers" (Sudnow, 2001: 79). Use of a typewriter, too, is characterised at one moment, during a short excursus within his monograph on playing jazz piano, as "gestural production . . . doing finger talking" (Sudnow, 2001: 80–81).

More centrally, though, Sudnow's concern in that extract is with Merleau-Ponty's notion of incorporation, and with what I discussed earlier in my chapter as an intimate tie between motricity and spatiality for Merleau-Ponty and Heidegger. Sudnow's awareness of the arguments advanced by these philosophers is evident in the comparable instances of incorporation that he mentions there (but without formally citing Merleau-Ponty or Heidegger), in relation to his point that the piano becomes, in the course of its skilled use, a genuine extension of the body. So, he writes of a joiner who is handling or manipulating "a hammer . . . to hit in the nail . . . to bond . . . two pieces of wood" (Sudnow, 1979: 7), clearly echoing the commentary on hammering in *Being and Time*, in which Heidegger (2010: 69) notes the necessity of the tool's withdrawal in the experience of skilful hammering, for it to become "really handy" in use. This notion that hammers withdraw in labour is likely to be, initially, somewhat puzzling. The plausible phenomenological claim, however, is that a hammer ceases to be, for the joiner, an external mass made up of its wooden handle and attached metal head, since "everyday dealings are . . . busy . . . not . . . with . . . tools themselves, but the work" (Heidegger, 2010: 69), just as Sudnow reports that the piano seems to dissolve in the practice of playing, when the pianist gets caught up in the flow of a music-making task. Other examples of sensuous extension are pulled from *Phenomenology of Perception*. In addition to that of the accomplished typist who incorporates the keyboard, which I have already outlined, Sudnow (1979: 6–7; and see Merleau-Ponty, 2012: 144–145) borrows Merleau-Ponty's example of "the blind man" who learns "to feel the pavement . . . through the tip of his cane", thereby increasing the "scope" of the hand's touch, and also that of the proficient car driver. Cane and car, then, are each appropriated as an "appendage of the body . . . no longer an object" (Merleau-Ponty, 2012: 154). An accomplished driver is able to "enter into a lane and see that 'I can pass'" (Merleau-Ponty, 2012: 144), without having to measure the width of the available road surface against that of the vehicle, precisely because the technology has been incorporated (and there is the emergence, in practice, of a meaningful lived or existential space, allowing for a skilful coping with or a negotiation of environments). Writing in a slightly different context, Sudnow (1979: 19) insists, quite rightly in my view, that the quantifying perspectives of the natural sciences, which often seek to "reduce movements . . . of the body . . . to objective terms . . . mathematical measurements", are still, as yet, "a long way from grasping the essence of such movements". Consequently, any adequate account of the structuring of action and interaction in social life must, he contends, "come to terms with . . . spatial appropriations" (Sudnow, 1979: 19).

Given Sudnow's interest in piano playing, I am surprised that he does not pull another, highly relevant illustration of incorporation from the pages of *Phenomenology of Perception*. What I am thinking of is the case of the "experienced organist" (Merleau-Ponty, 2012: 146) who is capable, while in rehearsal for a public performance, of adapting to an organ that is not "his customary instrument" and with which, at first, "he is unfamiliar". I express my surprise especially since this example of Merleau-Ponty's relates to music-making, serving to show how the acquired skills of instrumentalists have a degree of responsive flexibility, which can be conceptualised as (in a general sense of the term) improvisatory. In accounting for that musician's adaptability, Merleau-Ponty (2012: 146) adopts a strong anti-rationalist stance, dismissing the possibility that, during the rehearsal, the player "forms . . . a representation of the stops, pedals . . . and keyboards . . . behaves like someone who wants to draw up a plan". There is no mental plan or cognitive map that takes shape to guide the performance, and, furthermore, "the organist does not play within objective space" (Merleau-Ponty, 2012: 147). Instead, according to Merleau-Ponty (2012: 146), this proficient organ player "sizes up the instrument with his body . . . incorporates its . . . dimensions . . . settles into the organ as one settles into a house". It is therefore a matter of skilful bodily orientation, spatial appropriation and habitation, rather than one of intellection or representation. Although, as I have indicated, Sudnow does not mention this particular, pertinent example, he does suggest, even so, that he is aware of it, when writing in *Talk's Body* of a "piano-knowing hand" (Sudnow, 1979: 16) being positioned "over a child's-sized toy keyboard". Despite the difference in dimensions between a full-sized instrument and that miniature version, the experienced pianist is still able to adjust, in a generative fashion, to the smaller layout of keys, to display a "familiarity in moving about" (Sudnow, 1979: 17). Just as it leads to a questioning of rationalism, intellectualism or cognitivism, such responsiveness of the skilled practitioner to changing environmental conditions also helps to rule out what Merleau-Ponty (2012: 146) has named a "mechanistic theory" of human action – an idea of automatic bodily movement.

To round off this part of my discussion, I want to turn to *Breakout* and to fulfil a promise made at the outset (in explaining my quotidian cultural studies perspective) to consider something of the distinctiveness of new media technologies, with specific reference here to issues of incorporation (and see Richardson, 2008). Sudnow shows again, in his important book on videogaming, a familiarity with the work of Merleau-Ponty and Heidegger. For example, he writes of the on-screen "paddle . . . controlled by a steering wheel knob" (Sudnow, 2019: 24) on his game machine, as coming to feel, following concentrated practice, "properly in hand . . . an extension of . . . fingers" (Sudnow, 2019: 95), and he describes this finely differentiated paddle that "was divided into . . . discrete portions . . . which . . . each . . . angled the ball in a certain way" (Sudnow, 2019: 70) as "unthinkingly" operated (Sudnow, 2019: 93): "I am not thinking of its parts, but using it . . . to place a shot". In other words, the on-screen paddle, "through which . . . an aim in a . . . direction" (Sudnow, 2019: 96) is realised, withdraws in Sudnow's lived experience of play, as it is gradually "incorporated by the body" in the ongoing flow of the game task, along with the "steering wheel knob". However, Sudnow appears to accept that using an on-screen paddle to hit a pictured ball at bricks in a wall is not quite like handling a cane, driving a car or hammering a nail. As he marvels at an early stage in his book, and remember that *Breakout* was initially written at what was still the dawn of a so-called digital age, the computer is providing his agile fingers with "a territory for action whose potentials and richness are electronically enhanced beyond the wildest dream . . . each stroke and sequence of strokes can generate . . . effects . . . on screen . . . human hands . . . create sights, sounds . . . movements" (Sudnow, 2019: 19–21). Returning to, and rearticulating, a term from George Herbert Mead's pragmatist philosophy that I quoted a few pages back, this might be understood as the electronic expansion of a meaningful manipulatory area. At the same time, though, what I particularly value in Sudnow's pioneering study of videogaming is that while he clearly becomes absorbed in the software's microworld, as a committed player, he never loses sight, as an academic researcher, of his practical engagement with the technological hardware and, most notably, the development of his dexterous manual activity and motor intentionality at the controls. He attends to unfolding relations between "the body's motions, the equipment . . . and . . . the interface" (Sudnow, 2019: 27), and his claim is that bodily movement or motricity, in hands-on interaction with such a machine, can "make us . . . feel in . . . touch with things" in a dynamic digital setting. Many years before me, then, Sudnow was already offering a notion of "digital orientations" (Moores, 2014, 2018) in his account of the interconnected motions and directions of fingers, paddle and ball. Even if he did not employ precisely that concept (or my notion of doubly digital media) in *Breakout*, he does write, elsewhere in his work, of his moving hands and fingers in terms of their "digital flexibilities", "interdigitations" (Sudnow, 1979: 33, 99) and "digitational acuity" (Sudnow, 2001: 116).

To achieve places, everybody moves on time

Earlier, near to the end of the section before last in this chapter, I referred to *Ways of the Hand* as a detailed story of skill acquisition. The point of departure for such a story cannot, of course, be the kinds of incorporation or spatial appropriation that I have just identified, since the narrative needs to, and indeed it does, start with what Sudnow (2001: 12–13) describes as the "substantial initial awkwardness" of a "beginning pianist", who is "groping to put each finger into a good spot . . . trying to get a hold on chords properly . . . to get from one to the next, playing progressions". As he expresses it in *Talk's Body*, the manual activity of this "beginner" (Sudnow, 1979: 8–9), of the novice at the piano keyboard, is marked by a profound hesitation: "The finger starts to go somewhere, touches the edge of an adjacent key, and quickly backs off. . . . It is a looking finger". Following a considerable period of concentrated practice, however: "Looking's work load progressively lightens . . . places gradually become places toward which . . . appreciative fingers . . . are aimed" (Sudnow, 2001: 15). Crucially, when the pianist is finally "at home" and comfortable in handling the instrument:

> The fingers do not look. They feel . . . the edges of adjacent keys not as treacherous . . . places to be avoided but with almost the degree of intimacy with which the fingers feel each other. Keyboard places feel to this hand almost as if they are feeling the hand.
>
> (Sudnow, 1979: 9)

Similarly, the "beginning typist" (Sudnow, 2001: 136) is deliberately "spelling every word as he types, thinking of the spellings as a step-by-step search of looking fingers for correct places". Only after a lengthy period at the typewriter keyboard do the typist's hands "behave spellingly" (Sudnow, 2001: 136), so that, in "typing the letter 't'", fingers no longer have to "search for . . . where 't' is written", rather they reach "toward saying 'this' or 'that'" (Sudnow, 1979: 10), just as Sudnow's piano-playing hands eventually come to perform in a "jazzful" manner (Sudnow, 2001: 125) and are able, as Dreyfus (2001: x) puts it well in his foreword to *Ways of the Hand*, "to reach for the music".

Whether Sudnow's focus at any one moment, in his studies of manual activity, is on piano playing or typewriter use, or else, in *Breakout*, on his videogaming hands, he has an interest in what he calls the "intelligence of the . . . knowing hand" (Sudnow, 1979: 11; and see Merleau-Ponty, 2012: 145; Pallasmaa, 2009; Sennett, 2009: 149–178), researching the gradual formation of that "hand's knowledge". What I want to stress, though, in the current section of my chapter, is that he understands this formation not only via issues of motricity and spatiality – as the appropriation of technology as a bodily appendage – but also, equally and simultaneously, in terms of "sequential organization . . . linear . . . arrangement" (Sudnow, 1979: 104, 109) and "wayful . . . moves" (Sudnow, 2001: 72). It is in reflecting on sequentiality or ordering that Sudnow's academic debt is more to his old friend Sacks, and to conversation analysis, than to the phenomenological philosophies of Merleau-Ponty and Heidegger, notwithstanding the concerns that those philosophers both had with issues of temporality. As I will explain shortly, it is with these reflections, in addition, that Sudnow's work anticipates a number of Ingold's arguments in contemporary anthropology, including the case that Ingold (2015: 53) is now making for a perspective named "linealogy", where a key concern is with "processional . . . order . . . movement . . . from place to place" (Ingold, 2011: 53). Partly, Sudnow's discussions of sequential organisation have to do with what Suchman (2007: 125–168) conceptualises as "human-machine communication", and partly, too, they have to do with physically co-present, interpersonal (or, better, intercorporeal) relations, like those between the jazz musicians who are performing together in a group. I will be dealing first, in what follows, with his writings on order in body-technology interaction, and it is important for me to reiterate the point that, even where there is just a single user, such interaction still needs to be thought of as a social relationship of a sort, a specific instance of sociality. For Sudnow (1979: 109–110), this is evident in the example of piano playing because the instrument is "built by and for . . . hands . . . the keys . . . are . . . arranged . . . to fit the . . . hand", and so there is "a mutually influential interaction of . . . hands at the keyboard . . . and the hands of the . . . manufacturer". That interaction-at-a-distance is mediated, then, by the piano's materials, by its ivory, wood and steel, despite their seeming to dissolve in the lived experience of proficient play.

During the gradual formation of Sudnow's skilful, improvising, jazz piano hands, which was a lengthy and frequently a frustrating period for him, he reports the emergence of "up-a-little-down-a-little ways, rocking ways and every-other-finger ways . . . skipping ways, hopping ways, rippling ways . . . and more" (Sudnow, 2001: 59). Hands at the keyboard were slowly getting their "bearings" (Sudnow, 2001: 51), learning how "to be wayfully oriented" (Sudnow, 2001: 69), until, eventually, as Sudnow (2001: 125) declares, near to the end of his account in *Ways of the Hand*: "I'd gained handful command over myriad varieties of paths. . . . I don't think at all about where I'm going. My hands make it up as they go along". As he expresses this elsewhere, in *Talk's Body*, "ways of the hand now unfold . . . from place to place" (Sudnow, 1979: 71), or, as he comments on his videogaming hands, in a comparable fashion, toward the close of the story of skill acquisition told in *Breakout*, "at last . . . I felt my hand feel that it knew where to go" (Sudnow, 2019: 184). For me, and I am not aware of any other academic who has yet made this link, Sudnow's conceptual vocabulary there, in these quotes from the trilogy of monographs that I am considering in the present chapter, bears a striking resemblance to that employed in Ingold's more recent non-representational theorising. To be absolutely clear, I am not suggesting that Ingold has somehow taken particular ideas, or a broader theoretical perspective, from Sudnow, and, indeed, I can find no evidence of Ingold ever having read Sudnow's books. Rather, my intention here is to indicate an unseen, intergenerational connection in the social sciences. So, while it is certainly worth noting that Ingold (2000, 2013, 2015) is openly critical of terms such as language, technology, embodiment and interaction (preferring notions of song, craft, animacy and correspondence), and although this anthropologist is interested at least as much in walking and ways of the feet (Ingold and Vergunst, 2008) as in manual activities and ways of the hand, the overlaps with Sudnow's work appear undeniable. Especially when Sudnow (2001: 51) writes of the wayful orientation of hands "at home on the keyboard", which know how to improvise jazz "as they go along", and of a hand that learns where to go in playing a videogame, I think of Ingold's conception of an "alongly integrated . . . inhabitant knowledge" (Ingold, 2007: 89; and see Moores, 2015), and of Ingold's associated discussion of those "everyday skills of orientation and wayfinding" (Ingold, 2000: 219, 229) with which, in familiar environments, "we know as we go . . . without having to consult a . . . map . . . a . . . representation". Ingold's wider insistence, too, that "life is lived . . . along paths . . . and paths are lines of a sort" (Ingold, 2007: 2), or that beings are "instantiated in the world as a . . . movement along a way of life" (Ingold, 2011: 4), may be regarded as highly compatible with Sudnow's discovery of "customary routings and . . . manners of moving" (Sudnow, 1979: 121), across seemingly diverse practices including piano playing, typewriter use, videogaming and sociable talk (but also, tangentially, acts of reading and listening, on which there is not currently room for me to remark).

As promised, I come now to the matter of order in intercorporeal relations. For Sudnow (1979: 5), this ordering is about an ability "to achieve places" in interwoven, co-operative, "improvisationally choreographed courses of movement", and he contends that research on language should "begin with the description of that accomplishment". In *Talk's Body*, as the book's title suggests, he makes a basic point, yet one which is sometimes oddly overlooked in the discipline of linguistics, that speech is "bodily work" (Sudnow, 1979: 37), and, in *Ways of the Hand*, he observes that "when I . . . move my mouth to speak, I must say each next sound in a progressively unfolding way" (Sudnow, 2001: 80–81), as, also, the articulated gestures of piano playing or typewriter use "must be serialized". All such gestural production involves knowledgeable bodily movement that is, in other words, sequentially organised or, to borrow Ingold's term, alongly integrated. Importantly, this is the case not just for a single speaking, gesturing body. Talk-in-interaction, in co-present, face-to-face communications as well as in technologically mediated discourse, as conversation analysts have shown, is ordered in a progressively unfolding way. Maiwald and Suerig (2020: 24) offer a helpful general definition of sequentiality in talk-in-interaction as "the temporally structured character . . . of successive speech acts". Schegloff (1972: 76) explains, though, in an essay that was published in Sudnow's collection, *Studies in Social Interaction*, that sequential organisation needs to be understood in this context as more than simply " 'subsequent occurrence' in the sense of . . . successive positions". What is of particular concern, for conversation analysis, is how a turn in a sequence of talk projects a next action, or a range of possible next actions, for another speaker in the next turn (as when, for example, an initial greeting invites a further greeting in response to it). At one moment, Schegloff and Sacks (1974: 239) refer to this specific dynamic as the "sequential implicativeness of an utterance", and something that Sudnow seeks to do is to adopt the same kind of analytical principles, most clearly in *Talk's Body*, in describing and reflecting upon collaborative musical performances. So, writing of musicians in a "string quartet" (Sudnow, 1979: 90–92), a "corporate entity" of "participating bodies", he asserts that they "could be said to take turns . . . their variously directed . . . articulations into a coherently interweaving whole . . . engaged in . . . a continuous flow of movements". Similarly, in a closely related consideration of "turn-taking jazz play" (Sudnow, 1979: 106), his interest is in how "musicians allocate turns of talk" in co-ordinating the improvised solo parts, thereby extending "speaking rights" to one another, "each player always ready to be given the floor and come in at the right place". Their sociable utterances have a sequential implicativeness. There is "concerted action . . . the paced . . . integration of individual gestures . . . group choreography" (Sudnow, 1979: 108), and so, in straightforward terms: "Everybody moves . . . on time".

Methodology: description of the course of gestures

By way of a conclusion (in truth, what follows here is more of an opening out than a tying up), I want to return to matters of research "methodology" (Sudnow, 1979: 23) that were first aired at a far earlier stage in my chapter. I raised the question, briefly, of how empirical researchers might best proceed toward *investigating "communities of co-movers"* of the type that Sudnow's writings deal with, and I pointed to the real challenge of bridging, at least partially, a divide between pre-reflective "knowledge-in-practice" (Thrift, 2008: 122) and the what-can-be-said of discursive consciousness. For Sudnow (1979: 83), the answer to this question is to be found in a "description of the course of gestures", specifically of articulated gestures that he identifies in both body-technology and intercorporeal relations, ranging from the expressive keyboard routings of an experienced pianist or an accomplished typist to the turn-allocating interplay between members of a classical or jazz ensemble. In these concluding remarks, I will be running with Sudnow's recommended solution to the problem of researching sensuous, meaningful bodily movements and co-movements, proposing some possible new directions for descriptive inquiry. Before I get to this, though, I must acknowledge that there was, long ago, a further, unusual character at the edges of the social sciences, Georges Perec (1999, 2010), whose own rather eccentric work, which also advocates and demonstrates techniques of close, detailed description, has already been recovered and reviewed quite widely over recent years (Forsdick et al., 2019; Wilken and Clemens, 2017; and see Morley, 2007: 94–97; Sheringham, 2006: 248–291). Recognising the level of contemporary interest in Perec, let me say just a bit about how his research concerns compare with Sudnow's, because, while I value Perec's take on quotidian cultures, sharing the enthusiasm that many of my fellow academics have felt for it, I want to make a claim for the distinctiveness and ongoing relevance of the descriptive studies that were carried out by Sudnow.

Perec's work on the everyday is focused on what he chooses to call "the infra-ordinary" (Perec, 1999: 209–210), a concept he defines in opposition to the extraordinary events that are reported in the "daily" newspapers, which "talk of everything except the daily . . . what happens every day and recurs every day . . . the common . . . the habitual". As he puts this near to the beginning of his extended, book-length "attempt at exhausting a place" (a large public square in his home city), such an emphasis on the infra-ordinary involves describing "what happens when nothing happens . . . that which is generally not taken note of, that which is not noticed" (Perec, 2010: 3). It requires a commitment to documenting "non-events" (Ehn and Löfgren, 2010: 4, 217). Sociologist Les Back (2020: xv) praises Perec for having managed "to enchant the mundane through noticing detail", and, intriguingly from my perspective in this chapter (since it suggests a link of a sort with Sudnow), Back draws a parallel, too, with aspects of Goffman's sociology of everyday life, observing "the same quality" of "attentiveness" in the writings of those academic authors. A favoured method of Perec's, which is perhaps especially evident in his account of the non-events in an urban square over a three-day period, is one of fragmented listing, and although Sudnow prefers a narrative style with greater coherence, there is still a broad similarity between their techniques, with both of them aiming to offer exhaustive descriptions of action and interaction. Indeed, I would have to admit that reading books by Perec and Sudnow, from cover to cover, can itself be exhausting as much as it is enchanting, given the degree of detail they provide. In Sudnow's case, in *Ways of the Hand* and *Breakout*, printed words are occasionally supplemented by photographs or sketches of keyboard layouts, finger positionings, game controls and screen displays, adding to the undoubted complexity of these texts for readers. However, despite the considerable overlap between Perec and Sudnow, my argument is that Sudnow's research constitutes the more fruitful model for quotidian cultural studies, and certainly for investigations of the everyday that are to be informed by a non-representational theoretical approach. I make that claim not only because, as I have been at pains to stress, Sudnow gathers and combines important ideas about bodily movement, existential space and sequential organisation from different traditions of thought, but also since he concentrates his descriptive inquiries on gradual acquisitions of skill, on the formation of courses of movement or articulated gesture in particular spheres of social practice.

Sudnow's continued, relative marginality in the social sciences today, partly in light of the sympathetic discussions and appropriations of Perec's work in the past few years, does not help me in the task of imagining what future Sudnowian projects may look like, as there is so little to go on in terms of existing applications. I am aware of just a single example in the current literature, a monograph by music scholar Marko Aho (2016), of empirical research that has explicitly borrowed its methodology from Sudnow, and, more specifically, from *Ways of the Hand*. Aho (2016: 10) writes of "taking up an instrument previously unfamiliar to me . . . the kantele . . . a plucked string instrument . . . documenting and analysing the learning process". While Aho's descriptions of playing this folk instrument, in public as well as in private, do not feature quite the kind of close detail that is found in Sudnow's production account, they are still evidence of Sudnow-inspired work in the sub-field of "performative ethnomusicology" (Aho, 2016: 6, 137), and the potential for a further application of Sudnow's methods in studies of music is not difficult to see. In addition, in my most recent book (Moores, 2018: 172), I have pointed to another possible development out of "Sudnow's study of piano playing", by inviting readers to "imagine the value of a . . . description and phenomenological analysis . . . along . . . the same lines . . . of the acquired skills . . . which are bound up with learning . . . to use a new, digital media technology". It should be clear from things I say in the present chapter concerning *Breakout* that Sudnow's study of videogaming, which he associated directly with his earlier research on jazzful hand movements at the keyboard, was already three-and-a-half decades ahead of my proposal there – and yet there remains, in my view, significant scope for an empirical investigation of media use in that vein, to be conducted within a framework of everyday-life studies or quotidian cultural studies. Of course, any contemporary attempt to do research roughly along the same lines as *Breakout* would need to allow room for considering the later, substantial growth in online, multi-player interactions.

Looking beyond the areas of music-making and digital or manual media, I want to finish my chapter now with a last couple of suggestions for where future descriptive inquiries might be situated, reaching back to Sudnow's metaphorical notion of the group choreography in a jazz ensemble, or else in a spoken, multi-participant conversation. First, then, practices of collective dance are obvious instances of communities of co-movement, and, among the academic discussions of dance to date, there is some work that is compatible with Sudnow's writings on the performance of music or ordinary talk. For example, historian William McNeill (1995: 34, 156) contends, interestingly, that insufficient attention has been paid in his discipline to the importance of communal dance in the emergence of human societies and, more generally, to the "muscular, rhythmic dimension of . . . sociality", to "gestural . . . communication" and to "emotion . . . aroused by moving and giving voice together in time" (and see Fuchs, 2017). Philosopher and trained dance practitioner Maxine Sheets-Johnstone (2011: 420), too, is in step with Sudnow when, in reflecting on her own lived experience of dance, she states that "a dance improvisation is akin to a jazz jam session wherein a group . . . brings something into being", by means of concerted, responsive, co-operative actions. Without realising it, she is echoing the conversation analysts' conception of sequential implicativeness, insisting that "my sensings and movings are . . . not . . . discrete gestures . . . one after the next . . . but integrally entwined facets of a . . . dynamic form in-the-making" (Sheets-Johnstone, 2011: 423, 425). There is a collaborative achievement of places, in Sudnow's terms, with the members of her dance group, a corporate entity, knowing how to get from place to place, wayfully and interweavingly. Second here, so as to ensure that this notion of improvisationally choreographed movements is positioned in the context of quotidian cultures, I am advocating fresh, descriptively close accounts of what Jane Jacobs (2011 [1961]: 66), in the field of urban studies, once referred to as a city neighbourhood's "intricate sidewalk ballet", which human geographer David Seamon (2015 [1979]: 54–59), citing Jacobs, theorises as a life-world choreography.[2] Such a complex "spatial dance . . . of bodies and things" (Thrift, 2009: 92) is routinely and habitually performed, in ways that are patterned yet not entirely predictable, on street pavements but also in urban locations like "market places, transportation depots . . . cafés" (Seamon, 2015 [1979]: 56–57), and it may foster, over a period, feelings of "familiarity" for participants. If this final proposal of mine seems to lean toward a Perecian inventory of infra-ordinary happenings (Perec, 2010) or an attentive, Goffman-style description of public behaviour and social relationships (Goffman, 1963, 1971), or perhaps even toward a Lefebvrean "rhythmanalysis" of movements (Lefebvre, 2004), it can be given a distinctive Sudnowian twist by investigating the gradual acquisition of orientation skills in cities – a crucial element of the spatial dance. Yi-Fu Tuan (1977: 199), another geographer, outlines a broad experiential shift, from arriving "in a strange part of town" through to sensing that neighbourhood, "eventually", to be "thoroughly familiar", having learnt "where we are" and "routes" to "find our way". Describing the slow formation, within this shift, of particular courses of movement (of how a grasp of the world is gained, mainly by establishing footful command over paths) could be the aim of a Sudnow-inspired research project, and it would require an acumen for appreciating "order in the tiniest details" (Sudnow, 2001: 132).

Notes

1 Of course, I am aware of the sort of criticism that often comes from cultural studies, of work in phenomenology, conversation analysis and ethnomethodology (for example, the argument that these perspectives pay insufficient attention to matters of social difference and inequality). Indeed, Sudnow's research is open to precisely such a critique. As I have explained elsewhere (Krajina et al., 2014: 695; Moores, 2018: 14–16, in press), though, a non-representational theoretical approach does not necessarily deny the politics of quotidian cultures.

2 Tim Edensor and Caitlan Bowdler (2015), whose own points of reference include Jacobs and Seamon, have recently attempted to combine concerns with artistic dance practice and the choreography of urban public life.

References

Aho, M. (2016) *The Tangible in Music: The Tactile Learning of a Musical Instrument.* London: Routledge.

Back, L. (2020) "Foreword: Making the mundane remarkable", in H. Holmes and S. M. Hall (eds.) *Mundane Methods: Innovative Ways to Research the Everyday.* Manchester: Manchester University Press, xiv–xviii.

Dreyfus, H. (1991) *Being-in-the-World: A Commentary on Heidegger's Being and Time, Division I.* Cambridge, MA: MIT Press.

Dreyfus, H. (2001) "Foreword", in D. Sudnow, *Ways of the Hand: A Rewritten Account.* Cambridge, MA: MIT Press, ix–xiii.

Dreyfus, H. (2014) *Skillful Coping: Essays on the Phenomenology of Everyday Perception and Action.* New York: Oxford University Press.

Edensor, T. and Bowdler, C. (2015) "Site-specific dance: Revealing and contesting the ludic qualities, everyday rhythms and embodied habits of place", *Environment and Planning A*, 47(3): 709–726.

Ehn, B. and Löfgren, O. (2010) *The Secret World of Doing Nothing.* Berkeley, CA: University of California Press.

Forsdick, C., Leak, A. and Phillips, R. (eds.) (2019) *Georges Perec's Geographies: Material, Performative and Textual Spaces.* London: UCL Press.

Fuchs, T. (2017) "Collective body memories", in C. Durt, T. Fuchs and C. Tewes (eds.) *Embodiment, Enaction, and Culture: Investigating the Constitution of the Shared World.* Cambridge, MA: MIT Press, 333–352.

Gallagher, S. (2017) *Enactivist Interventions: Rethinking the Mind.* Oxford: Oxford University Press.

Garfinkel, H. (1984 [1967]) *Studies in Ethnomethodology*. Cambridge: Polity Press.

Giddens, A. (1984) *The Constitution of Society: Outline of the Theory of Structuration*. Cambridge: Polity Press.

Goffman, E. (1963) *Behavior in Public Places: Notes on the Social Organization of Gatherings*. New York: The Free Press.

Goffman, E. (1971) *Relations in Public: Microstudies of the Public Order*. New York: Basic Books.

Goffman, E. (1990 [1959]) *The Presentation of Self in Everyday Life*. London: Penguin.

Heidegger, M. (2010) *Being and Time*. Albany, NY: SUNY Press.

Hoggart, R. (1990 [1957]) *The Uses of Literacy: Aspects of Working-Class Life with Special Reference to Publications and Entertainments*. London: Penguin.

Ingold, T. (2000) *The Perception of the Environment: Essays on Livelihood, Dwelling and Skill*. London: Routledge.

Ingold, T. (2007) *Lines: A Brief History*. London: Routledge.

Ingold, T. (2011) *Being Alive: Essays on Movement, Knowledge and Description*. London: Routledge.

Ingold, T. (2013) *Making: Anthropology, Archaeology, Art and Architecture*. London: Routledge.

Ingold, T. (2015) *The Life of Lines*. London: Routledge.

Ingold, T. and Vergunst, J. L. (eds.) (2008) *Ways of Walking: Ethnography and Practice on Foot*. Aldershot: Ashgate.

Jacobs, J. (2011 [1961]) *The Death and Life of Great American Cities*. New York: Modern Library.

Krajina, Z., Moores, S. and Morley, D. (2014) "Non-media-centric media studies: A cross-generational conversation", *European Journal of Cultural Studies*, 17(6): 682–700.

Landes, D. (2012) "Translator's introduction", in M. Merleau-Ponty, *Phenomenology of Perception*. London: Routledge, xxx–li.

Lefebvre, H. (2004) *Rhythmanalysis: Space, Time and Everyday Life*. London: Continuum.

Maiwald, K.-O. and Suerig, I. (2020) *Microsociology: A Tool Kit for Interaction Analysis*. London: Routledge.

McNeill, W. (1995) *Keeping Together in Time: Dance and Drill in Human History*. Cambridge, MA: Harvard University Press.

Mead, G. H. (1972 [1938]) *The Philosophy of the Act*. Chicago, IL: University of Chicago Press.

Merleau-Ponty, M. (2012) *Phenomenology of Perception*. London: Routledge.

Moores, S. (2005) *Media/Theory: Thinking about Media and Communications*. London: Routledge.

Moores, S. (2009) "That familiarity with the world born of habit: A phenomenological approach to the study of media uses in daily living", *Interactions*, 1(3): 301–312.

Moores, S. (2012) *Media, Place and Mobility*. Basingstoke: Palgrave Macmillan.

Moores, S. (2014) "Digital orientations: 'Ways of the hand' and practical knowing in media uses and other manual activities", *Mobile Media and Communication*, 2(2): 196–208.

Moores, S. (2015) "We find our way about: Everyday media use and 'inhabitant knowledge'", *Mobilities*, 10(1): 17–35.

Moores, S. (2017a) "Arguments for a non-media-centric, non-representational approach to media and place", in P. Adams, J. Cupples, K. Glynn, A. Jansson and S. Moores, *Communications/Media/Geographies*. New York: Routledge, 132–159.

Moores, S. (2017b) "For everyday-life studies", in P. Adams, J. Cupples, K. Glynn, A. Jansson and S. Moores, *Communications/Media/Geographies*. New York: Routledge, 183–188.

Moores, S. (2018) *Digital Orientations: Non-Media-Centric Media Studies and Non-Representational Theories of Practice*. New York: Peter Lang.

Moores, S. (2020) "Cities of feet and hands: Urban habitations", in Z. Krajina and D. Stevenson (eds.) *The Routledge Companion to Urban Media and Communication*. New York: Routledge, 107–115.

Moores, S. (in press) "The everyday skills that get us by: Non-representational theories for a linealogy of quotidian cultures", *European Journal of Cultural Studies*.

Moores, S. and Metykova, M. (2010) "'I didn't realize how attached I am': On the environmental experiences of trans-European migrants", *European Journal of Cultural Studies*, 13(2): 171–189.

Morley, D. (2007) *Media, Modernity and Technology: The Geography of the New*. London: Routledge.

Morley, D. (2017) *Communications and Mobility: The Migrant, the Mobile Phone, and the Container Box*. Hoboken, NJ: Wiley-Blackwell.

Morris, D. (2004) *The Sense of Space*. Albany, NY: SUNY Press.

Pallasmaa, J. (2009) *The Thinking Hand: Existential and Embodied Wisdom in Architecture*. Chichester: Wiley-Blackwell.

Perec, G. (1999) *Species of Spaces and Other Pieces*. Revised edition. London: Penguin.

Perec, G. (2010) *An Attempt at Exhausting a Place in Paris*. Cambridge, MA: Wakefield Press.

Richardson, I. (2008) "Pocket technospaces: The bodily incorporation of mobile media", in G. Goggin (ed.) *Mobile Phone Cultures*. London: Routledge, 66–76.
Sacks, H. (1984) "Notes on methodology", in J. M. Atkinson and J. Heritage (eds.) *Structures of Social Action: Studies in Conversation Analysis*. Cambridge: Cambridge University Press, 21–27.
Sacks, H. (1995) *Lectures on Conversation*. Oxford: Wiley-Blackwell.
Sacks, H., Schegloff, E. and Jefferson, G. (1978) "A simplest systematics for the organization of turn taking for conversation", in J. Schenkein (ed.) *Studies in the Organization of Conversational Interaction*. New York: Academic Press, 7–55.
Schegloff, E. (1972) "Notes on a conversational practice: Formulating place", in D. Sudnow (ed.) *Studies in Social Interaction*. New York: The Free Press, 75–119.
Schegloff, E. (2007) *Sequence Organization in Interaction: A Primer in Conversation Analysis*. Cambridge: Cambridge University Press.
Schegloff, E. and Sacks, H. (1974) "Opening up closings", in R. Turner (ed.) *Ethnomethodology: Selected Readings*. London: Penguin, 233–264.
Seamon, D. (2015 [1979]) *A Geography of the Lifeworld: Movement, Rest and Encounter*. London: Routledge.
Sennett, R. (2009) *The Craftsman*. London: Penguin.
Sheets-Johnstone, M. (2011) *The Primacy of Movement*. Expanded edition. Amsterdam: John Benjamins.
Sheller, M. and Urry, J. (2006) "The new mobilities paradigm", *Environment and Planning A*, 38(2): 207–226.
Sheringham, M. (2006) *Everyday Life: Theories and Practices from Surrealism to the Present*. Oxford: Oxford University Press.
Suchman, L. (2007) *Human-Machine Reconfigurations: Plans and Situated Actions*. New York: Cambridge University Press.
Sudnow, D. (ed.) (1972) *Studies in Social Interaction*. New York: The Free Press.
Sudnow, D. (1979) *Talk's Body: A Meditation between Two Keyboards*. New York: Alfred Knopf.
Sudnow, D. (2001) *Ways of the Hand: A Rewritten Account*. Cambridge, MA: MIT Press.
Sudnow, D. (2019) *Breakout: Pilgrim in the Microworld*. Los Angeles, CA: Boss Fight Books.
Thrift, N. (1996) *Spatial Formations*. London: Sage.
Thrift, N. (1999) "Steps to an ecology of place", in D. Massey, J. Allen and P. Sarre (eds.) *Human Geography Today*. Cambridge: Polity Press, 295–322.
Thrift, N. (2008) *Non-Representational Theory: Space/Politics/Affect*. London: Routledge.
Thrift, N. (2009) "Space: The fundamental stuff of geography", in N. Clifford, S. Holloway, S. Rice and G. Valentine (eds.) *Key Concepts in Geography*, 2nd edition. London: Sage, 85–96.
Todes, S. (2001) *Body and World*. Cambridge, MA: MIT Press.
Tuan, Y.-F. (1977) *Space and Place: The Perspective of Experience*. Minneapolis, MN: University of Minnesota Press.
Urry, J. (2007) *Mobilities*. Cambridge: Polity Press.
Wilken, R. and Clemens, J. (eds.) (2017) *The Afterlives of Georges Perec*. Edinburgh: Edinburgh University Press.
Williams, R. (1989 [1958]) "Culture is ordinary", in R. Williams, *Resources of Hope: Culture, Democracy, Socialism*. London: Verso, 3–18.
Willis, P. (2000) *The Ethnographic Imagination*. Cambridge: Polity Press.

4
SOCIALITY ON THE MOVE

Leopoldina Fortunati

Introduction

This chapter deals with the attempt to reconstruct the shift from a debate considering mobility and sociality as two separate dimensions to consideration of a unique and integrated dimension of mobile sociality. Although this debate has accumulated many interesting theoretical perspectives over time, such as those by Urry (2000, 2011a, 2011b) and Sheller (2004, 2011), and although it has put in motion a considerable corpus of empirical research (e.g. Elliot and Urry, 2010; Larsen et al., 2006), it so far has never engaged directly with the theoretical analysis of the main notions implicated. Exceptions include the efforts made by Fortunati and Taipale (2017a, 2017b) and Fortunati et al. (2013). My final aim is to develop further the theoretical approach proposed by these works we carried out in the last decade and to advance the theoretical analysis of the various notions that come into play within this shift. These concepts are mobility – understood as work mobility and mobility in the reproductive sphere – and the older notions of social mobility meaning a move in individuals' social position and/or professional roles, which can be horizontal and/or vertical. Social mobility is not actually connected to a spatial move, but rather to a symbolic one that should be read more correctly as a change in social circumstances.

The concept of sociality is key to my argument. The English use of the term sociality is a value-neutral term. It contrasts with the term sociability which refers to someone who has positive social skills, and is meant as a compliment; compare this with the term unsociable, which has negative associations. As we shall see in my argument, the value-neutral term of sociality brings to the foreground individuals' activities of meeting, speaking, spending time and doing things with other people. Sociality is thus the ensemble of relations or interactions among the individuals who are part of a given society, and it becomes all the more intense the higher the degree to which individuals associate in social circles and groups (gregariousness) and cooperate together.

The outline of my chapter includes a reconstruction of the sociological meaning of mobility by recurring to Marxian categories but going beyond them. The following section will analyse the classic notion of social mobility (which is not to be confounded with mobile sociality) and shall explain that mobility has registered so wide an increase just because the mechanisms of social mobility have jammed. In the third section, I will analyse the meaning of sociality, as well as its implications for mobility. The fourth section will be dedicated to weaving the notions of mobility and sociality as two interconnected dimensions and signals a fruitful new area of future research.

Theoretical advancements in mobility

It is necessary to dedicate a section to the analysis of the concept of mobility because we will need to integrate the main features that will emerge here about mobility in the elaboration of the new concept of mobile sociability, which is the aim of this chapter. This passage is, in a certain sense, obligatory because the new concept of mobile sociality is composed of both the notion of mobility and that of sociality. The reflections contained in this chapter about mobility represent hopefully an advancement on a long path of research and debate on this theme that I have undertaken with some colleagues, first of all Sakari Taipale.

Among scholars, there is a wide agreement that spatial mobility represents a key element of societal structure (Urry, 2007; Adey et al., 2014), and as such it has the capacity to produce relevant social change, e.g. in the labour market, in the family and in sociality. In the sociological debate on mobility, however, Urry (2000) was the first to denounce the lack of a solid, sociological understanding of this category. Recently, Taipale and I have proposed to define mobility as a productive force of social labour, as we will see in the next section (Fortunati and Taipale, 2017a). Scholars have talked about mobility generically as a social phenomenon, or they have taken the sociological significance of this category for granted. Despite the insufficient theoretical elaboration of this concept, during recent decades this has not stopped many researchers from investigating several types of mobilities at the empirical level and mapping the social practices, as well as the implications of the increase of mobility on the entire society (e.g. Kellerman, 2006, 2012; Oishi, 2010; Peters et al., 2010; Newman and Giardina, 2014). This work has been very precious for making sense of the qualitative and quantitative dimensions of mobilities, written in the plural as Urry suggested (2007). A further reflection on the definition of mobility, however, may advance our understanding of the social meaning of mobility. How can we arrive at grasping a profound socio-political definition of what mobility is with respect to society that resonates with the fundamental sociological categories?

Taipale and I thought to define mobility starting from Marxian categories, but going beyond them. It became clear to us that the key to understanding the social meaning of mobility was in the Marxian analysis, precisely in the part of *Capital* in which Marx (1964) talks of social forces. There Marx identifies only two productive forces of social labour: cooperation and division of labour. These productive forces, he argues, increase the productivity of labour, and because they do not cost anything to capital, he also defined them as natural forces. We proposed considering mobility as the third natural force of social labour, as we explain in detail in what follows. Here, to be as articulate as possible, in our analysis we will distinguish between work mobility from mobility in the reproductive sphere (Fortunati, 1981/1995), because these two forms of mobility assume different meanings.

Let us start with work mobility. Drawing on the first volume of *Capital* (Marx, 1964), Taipale and I proposed seeing work spatial mobility as a productive force of social labour. Mobility contributes highly to the production of value. Furthermore, we also proposed seeing it as a natural force of social labour: mobility makes the workers available where they are needed: for their employers, they are commodities in the circulation phase (Marx, 1965a: 305). Their mobility does not present any adjunct cost to capital, although it is fundamental to enable the production process. It is incorporated inside production as an agent of the process; that is, the value produced by mobility is transferred in the production process from labour to capital

(Marx, 1965b: 940) in a "natural way". However, the cycle of value connected to mobility does not stop here, as it involves a further production of value. Workers, in order to move, have to take trains, airplanes or buses, or drive cars, so that they also become at the same time commodities of the transportation industry. The value produced by the means of transportation ends to also be incorporated in individual consumption (Marx, 1965a: 163), because, as Fortunati (1981/1995) pointed out, individual consumption should be considered as being part of the production process since it enables labour's participation in production. In any case, this value incorporated in the labour force will be exploited by capital in the production process without shelling out specific money for this.

If we share this perspective, it becomes clear that mobility valorises capital on two levels: directly in the production sphere and indirectly within the reproductive sphere, where the labour force is sustained and reproduced as a resource for the production process. Conversely, it is also clear that it devalorises the labour force since mobility as a natural force of social labour means, on the one hand, the prolonging of the working day for millions of workers, and, on the other hand, the fact that the social costs of work mobility in terms of wasted time, stress and money are paid by workers, not by entrepreneurs or society. This is also true for migrants.

At this point in the discussion, it is interesting to understand whether something has changed in contemporary societies regarding mobility in respect to the past. With industrialisation, the ruling class forced peasants to move from the countryside to factories (Marx, 1964). As for contemporary societies, we can broadly say that at least two aspects have changed. First, this push for mobility has never been as strong as it is today, because workers are pushed to move not only by employers but also by governments that implement labour-market policies that often presume or cause commuting (Ludwig-Mayerhofer and Behrend, 2014). A large part of this mobility is commuting that occurs *at a local or regional level*, but there is a significant number of workers, managers, students and professionals who regularly commute between two countries on the same or even on different continents (see the analysis of "globals" in Elliot and Urry, 2010; Polson, 2011). Second, this push has substantially modified the concept of labour force itself. For many workers, it is no longer sufficient to sell their work capacity, as stated by Marx, but now they also have to sell their capacity to move. Therefore, the recent development of mobility has strong implications also for the concept of labour force itself.

Let us move now to discuss spatial mobility in the reproduction sphere. From an analytic perspective, mobility in everyday life becomes, on the one hand, a good that enters into the basket of goods increasingly consumed by people (Ferrarotti, 1999). Long-distance mobility in the forms of tourism has become a must and often a status symbol for many individuals. But as previously outlined, individual consumption should be considered as a part of the reproduction process of the labour force since it enables labour's participation in production. As such, this type of mobility also has to be considered a natural force of reproductive social labour and as contributing to the process of capital valorisation. On the other hand, as happened to workers, people on the move in their everyday life also become commodities of the transportation industry. The value produced by the means of transportation ends up also being incorporated into their individual consumption (Marx, 1965a: 163). As we shall see, the individuals who travel make the labour force potentially more complex and sophisticated.

Both these forms of mobility – work mobility and mobility in the reproductive sphere – have been enhanced by two factors: first, the development of public means of transportation and the personalisation of transportation means, such as cars, motorcycles and bicycles (Williams, 1974); second, by the massive diffusion and adoption of information and communication technologies, which, with mobility and space, concur to build together the same socio-technical system (Sheller, 2011: 1). These technologies include both mobile media, such as tablet computers, laptops and mobile phones, and fixed technologies, such as desktop computers and landline phones. All of them concur in sustaining the same purposes: communicating, searching and creating information, and being in touch with and supporting many different tasks, operations and activities of everyday life. In this network, the mobile phone, and especially, the smartphone (Fortunati et al., 2013), are at the front edge. Taipale and I defined these digital technologies as the network of personal technologies that forms a kind of meta-medium, a continuum in which old and new media converge around every individual (Fortunati and Taipale, 2017b). This has the implication that for many workers, it is no longer sufficient to sell their work capacity and their capacity to move, but also their capacity to mediate mobility with the network of personal technologies.

Coming back to the general notion of mobility, it is worth noticing that mobility in itself is also changing. As Taipale and I pointed out (2017b, p. 563), "first, many more people move today than in the past; second, the distance travelled has lengthened; and third, the speeds reached by mobility is much higher than before". Hence, current long-distance mobility is characterised by larger-scale and faster displacements, in which the distance travelled is longer than in the past. According to Urry (2011a: XV), if in the nineteenth century, the daily distance travelled by people in the United States was 50 metres, today this distance is 50 kilometres. At the global level, people now travel 23 billion kilometres.

If we speculate on why mobility in the reproductive sphere has increased so much, one reflection is that workers are required today to be mobile as this fulfils capital's need in terms of valorisation; another reflection is that capital has a vested interest in developing macro mobility up to this point, as this increases valorisation. It is less clear why people adhered so enthusiastically to mobility in their daily lives. This will become clear in the next section in which we will analyse the old sociological concept of social mobility that will be able to clarify both the logics of current mobility and the logics that inspire it with respect to mobile sociality.

Theoretical advancements in social mobility

Social mobility is a concept which belongs to the classic sociological tradition and which indicates the movement of individuals, families, households or other categories of people within or between social strata in a society (Sorokin, 1927). It represents a change in individuals' social status caused by their movement within and between social strata in an open system of social stratification. Social mobility means any change in a downward or upward direction or in a horizontal direction (e.g. from plumber to carpenter). It is clear from this description that the mobility put in motion here is not necessarily connected to a spatial move but to a move in individuals' social position and/or professional roles. Social mobility, therefore, implies a symbolic move that should be read more correctly as a change (Savage, 1988). Upward mobility is particularly central to the capitalist system because it justifies its most important ideology, which is based on the fact that each individual can potentially move to a superior position. When the mechanism of upward mobility does not work efficiently (it is not capable of selecting the best human resources for the highest positions), it tends to have depressive effects on society. Recent empirical studies on vertical social mobility point toward disappointing relative rates of social mobility across advanced countries (Breen, 2004; Goldthorpe, 2013).

Fortunati and Taipale (2017a) proposed explaining the growth of spatial mobility as a reaction against the problems and barriers of upward social mobility in postmodern industrialised countries (Breen and Luijkx, 2004). In other words, the increase of spatial mobility can be interpreted as a reaction against the sense of social immobilism brought about by the stagnation of social mobility. A society that remains motionless for too long risks losing vitality. Thus, the development of spatial mobility can be interpreted as a means to compensate and moderate frustrations arising from a lack of social mobility. The increasing recurrence of spatial mobility has been a powerful expedient that has helped people cope with the rigidity of social mobility. People started to travel more intensely for the most varied reasons, from amusement to sightseeing, from meeting different people to increasing their knowledge and understanding of the world, or just for the pleasure of being elsewhere. This behaviour, which seems to increase year after year, can be understood as a grassroots political and social initiative through which people have tried to enrich their lives. It represents for an increasing number of people a great leap forward in expanding the mind, observing other ways of life and experiencing other mentalities and languages. Through the growth of mobility and experiences of the wider world, people have an opportunity to become less parochial.

Obviously, this reaction is not only connected with the barriers to social mobility and is not without inherent problems. As Simmel argues (1900/1991), the expansion of spatial mobility is connected also at the psychological level to the general transformation of the self with modernity, which becomes devoid of a psychologically stable centre. Simmel takes from Marx the concept of alienation and extends it to modern life that, according to him, makes people nervous and thus unstable in their attitudes and behaviours. Simmel develops this concept by writing:

> The lack of something definite at the center of the soul impels us to search for the . . . momentary satisfaction in ever-new stimulations, sensations and external activities. Thus it is that we become entangled in the insatiability and helplessness that manifests itself as the tumult of the metropolis, as the mania for traveling, as the wild pursuit of competition, and as the typical modern disloyalty with regard to taste, style, opinions and personal relations.
>
> (Simmel, 1900/1991: 41)

In this sense, the increased spatial mobility in the reproductive sphere seems to have also accentuated the fragility of the self that has become increasingly uprooted from a space tied to temporal continuity.

In this section, we have examined the concept of social mobility, barriers for social mobility and a reduction in opportunities to change an individuals' social position and/or professional roles; these barriers are also a means of understanding the reasons behind the growth of spatial mobility. The next section will unpack the meaning of sociality and map its main features in social theory.

Theoretical advancements in sociality

As discussed at the beginning of this chapter, the English meaning of the term sociality underpins my argument of the term as value neutral. The related term sociability, while seemingly similar, is in fact a value-laden term meant as a compliment to someone's skills in positive social interactions, for example to say someone is sociable is a compliment to their conversational skills, or hosting abilities. In classic studies of sociality and social relations (see for example Simmel [1908/1998]), sociality is the ensemble of relations or interactions among the individuals who are part of a given society and becomes all the more intense the higher the degree to which individuals associate in social circles and groups (gregariousness) and cooperate together. At the basis of the civilisation process, there is this process of becoming human through the development of sociality with other humans. The tendency to unite with other humans is made even more compelling by the fact that humans are, as Aristotle says in his work *Politics* (Aristotle et al., 1920), also naturally political animals (EN I 5.1097b11; pol. I 2.1253a2-3). Living in a society, they have to develop strategies to manage the common good. From another perspective, sociality can be seen also as a survival response to evolutionary pressures (Smelser and Baltes, 2001; Gallino, 1993: 592).

In German sociology, the discussion about sociability and sociality was furthered by the scholars such as Tönnies (1887), who introduced the dichotomy between community and society. He aimed at distinguishing relationships understood, on the one hand, as forms of immediate and personal interactions based on values, roles and beliefs, which he described as community (Gemeinschaft), and, on the other hand, as indirect interactions based on impersonal roles, formal values and less shared beliefs, which he described as society (Gesellschaft). Weber (1921/2013), answering Tönnies, proposed that while the relationships based on Gemeinschaft can be attributable to subjective attitudes, those based on Gesellschaft are rooted on the rational agreement of mutual consensus. Simmel (1910/1949) wrote an essay specifically on the sociology of sociability, and Von Wiese (1933/1968) stressed the need to see social relationships as processes. Thus, the meanings of sociability and sociality, including the differences, similarities and values, became part of social theory.

In the French world, the contribution of Gabriel Tarde (1890, 1895, 1897, 1898) was fundamental in examining the historical transformation of forms of sociality and in identifying three social laws: imitation, opposition and adaptation. Fundamental also are his reflections on friendship and its change after the development of urbanisation. A few years later, Bouglé (1902) broadened the idea of sociality by proposing to also include in this concept the notions of association and cooperation. By studying pauperism and the life of the working classes, social economy identified education and leisure time as areas where it was possible to impose a model of a controllable and ordered sociality (Gemelli and Malatesta, 1982: 19).

In particular, Simmel's (1910/1949) contribution to the elaboration of sociability is generally considered fundamental, although it focuses on individual interaction. On the contrary, Halbwachs (1933) explored the forms of sociality by analysing those connected to the lifestyle and mentality of the working class. Five years later, Gurvitch (1938) argued that the study of the forms of social interactions should be considered as the starting point for understanding the entire society. He thus observed that they are neither totally spontaneous nor totally organised, and that they are not stable collective units but rather particular ways of being of these units. He concludes by saying that this relative evanescence of the forms of sociality has been the reason that explains, in general, their elusive analysis.

Also relevant to our theme was the experience of Les Annales, which developed a broad vision of everyday life practices, ritualisation and ceremonies. Fortunati et al. (2013) argued that in the forms of sociality, termed by Goffman (1971) as forms of focalised interaction, many social dimensions converge, such as politics (strikes, social movements, forms of resistance, political parties, unions), religion (masses, ceremonies, processions, pilgrimages, feasts, charity institutions), market (markets, fairs, shopping centres, shops), prostitution, playfulness (carnivals, secular and public feasts, games, pubs, taverns, cafés), communication (direct or mediated by analogue and digital technologies), culture and performing arts (from theatre to cinema, from concerts to singing and dancing), and family (visits, celebrations and ceremonies). It is inside these forms of social interactions that fundamental events of daily life take place or situations arise that can evolve in various directions: for example, men and women can meet each other, choose their partner and date him/her; family, relatives and friends can celebrate the birth of a child and rites related to birth, participate in funerals or in weddings, or assist their dearest during illnesses.

One of the key arguments in my chapter is to understand the meaning of sociality in terms of a critique informed by political economy. Sociality, seen as a practice of building and managing social relationships, is characterised by a twofold aspect: immaterial (communication, affects, entertainment) and material (the forms of social interactions based on doing things together with others) and is part of the wide range of typologies of reproductive labour (Fortunati, 1981/1995). The reproductive labour is the labour that all individuals (but especially women) perform in the domestic and social sphere in order to reconstruct their own and others' labour force. Even if this seems to have nothing to do with the economic system, in reality this labour gives rise to very important, even if invisible, valorisation processes. The value produced is incorporated in the labour force of those who cultivate and manage social relationships: not only does the labour force need to be socialised, but it also needs to keep its social being and the inherent abilities well-oiled day by day. As for mobility, we are at the heart of the fundamental mechanisms of value creation.

To develop the notion of – what we then still called – sociability further, Fortunati and colleagues (2013) argued that it is closely interconnected to three elements central to the functioning of every society: communication, labour and mobility. The linkage between sociality and *communication* is evident since it is very difficult – if not impossible – to build social relationships without communicating. The communicative sphere is fundamental for any type of relationship, from intimate relationships to work relationships. Even the primary socialisation processes show this close interrelation between sociality and communication: how would it be possible to accompany a child during their growth without talking to him/her and teaching him/her communication skills? On the other hand, communication is not only needed to elicit sociality, but sociality also promotes communication, as emerged in another research project carried out by Fortunati and colleagues (2013: 34). We argued that communication, as an action with a low output of energy (Braga, 1969), has to be embedded in social activities to be effective and to last over time. For example, doing social activities – such as going to visit friends or relatives at home or receiving them at home, walking, shopping and watching sporting events – helps to keep "communication" alive. In other words, we need to do things with others to nourish the communicative process.

In this work, we have also addressed an important question that was pending in the debate on sociality: has the use of this network of personal technologies increased or decreased sociality? There are two different positions among scholars: some have found that the network of personal technologies has brought an increase in the sphere of social relationships (Wellman, 2001; Kraut et al., 2002; Ling, 2004, 2008; Fischer, 2010). Other scholars have found the contrary, that digital society has brought about a decrease in sociality (Kraut et al., 1998; Nie and Erbing, 2000; Putnam, 2000; Turkle, 2011). The results of our research in 2013 unexpectedly showed that, although there was an increasing use of digital technologies, the amount of face-to-face sociality remained practically unchanged in Europe from 1996–2009. On the basis of the statistical analyses carried out, this finding suggested that the increasing use of digital technologies halted a decline in face-to-face social relations, thus enabling rather than disabling social interaction.

The second element that is interconnected with sociality is *labour* in general and care work in particular or, in more technical terms, reproductive labour (Fortunati, 1981/1995). This is the work needed to reproduce the labour force, both in the sense of giving birth to children and growing future generations and of feeding and clothing workers when not at work. For example, family relationships, which normally are an important part of individuals' sociality, are based on and nourished by reproductive labour. This requires a large share of immaterial labour in which many different tasks converge, such as affect, love, sex, knowledge sharing, entertainment, information and psychological support. All these tasks are conveyed by communication and contribute to constituting family sociality. This labour, which still involves women much more than men, is the spine of the value production in the domestic sphere and of basic sociality for any individual (Fortunati, 1981/1995; Hochschild, 1983). In sum, to build concrete forms of sociality, we need to work in coordination with others and to handle the organisational and logistic features of this work.

The other, third structural element which closely intersects with sociality is *mobility*. To develop our social relations and become part of a society, it is necessary to move beyond family relationships; while family bonds are fundamental, they are not sufficient for the balanced development of people in society. We need to go out and move about in public spaces in order to encounter friends or to meet new people. In this respect, sociality structurally involves mobility. In fact, staying indoors for too long isolates an individual from others. The four walls of the house can turn from a reassuring space into a prison cell.

On the basis of this argumentation, it is evident that sociality is a dynamic sphere in which an intensified logic is applied: individuals feel more reassured and are more reassuring if they are in the conditions and are able to practise social interaction. At the same time, it is evident that the mobility set in motion by sociality in these cases is short-range or, at maximum, medium-range mobility. In general, the previously mentioned forms of mobility are compatible and well supported, and in turn support individuals' sedentariness. These serve to mitigate the substantial meaning of sedentariness, which is the fundamental acceptance of the separation from others and rooting in a well-known place, family and tradition.

What happens with long-range mobility and social interactions? From the research we carried out in Europe (Fortunati et al., 2013), no major innovations emerged in the most common forms of sociality from this research, but there are some changes that are good to acknowledge because they have implications for the present musings. The first big change is that, while in the past, visiting and receiving friends and relatives at home was a frequent social practice, today this kind of sociality tends to move outside the home towards public places, with the consequence that meeting friends and relatives in public spaces doubles mobility because now all the people involved in the meeting have to move. The second relevant change is that the other social practices, such as walking, shopping and watching sporting events, going to restaurants, pubs and dancing, the cinema, theatre and museums, have actually become more widespread in the social body (meaning that more people experience them and enjoy a more varied sociality). They have, however, become less frequent compared to the past: what was previously considered a weekly practice has now become monthly (Fortunati et al., 2013: 902–903). Hence, it can be concluded that the social body is more broadly but less intensively socialised and probably also more homogenised with respect to social styles. On the one hand, sociality seems to have become a less intimate, family-centred and home-based practice, whereas, on the other, mobility receives contrasting pressures towards its contraction and expansion, whereby it is difficult to grasp if in the end mobility connected to the maintenance of social relationships in everyday life has decreased or increased.

Theoretical advancements in mobile sociality

At this point, it is time to understand which features of mobility and society the new concept of mobile sociality integrates. First of all, the concept of mobile sociality includes – above all – the notion of long-distance mobility that consists of long-range displacements (migration and tourism). It does not include short-range mobility because we saw that this type of mobility is presumed and is also included structurally in the notion itself of sociality, along with communication and labour. The analysis of the forms of sociality has revealed an important connection with local mobility. If mobility is short range, it coexists well, or, even better, it is functional to the construction and maintenance of those social relationships that are usually rooted in the place where we live. A large part of our social relations take place in our local and daily lives.

On the contrary, long-distance mobility usually puts in crisis the maintenance of one's own social relationships and the construction of new ones. It characteristically disconnects people from the local space and obliges sociality to deal with spatial distance. This creates wounds and traumas in many cases, because social relationships need spatial continuity to become temporally frequent. Sociality, when it is accompanied by mobility, becomes separation, discontinuity and distance from the social relationships of origin, and it introduces the need to identify strategies in order to cope with these conflicting dimensions. Much research on migrants and new globals has identified how locals are typically hostile to expatriates, so these are often obliged, at least initally, to socialise with other expats (Fortunati et al., 2012). In this regard, mobile sociality is certainly a concept full of tension, given that sociality needs a certain degree of sedentary life-style practices – and when the sedentariness is put into crisis, social relations suffer.

Second, long-distance mobility in the form of tourism, migration and mobile globality has greatly increased, involving a quarter of humanity (Fortunati et al., 2012). This means that we do not talk of a residual phenomenon but of a reality that involves societies all over the world. In addition, long-distance mobility is generative of further long-distance mobility. Empirical research first showed that those who migrate, or the so-called globals, tend to attract their social relations of origins in the new places to which they went (Fortunati et al., forthcoming). Family, relatives and friends often travel to visit their dearest in distant places. This happens, even in other terms, with tourism when tourists play up the places they have visited to their friends and relatives so much that often these individuals are stimulated to visit those places. Second, long-distance mobility is not only the movement from, but risks becoming a perennial movement to – a kind of compulsion to reiterate mobility in other places. Becoming stably mobile is the risk that many globals take. Coming back home or settling stably somewhere often becomes a mirage, a dream, because the interruption especially of the weak ties with a specific territory makes it difficult for them retie those threads, those relationships and that information that, for example, allow you to find a job (Granovetter, 1973). In this context, long-distance mobility becomes a condition of life, since these people are destined to stay in a mobile modality all their life.

Third, with the concept of mobile sociality, we are at the heart of the capital valorisation process. We are talking of a concept which is crucial with respect to the most important sociological categories because it incorporates two elements – mobility and sociality – that are central to the functioning of contemporary societies since they act as productive forces of social labour. As such, they valorise capital both in the production sphere (industrial sectors and advanced services) and in the reproduction sphere (reintegration of the energy of the labour force). This new concept is, however, not only crucial but ambivalent, because it incorporates, as we saw previously, a very negative implication: the consequent devalorisation of the labour force. Furthermore, both these elements – long-distance mobility and sociality – are part of the grassroots political and social initiatives carried out by people to express self-determination in enriching their lives. Thus, mobile sociality simultaneously devalorises and valorises people. This ambivalence should be explored also at the empirical level.

Fourth, both mobility and sociality are supported today by networks of personal technologies. Many research projects have been published about the aspects, both positive and negative, of the use of digital media with respect to social interactions. But the implications of technology use are typically weighed in comparison with combinations of face-to-face and mediated sociality. Along with long-distance mobility, the use of mobile phones, computers and social media does not represent an alternative to in-person social relationships, but the only possibility to stay in touch with family, relatives and friends of origin. Digital technologies such as Facebook, Instagram, TikTok, WhatsApp, FaceTime, Skype, emails, texts and calls offer primary ways to maintain one's social relationships when on the move. However, it is worth noticing that while digital media are fundamental to heal the wounds related to mobility understood as uprooting and geographical distance, they cannot do anything to expand the possibility of socialising beyond verbal and visual communication. On the contrary, digital media compete with in-person social interaction with locals in the new country/city, and in these situations, they show all the pros and cons of their use regarding sociality. Furthermore, they cannot really help in overcoming the cultural distance between migrants, tourists, globals and locals.

Fifth, the study of the forms of mobile sociality is also needed. Everyday social interactions involve going to visit friends or relatives at home or receiving them at home, walking, shopping and watching sporting events, going to restaurants, pubs and dancing, the cinema, theatre and museums, all of which are related to mobility, although to different degrees. There are also social dimensions such as politics, religion, market, prostitution, playfulness, communication, culture and performing arts that flow together. We may ask which social dimensions converge in the forms of mobile sociality. This question also needs empirical research, understanding the material and immaterial aspects of the forms of mobility and sociality.

Discussion

The aim of this chapter was to advance the theoretical reflection on mobile sociality by examining closely the main concepts arising from social theory that concur to define it. The chapter first offered an emphasis on spatial mobility in relation to both work and the reproductive sphere, managing to show that mobility has increasingly become an unacknowledged part of labour – nowadays including digital media use. In another part, I argued that the increasing mobility is also related to a growing lack of social mobility. Last, but not least, I emphasised that the analysis of sociality on the move addresses three elements: communication, labour and mobility. In the concept of mobile sociality, I hope to have shown that these different notions are integrated, allowing not only mobility and sociality to merge, but also the whole range of developments to be included, since the value-laden term sociabilities has been replaced with socialities. We hope that future studies can build on this work.

References

Adey, D. et al. (eds.) (2014) *The Routledge Handbook of Mobilities*. London: Routledge.

Aristotle, Jowett, B., & Davis, H. W. C. (1920) *Aristotle's Politics*. Oxford: Clarendon Press.

Bouglé, C. (1902) *Vie Spirituelle et Action Sociale*. Paris: É. Cornély.

Braga, G. (1969) *La comunicazione sociale*. Roma: Eri.

Breen, R. (ed.) (2004) *Social Mobility in Europe*. Oxford: Oxford University Press.

Breen, R. and Luijkx, R. (2004) "Conclusions", in R. Breen (ed.) *Social Mobility in Europe*. Oxford: Oxford University Press, 383–410.

Elliot, A. and Urry, J. (2010) *Mobile Lives*. London: Routledge.

Ferrarotti, F. (1999) *Partire, tornare. Viaggiatori e pellegrini alla fine del millennio*. Roma: Donzelli.

Fischer, C. S. (2010) *Made in America: A Social History of American Culture and Character*. Chicago, IL: University of Chicago Press.

Fortunati, L. (1981/1995) *The Arcane of Reproduction. Housework, Prostitution, Labour and Capital*. New York: Autonomedia.

Fortunati, L., de Luca, F., and Farinosi, M. (forthcoming) "Mobile sociability and the use of the network of personal technologies". Paper submitted to the journal Information, Communication & Society.
Fortunati, L., Pertierra, R. and Vincent, J. (eds.) (2012) *Migration, Diaspora and Information Technology in Global Societies*. New York: Routledge.
Fortunati, L. and Taipale, S. (2017a) "A different glimpse into mobilities: On the interrelations between daily special mobility and social mobility", *The Information Society*, 33(5): 261–270.
Fortunati, L. and Taipale, S. (2017b) "Mobilities and the network of personal technologies: Refining the understanding of mobility structure", *Telematics & Informatics*, 34(2): 560–568.
Fortunati, L., Taipale, S. and de Luca, F. (2013) "What happened to body-to-body sociability?", *Social Science Research*, 42(3): 893–905.
Gallino, L. (1993) *Sociabilità, Dizionario di Sociologia*. Torino: Utet, 592–593.
Gemelli, G. and Malatesta, M. (1982) *Forme di sociabilità nella storiografia francese contemporanea*. Milan: Feltrinelli.
Goffman, E. (1971) *Relations in Public: Micro Studies of the Public Order*. New York: Basic Books.
Goldthorpe, J. H. (2013) "Understanding – and misunderstanding – social mobility in Britain: The entry of the economists, the confusion of politicians and the limits of educational policy", *Journal of Social Policy*, 42(3): 431–450.
Granovetter, M. S. (1973) "The strength of weak ties", *American Journal of Sociology*, 78(6): 1360–1380.
Gurvitch, G. (1938) *Les formes de la sociabilité. In: Essais de Sociologie*. Paris: Sirey.
Halbwachs, M. (1933) *L'Évolution des Besoins dans les Classes Ouvrières* [The Evolution of Needs in the Working Classes]. Paris: Alcan.
Hochschild, A. R. (1983) *The Managed Heart: Commercialization of Human Feeling*. Berkeley, CA: University of California Press.
Kellerman, A. (2006) *Personal Mobilities*. London: Routledge,.
Kellerman, A. (2012) *Daily Spatial Mobilities. Physical and Virtual*. Farnham: Ashgate.
Kraut, R., Kiesler, S., Boneva, B., Cummings, J., Helgeson, V. and Crawford, A. (2002) "Internet paradox revisited", *Journal of Social Issues*, 58(1): 49–74.
Kraut, R., Patterson, M., Lundmark, V., Kiesler, S., Mukhopadhyay, T. and Scherlis, W. (1998) "Internet paradox: A social technology that reduces social involvement and psychological well-being?", *American Psychologist*, 53(9): 1017–1031.
Larsen, J., Urry, J. and Axhausen, K. (2006) *Mobilities, Networks, Geographies*. Aldershot: Ashgate.
Ling, R. (2004) *The Mobile Connection: The Cell Phone's Impact on Society*. San Francisco: Morgan Kaufmann.
Ling, R. (2008) *New Tech, New Ties: How Mobile Communication is Reshaping Social Cohesion*. Cambridge: MIT Press.
Ludwig-Mayerhofer, W. and Behrend, O. (2014) "Enforcing mobility: Spatial mobility under the regime of activation", *Mobilities*, 10(2): 326–343.
Marx, K. (1964) *Il Capitale* [Capital], Vol. I. Rome: Editori Riuniti.
Marx, K. (1965a) *Il Capitale* [Capital], Vol. II. Rome: Editori Riuniti.
Marx, K. (1965b) *Il Capitale* [Capital], Vol. III. Rome: Editori Riuniti.
Newman, J. and Giardina, M. D. (2014) "Moving biopolitics", *Cultural Studies – Critical Methodologies*, 14: 419–424.
Nie, N. and Erbing, L. (2000) "Internet and society: A preliminary report", *IT and Society*, 1(1): 275–283.
Oishi, S. (2010) "The psychology of residential mobility implications for the self, social relationships, and well-being", *Perspectives on Psychological Science*, *5*: 5–21.
Peters, P., Kloppenburg, S. and Wyatt, S. (2010) "Co-ordinating passages: Understanding the resources needed for everyday mobility", *Mobility*, 5: 349–368.
Polson, E. (2011) "Belonging to the network society: Social media and the production of a new global middle class", *Communication, Culture & Critique*, 4: 144–163.
Putnam, R. D. (2000) *Bowling Alone*. New York: Simon and Schuster.
Savage, M. (1988) "The missing link? The relationship between spatial mobility and social mobility", *British Journal of Sociology*, 39(4): 554–577.
Sheller, M. (2004) "Automotive emotions: Feeling the car", *Theory, Culture and Society*, 21: 221–242.
Sheller, M. (2011) "Mobility", *Sociopedia.isa*. Available at: www.sagepub.net/isa/resources/pdf/Mobility.pdf.
Simmel, G. (1900/1991) "Il dominio della tecnica", in T. Maldonado (ed.) *Tecnica e cultura. Il dibattito tedesco fra Bismark e Weimar*. Milano: Feltrinelli, 37–46.
Simmel, G. (1908/1998) *Soziologie*. Leipzig: Duncker & Humblot (trans. *Sociologia*. Torino: Edizioni di Comunità).

Simmel, G. (1910/1949) "The sociology of sociability", *American Journal of Sociology*, 55(3): 254–261.
Smelser, N. J. and Baltes, P. B. (eds.) (2001) "Evolution of sociality", in *International Encyclopedia of the Social & Behavioral Sciences*. New York: Elsevier, 14506.
Sorokin, P. A. (1927) *Social Mobility*. New York: Harper & Brothers.
Tarde, G. (1890) "Les lois de l'imitation: étude sociologique". Available at: http://classiques.uqac.ca/classiques/tarde_gabriel/lois_imitation/tarde_lois_imitation.pdf
Tarde, G. (1895) "La logique sociale". Available at: http://classiques.uqac.ca/classiques/tarde_gabriel/la_logique_sociale/la_logique_sociale.html
Tarde, G. (1897) "L'opposition universelle: essai d'une théorie des contraires". Available at: http://classiques.uqac.ca/classiques/tarde_gabriel/opposition_universelle/opposition_universelle.html
Tarde, G. (1898) "Les lois sociales: esquisse d'une sociologie". Available at: http://classiques.uqac.ca/classiques/tarde_gabriel/les_lois_sociales/les_lois_sociales.html
Tönnies, F. (1887) *Gemeinschaft und Gesellschaft*. Lipsia: Verlag von Fues.
Turkle, S. (2011) *Alone Together: Why We Expect More from Technology and Less from Each Other*. New York: Basic Books.
Urry, J. (2000) *Sociology beyond Societies. Mobilities for the Twenty-First Century*. London: Routledge.
Urry, J. (2007) *Mobilities*. Cambridge: Polity Press.
Urry, J. (2011a) "Foreword", in G. Pellegrino (ed.) *The Politics of Proximity. Mobility and Immobility in Practice*. Farnham: Ashgate, XV–VII.
Urry, J. (2011b) *Climate Change and Society*. Cambridge: Polity Press.
Von Wiese, L. (1968) *Sistema di sociologia generale*. Translated by Mario Digilio. Torino: UTET. [1933. System der Allgemeinen Soziologie als Lehre von den sozialen Prozessen und den sozialen Gebilden der Menschen (Beziehungslehre)].
Weber, M. (1921/2013) *Economy and Society*. 2 vols. Edited by G. Roth and C. Wittich. Oakland: University of California Press.
Wellman, B. (2001) "Physical place and cyberplace. The rise of personalized networking", *International Journal of Urban and Regional Research*, 25(2): 227–252.
Williams, R. (1974) *Television: Technology and Cultural Form*. London: Collins.

PART II

Valuing mobile socialities

INTRODUCTION PART II

Lynn Schofield Clark

The chapters in Part II of the text address broad questions of *value*. As authors in this section, we reflect on where, when, why or how the relationships between communication technologies, people and places – or the relationships of mobility and sociality – are rendered valuable, and for whom. We explore questions of how socialities are changed for better or worse through our mobile phones or through platforms such as WeChat when mobility is either enacted, forced or curtailed. We consider the extent to which socialities, leveraged through the affordances of digital media platforms, can support social, economic and political developments. Previous scholarship has explored the varied ways that people attach value to their mobile phones – the cult of the iPhone, for instance, or the ways that people personalize their devices through bejewelled phone cases – but rather than media devices, these chapters place socialities and social relations at the centre when considering how people and societies come to attach value within situations of mobile socialities.

How we ascribe value to certain actions, practices or policies, and questions of what can or should happen when those values are undermined, are often raised in the realm of ethics. The chapters in this section consider how individuals are currently making determinations about their own practices related to mobile socialities. As scholars, we need to know more about these 'on the ground' practices and valuations, so that we can then step back and consider how those valuations relate to the values of the common good. We can then better understand, and evaluate, how these may come into conflict, and can therefore raise important questions about the ways that the smooth efficiencies of the platforms are also inadvertently promoting or allowing what we value as important for a functioning, democratic society.

Some of the chapters in this section highlight situations of unequal access, either to resources or to control over one's own mobility. Questions of conflict and contradiction often emerge in these discussions, as does the contingent nature of socialities emerging in relation to situations of mobility. What is of particular interest in these chapters is this: what are the possibilities for well-being, empowerment and the fulfilment of social obligations that emerge within the contexts of mobile socialities? How do new opportunities emerge in mobile socialities that reinforce the value of human life and of meaningful relationships? How is democratic decision-making valued? And how do mobile socialities open new spaces for articulating the value of justice and the reparation of wrongs?

Schofield Clark's Chapter 5 begins Part II by placing at the centre of inquiry questions of how migrants bring value to their new locations by introducing new imaginaries. These new imaginaries come to be as migrants articulate their lived experiences in ways that challenge existing narratives, and thus can call into question the policies and situations that exacerbate the struggles of mobility and immobility. Mobile media allow these narratives to become portable and thus accessible to key audiences of varying sizes, as this chapter explores, thus situating these personalised anchoring narratives as catalysts for both personal and social transformation.

Chapter 6 by Deborah Chambers is similarly concerned with how people from groups experiencing marginalisation place value on relationships. This chapter explores how algorithmic systems ascribe value to prospective intimate partners through dating applications (apps) that utilise geolocative information, and considers how users then negotiate the resulting data about prospective partners in relation to their own value judgements. Chambers further considers some moral dilemmas that emerge in relation to issues of location, introducing a notion of the geo-social flâneur and what she terms geo-enabled intimacies.

Chapter 7 by Jamie Coates then explores questions of value as they emerge and are negotiated in relation to Sino-Japanese informal economies. Like the first and second chapters in Part II, this chapter also explores meanings and imaginings ascribed to the emergent practices of mobile socialities. But this chapter also understands value in relation to the commercial marketplace, considering how the relationships and mobilities play out in relation to Daigu: the informal digital economies that are emplaced between interpersonal, organisational, national and transnational levels of socialities.

Next, Chapter 8 by Elisabetta Costa explores the ways that young Muslim women, valuing both harmony with family members and relationships that take place outside the family circle, leverage digital and mobile communication to negotiate their relationships in ways that reduce conflict with traditional norms and familial expectations. Costa argues that the ways that media create these openings constitute new spaces for what she terms "immobile mobilities": ways for women to claim spaces for agency while remaining within the bounds of what is, for them, largely a setting of immobility.

The final chapter in this section, Chapter 9 by Shuhan Chen and Peter Lunt, considers research into the ways that Chinese university students ascribe value both to higher education and to WeChat (WeXin in Chinese). They find that students leverage the affordances of WeChat as they fulfil traditional social obligations of their home region, and also while they experiment with new ways to enhance relationships with professors and employers in their new settings. WeChat thus mediates and gives shape to the experiences of mobile socialities as these students are able to manage both the relationships here and there that they value.

Each of these chapters foregrounds in various ways the relationships between lived mobile socialities and shared values, including the value of human life and relationships, the value of justice and the reparation of wrongs, and the value of democratic systems that protect the rights of all persons to lead lives with dignity. And as these chapters explore the benefits and costs at play as values clash in settings increasingly understood as contentious, they also help to carve out spaces for future research regarding the moral and ethical dilemmas of new media socialities and mobility.

5

ANCHORING NARRATIVES

Placing narrative in dialogue with the mobile socialities framework

Lynn Schofield Clark

Counsel and advise him, and if he does not listen, then let adversity teach him.

Ishmael opened his digital story by voicing this Ethiopian proverb that his grandmother had often shared with him. Speaking over photos of Ethiopian places that had been meaningful to him, as well as photos of his grandmother and family members gathered there, he described his grandmother as open-minded and as someone who "let me be who I am." "She was the person who listened to me," he said in his story. "She was the person that taught me what was wrong and what's right." Remembering how she smiled and "how that made me feel like I had accomplished something" was a source of continued motivation for Ishmael. His grandmother had wanted him to move to the U.S. in order to pursue education and to achieve his dreams. And so at age 11, Ishmael had left a war-torn situation in his homeland, and, with few economic or social resources, he had begun a new chapter in his life in a new location in the United States.

This chapter places stories like Ishmael's in conversation with *mobile socialities*, a theoretical framework within media studies that foregrounds issues of mobility, social relationships, and communication technologies. In this chapter, I focus on the role of narrative in relation to mobile socialities in order to explore the role that stories can play in healing individuals and societies that have been set adrift in relation to mobility.

The mobile socialities approach explores how mobilities, social relationships, and communication technologies are mediating and reshaping one another as the world's cultures, economies, and populations become increasingly interdependent, and at an ever-greater pace all across the world (Hartmann, 2018). The term *mobile* draws on Tim Cresswell's definition of mobility as "a geographical fact that lies at the centre of constellations of power, the creation of identities, and the microgeographies of everyday life" (2010: 551). Mobility refers to our ability to move through space, and also references the physiological, psychological, emotional, and social costs some must endure in order to make such moves (Fortunati and Taipale, 2017; Morley, 2017). These costs constitute the "darker side of hypermobility" (Cohen and Gössling, 2015), drawing attention to the fears expressed, as well as the complicated workarounds required in order to manage the demands of everyday life in societies in which mobility has become the norm rather than the exception.

How we navigate mobilities and immobilities has, of course, a great deal to do with our social and economic relationships and our lived experiences of connectivity. And thus the concept of *sociality*, as Magnus Andersson develops it, highlights the human desire for relationship and connectivity that, as Jose Van Dijck (2013) has pointed out, is increasingly mediated by technological platforms. A great deal of research has demonstrated that the mobile phone and other forms of digital communication are reshaping migrant and diaspora communities as people on the move maintain connections with their families of origin and create networks of compatriots and locals in the co-construction of new communities (Fortunati et al., 2013; Horst and Miller, 2006; Madianou and Miller, 2013; Pertierra, 2013). Immediate social relations among such communities are vitally important, not only for their emotional value but also for the access such relations provide to employment, housing, child care, transportation, translation, and support for education, health, food, and nutrition; these are the "weak ties" that provide the best opportunities for individual, familial, and societal vitality (Granovetter, 1983).

Narrative is also understood as inherently social. The "narrative turn" arose in interdisciplinary social thought in the 1980s, largely in response to what Jerome Bruner (1991) has characterized as a simplistic approach to the processing of information known as the computational model of the mind. Highlighting the role of meaning as central to the individual mind and to social interaction, Bruner argued that narrative must be understood in relation to communities of practice, as narrative is a form of situated knowledge that emphasizes the agency of social actors. We organize our experience and memory mainly in the form of narrative, and it is through narrative that we come to understand our experiences. The construction of narrative, then, is not merely a matter of selecting events that are worth remembering, or of manufacturing events from fantasy; narrative is about constituting and framing stories in a broader perspective and in relation to specific aims shaped by those with whom we are interacting. "We constantly construct and reconstruct ourselves to meet the needs of the situations we encounter," wrote Bruner (2002: 64). In our darkest days, a narrative can remind us of our best selves: who we are, who loves us, and who we aspire to be. Through narrative, therefore, we are able to transform ourselves and our societies.

In this chapter, I propose the concept of *anchoring narratives* as a way of thinking about the work that certain narratives do to provide security and stability for individuals and communities experiencing the traumas of mobility, immobility, and anti-mobility. In the same way that an anchor secures stability for a ship that is tossed about for reasons beyond its control, anchoring narratives are put to work to provide security and stability as people experience precarity and uncertainty. They are the stories that we tell ourselves about who we are based on who we have been in the past that enable us to conceive of who we can be in the future. They are positive stories of identity that migrants can refer to as they negotiate their social worlds and their new homes, and may include stories of how they have successfully made such negotiations. They are also narratives that may be electronically mediated; they may be, like Ishmael's story, personal stories that have been encoded as media products in the form of videos, audio and music recordings, or photo essays. Assisting young people in the articulation of their own anchoring narratives, it is hoped by those working with them, will give young people a means to reflect on their own struggles as they build resilience and form strong community bonds that will support them in their efforts to address the economic, geographic, and other structural limitations that they face now, have faced in the past, and may encounter in the future. Anchoring narratives, when recorded and shared, then, may help young people to draw connections between their own suffering and that of others so that they can understand themselves as part of a collective that critiques the causes of that suffering, enabling them to work jointly to bring about change.

This chapter will argue that the mobile socialities framework opens ways for media studies to consider aspects of narrative that emphasize resilience, healing, and movement toward positive democratic engagement, which is important in a time of migratory societies. I begin by tracing a few stories of how young people have engaged in the development of anchoring narratives in response to their experiences of mobility and its consequences. I situate these stories in relation to contemporary theories of narrative in order to draw connections between narrative as a means by which people grapple with their lived experiences of the traumas of mobility, and narrative as a means by which newcomers and others who are disenfranchised in relation to mobility recreate political systems and reimagine processes of democratic decision making. I thus consider the ways that these narratives, once technologically mediated and recorded, can serve as catalysts for remembering – and for collective action.

Creating mediated narratives with populations made vulnerable due to mobility

For more than a decade, I have been engaging in work with critical race activists, social workers, public health professionals, and educators in developing media production programs that support leadership development among young people like Ishmael who live in neighborhoods of economic precarity. Conversations about mobility in such settings are fraught, as people living in these neighborhoods have experienced intergenerational trauma as a result of land dispossession, forced refugee resettlement, incarceration, racism and the legacies of U.S. slavery, as well as other forms of systemic oppression. Although those living in and working with these communities often emphasize the remarkable resilience of those living in such situations, it is important to recognize that in general, the young people in these communities have experienced geographic mobility, as well as immobility for that matter, as disruptive and disheartening. Pride in one's current locale is laced with frustrations at the appalling lack of services available in most areas of affordable or publicly available housing, inadequacies that only became more pronounced in the pandemic. And leaving one's home neighborhood, whether because of ever-increasing rent and expenses for the family or for the individual young person's advanced schooling or enlistment in the military, can be disorienting, lonely, and also traumatic.

Ishmael was one of six young people to produce digital stories about resettlement and resilience as part of an after-school digital storytelling program that I co-led with several other people, including Badiah Haffejee, a then-Ph.D. student in social work who herself had relocated to the U.S. from South Africa. Haffejee had special expertise in the integration of newly resettled refugee women and children in the U.S., and had helped to shape the after-school project through her interest in the ways that young people could utilize narrative to develop resiliency and healing from trauma (Haffejee, 2015; Haffejee and East, 2016).

Digital storytelling, as Burgess (2006) has explained, arose as part of a movement that utilizes the tools of scripting, selecting illustrations and music, recording original audio and visual elements, and editing together a brief video as a means of elevating ordinary voices (Lambert, 2002). When utilized in contexts among people who have experienced societal marginalization or dispossession, processes of digital storytelling must be adapted to meet the needs of that particular community (Beltrán et al., 2014; Beltrán and Begun, 2014; Castleden et al., 2008; Dreher et al., 2016; Mallapragada, 2017). Many such communities have long prioritized storytelling as a means of constructing and sharing knowledge grounded in the community (Beltrán and Begun, 2014; Haas, 2015; Kelly, 2018; Njeru et al., 2015). Stories, as Indigenous writers remind us, have been important to "survivance," a word used in Indigenous communities to refer to the forms of poetic and aesthetic expression that speak to the collective experience and that express both survival and resistance (Haas, 2015; Vizenor, 1994). Critical youth scholars in education Shawn Ginwright and Taj James (2002) have similarly argued for the importance of narrative as developed and understood in context, recognizing the need to analyze both internalized and external oppression in the creation of narrative as a means of healing the wounds of oppression (see also Bailey et al., 2019; Barman-Adhikari et al., 2019).

Ishmael and his fellow film creators had chosen to produce their stories as a way of sharing their lived experiences with those in their school who, they believed, harbored many misunderstandings about migration, particularly in relation to the places they had left. They believed that their peers also failed to understand either their reasons for relocating, or their views of and visions for their new settings. And thus, they had wanted to create a "legacy" for the school and for its future immigrant population, as they often explained. When their stories were completed and they were invited to screen their films as part of their high school's end-of-the-year art show, Ishmael and his peers were thrilled. At the event, which took place in the school building and was attended by nearly one hundred of their peers and teachers, Ishmael and his friends celebrated the moment with orange soda and supermarket cookies. They texted their fellow refugee and immigrant peers who soon arrived to congratulate the filmmakers who had been affirmed as agentive artists and creators who successfully gave voice to the immigrant and refugee experience. The filmmakers and their friends sat on top of the school desks as the warm spring air wafted through the school's open windows, screening their stories over and over as peers and teachers asked to view them, laughing freely together and speaking happily about their plans for upcoming graduation parties and the colleges they would be attending in the fall.

The digital stories that Ishmael and his friends chose to create were part of a larger narrative. They sought to resist the regressive narratives too often imposed on them by those who viewed immigrants as victims in need of a "white savior," or who thought of immigrants as freeloaders seeking benefits to which they were not entitled. The students instead shared with their peers and teachers the stories of themselves as cared for by others elsewhere in the world, and stories of being torn from those loved ones largely due to circumstances beyond their control and because of hopes that they and those loved ones held for something better in their lives. They articulated a narrative of aspiration, hope, and achievement, giving voice to the desires they had to make their families and communities proud of them. And they also articulated stories of the roles they wanted to play in their new locations. The stories evoked praise and sparked positive conversations about the experiences of migration that would not have happened otherwise. The stories, and stories about the context in which they were shared, were welded together into anchoring narratives, providing a sense of stability and security as they were put to work for the young people, both in that moment and as these young people experienced precarity and uncertainty in the years that followed. The stories themselves were deeply personal and protective, as was the experience of sharing them in a broader context.

Rahma's story

> Today I'm going to tell you about my neighborhood. I see people often drunk and high. I lived in this neighborhood my whole life, where sometimes I don't feel safe. I wish the police would watch my neighborhood more so I would see less lights and sirens. I wish I could feel more safe, and keep my windows open to feel the fresh air. My neighborhood smells like trash, weed, beer, and skunk. I wish I smelled the fresh flowers that surround the park, or when I walk next door to my grandma's house and she's cooking *boerewors* [fragrant sausage]. Often, I hear the sounds not just of little kids but of drunk people wandering around and parties going late into the night. I wish I could hear the quiet sounds of the early mornings throughout the day.

At 13 years old, Rahma, who was the middle sister in a large family of Somali migrants, was involved in a different after-school program that, like Ishmael's, encouraged young people to tell a story of their lived experiences in order to share it with the communities in which they wanted to make a difference. Through such programs, young people learn to "read and write their worlds," in the words of youth educator and activist Johnny Ramirez (2020). By focusing on her neighborhood, Rahma highlighted the ways that forced mobilities and immobilities and a lack of social services combine to create crowded neighborhoods in which families of resettlement and immigration live alongside people suffering from substance abuse, homelessness, and mental illness. These problems of crowding are exacerbated as nearby gentrifying neighborhoods grow and the stock of affordable housing shrinks.

Rahma's video was part of a one-day exhibition mounted at a city municipal building that brought young media producers into the realm of policymakers and politicians, with the goal of introducing policymakers to Rahma and a group of close to twenty of her peers who wanted their stories to inform local and state policy. During the event, the young people met with their state house representative and a state senator, and close to one hundred people visited the exhibit. A local television station covered the event, highlighting the themes of gentrification, displacement, and entrenched poverty that were prevalent in many of the youth-created representations. The state house representative spoke about the unacceptable conditions in the neighborhoods like Rahma's and, noting that he had grown up in a similarly disadvantaged local community, he mentioned several legislative initiatives he had led that had measurably improved the quality of life for those living in the community he represented.

Rahma's story was aspirational, like Ishmael's, in the sense that it contained her memories and the longing she had for quiet, fresh air, and the good smells of flowers and her favorite foods. The story itself, focused on current problems rather than past remembrances, was not protective in the way that Ishmael's had been. However, when she shared the story in a public place among peers and with powerful people who recognized and shared her concerns, her story was placed into larger contexts of community and of policymaking. Her story, and the context in which it was told, drew attention to the connections between personal experiences of immobility, and larger contexts of the need for political advocates and regulations that safeguard the rights of those whose communities become casualties of the forced mobility of others. When her digital story was recognized by others as meaningful in its expression of shared concerns and as a catalyst for further conversations about legislation and proposed changes, Rahma then had a story about her digital story. The story of the video's screening and her interaction with policymakers became an anchoring narrative of herself as agentive in and capable of articulating and addressing concerns of her community. Like the anchoring narratives of Ishmael and his peers that provided them with a sense of hope and security and a sense that their experiences were heard and valued, Rahma's anchoring narrative gave her a sense of hope that she had played a role, and could continue to play a role, in giving voice to shared experiences in order to bring about positive change for both herself and for many others affected by mobility and immobility.

The role of mediation in anchoring narratives

A video, a recorded song, or even a photo becomes a material object that is separate from its creator and is available for reflection and critique, both over the course of the act of its creation and afterwards when it is viewed in differing contexts and at different points in time. Jose Van Dijck (2007) has argued that we need to rethink memory in light of opportunities for reflection and sharing that have emerged in relation to these technological developments. The digital realm has made stories of existential questions of time, space, identity, and community more accessible than was the case in the past, as McIlwain (2004) has pointed out. And because the memories in our personal archives are often digitized, the past can return to "puncture" the present by means of technological mediation, as Roland Barthes (1981) pointed out about photographs and as Pinchevski (2011) notes of the archived testimonies of Holocaust survivors. Such encounters with mediated representations of traumatic experiences can change not only one's individual response to trauma, but can also change responses on a societal level.

I am arguing that whereas there are many important narratives that migrants can articulate of their shared experiences, it is vitally important for those stories to find an audience, and this finding of an audience is often dependent upon the mediation of migrants' stories. This is because, as Paul Ricoeur (1988: 249) has noted, stories can be "a provocation to be and act differently." Narratives organize experiences and ideas into a meaningful framework that can come to operate as a cultural anchor that provides a "conceptual handle or peg," as Benford and Snow (2000) have argued. In this way, such narratives can connect related ideas to societal responses, playing an important role in what sociologist Ann Swidler (2001: 206) has argued is "the biggest unanswered question in the sociology of culture": "whether and how some cultural elements control, anchor or organize others." Activists create and leverage cultural anchors to promote stability in the midst of changing circumstances in order to provide coherence to social movements, as Ghaziani and Baldassarri (2011) have pointed out. In a similar way, stories of migration and of individual anchoring narratives can become cultural anchors for how societies understand migration and its effects on individuals. These cultural anchors articulate and draw into coherence the concerns about interrelated policies that have created the traumas to begin with. As activists leverage these cultural anchors, they promote stability among movements that articulate how those policies could and need to be changed in order to ensure the human rights of migrants and to provide the social, economic, and political resources necessary for their integration into their new locations.

Ishmael, Rahma, and many of the thousands of other young people who take part every year in programs that teach mediated storytelling skills do so to participate in what Akom et al. (2008) have termed "youthtopias": spaces in which young people can explore and leverage their own stories as a means of highlighting tensions between what is and ought to be as they participate in naming and seeking to eradicate systemic racism and injustice (Chávez and Soep, 2005; Clark and Marchi, 2017; Maira and Soep, 2005). There is thus a normative and ethical dimension to such programs. As Strack et al. (2004: 49) note, "Youth should and need to be given the opportunity to build and confirm their abilities, to comment on their experiences and insights, and to develop a social morality for becoming a positive agent within their communities and society."

Such efforts are rooted in intersectional feminist and critical race theory, approaches committed to the idea that reconceptualizing knowledge requires the expertise of those who, due to interlocking systems of racism and subordination, are traditionally excluded from public life (Bailey et al., 2019; Collins, 1986; Crenshaw, 1990). Whereas most existing research into such programs has identified positive youth development of agency and leadership as primary outcomes (Anyon et al., 2018), such storytelling can have benefits on several levels and for both those creating and those observing the stories, as both Ishmael's and Rahma's stories illustrate.

Narrative is potentially transformative, both for the person whose story becomes part of a narrative and for those for whom the narrative becomes meaningful. Narrative imagination, as Martha Nussbaum (1992) argued, is what enables us to place ourselves in the shoes of another, and therefore acts as a counter to anti-democratic forces in the world. Building on this, Hannah Meretoja (2017) has argued that narratives enlarge (or diminish) the spaces of possibilities in which we can act and reimagine the world with others. Audience is key to the construction of an anchoring narrative, then, because the migrant's story becomes part of a shared experience that broadens out from the protection of the individual to the possibility of shared imagination for a different and better future, both for oneself and for others.

Sometimes, a small audience is all that is needed for a mediated anchoring narrative to serve a protective and catalytic role.

Ishmael lived in a neighborhood very much like Rahma's. And unfortunately, after the end-of-the-year art show which took place during his last year in high school, Ishmael ran into several problems. First, he had received some advice from his track coach that he had misunderstood, and because he had not shared the goal he had formed as a result of that advice, no one was able to help him to realize his misunderstanding until it was too late. And so on his own, Ishmael had worked long into the month of April in order to achieve a race time that would make him eligible for a scholarship to a school in California, only to learn once he had achieved that time that the deadline for the scholarship had long since passed. Several people in his school then scrambled to help him gain admission and a track scholarship at a community college in a neighboring state. But then, in the final weeks of the school year, he stopped attending school altogether. He did not attend the final after-school club's graduation party or the track team's end-of-the-year banquet. Ishmael had loved high school, and his absence at these events was puzzling. One of his school friends learned that his grandmother had become very ill, and his father had returned to Ethiopia, leaving Ishmael alone in their apartment with his disappointment and grief during those final weeks of school. He had taken to smoking marijuana every day, refusing to answer the phone calls of his school friends. The adults in the club also learned that his lack of attendance at school had made him ineligible to participate in the graduation ceremony he had been looking forward to for a long time. When he did not attend his close school friend's graduation party the day after graduation, the adults who attended the party and had been involved with Ishmael in the after-school club discussed what we could do. Among those adults was Haffejee, who, as noted earlier, had personal experience with disruptive migration. With her training in social work and trauma, Haffejee suggested that we should go directly to Ishmael's apartment. Two carloads full of young people and adults headed to the apartment complex where Ishmael lived. When we got to his door, Ishmael did not want to let us in because, as he told his friend in the Ethiopian language of Amharic, he was high. Haffejee then suggested that the younger people with us should go back to the party, and she and I stayed, speaking with Ishmael through the closed door of his apartment. After nearly an hour he finally let us in, and we sat with him in his sadness and disappointment as he told us of his grandmother's declining health and his shame both about his lost opportunity to move to California and his failures at school. Haffejee gently reminded him of the importance of what he had said in his video about his grandmother's role in his life. His grandmother had loved him a great deal, she said, and would want him to keep his promise to her that he would finish his schooling and achieve his goals. He could still complete his degree and go to college, she reminded him. He was encouraged, and calmed as she recalled the aspirations that both he and his grandmother had had for him. And any time he felt like he was wavering, she continued, he could watch that video and be reminded of who he was and what it was that he wanted.

In that moment, Haffejee recalled *as* an anchoring narrative both the video story and Ishmael's experience of having shown it in a supportive community. Whereas the digital story had been a project that had made him proud and had reinforced his positive experiences with his peers, the fact that it was in video form would, she reminded him, enable him to reflect on and return again to a memory of the gravitas of his relationship with his grandmother and the aspirations that both he and his grandmother had held for him.

Rahma's anchoring narrative as an agentive young person had been sparked by the video she created, but it was given weight due to the context in which her story had been heard. Like Ishmael's story, hers too had been created in the context of an after-school environment led by people with both expertise in and lived experiences with trauma. Her video had been viewed as aligned with the wider concerns of members of her community, including activists and agitators as well as elected officials, who had identified needed change and who were working toward greater equity and justice in the neighborhoods of her new city. By watching others watch her video, she was affirmed in the sense that her experience was shared and that she could play a role in naming the need for it to change. "Personal narrative is borne out of experience and gives shape to experience. . . . Narrative and self are inseparable," as Elinor Ochs and Lisa Capps (1996) have observed. The anchoring narratives and the fact that they were mediated thus allowed for the illumination of and reflection upon past observations, present experiences, and future aspirations for the self. As such, both the process of evoking anchoring narratives and the media products that made those narratives available for others became part of the project of healing, resilience, and resistance necessary in relation to the traumas of mobility for both the individual and, potentially, for the larger community.

Anchoring narratives as a way of rethinking relationships between media and trauma

Narrative, I have argued, holds a potentially catalytic role for individuals and for society. And yet, narrative – and in particular, narrative therapy – seems to have fallen out of favor in recent years. In his work on media and the mediation of trauma, Pinchevski (2019) has observed that with the enthusiasm for creating technological efficiencies to solve social problems, we are currently seeing a shift away from narrative in favor of cognitive behavioral therapies and emergent approaches employing virtual reality. And indeed, in recent times, some have argued that there is great potential for mobile apps to advance well-being, particularly in relation to what has been termed internet or cognitive behavioral therapy (iCBT or cCBT) that encourages young people to engage in relaxation techniques and other behavioral modifications to address themselves to the stresses of everyday life (Ito et al., 2020). Technologies can create efficiencies, which may seem attractive in situations like those of the U.S. where organizations employing persons trained in trauma-informed practices are usually overburdened and underfunded. But several reviews of the existing research into such iCBT interventions have noted that the research into mental health intervention apps is at present inconclusive, and such interventions may not be effective for persons with more complex mental health needs such as those related to trauma (Hill et al., 2018; Hollis et al., 2017). Moreover, scholars have underscored the importance of community involvement in resilience building and well-being, emphasizing the role played by relationships that are formed with trained and empathic mentors and intermediaries, whether in person or via social media platforms (Hendry et al., 2017; McCosker, 2018). Despite the enthusiasm for technologically mediated solutions to trauma, then, studies suggest the need for a more fulsome understanding of the contexts in which interventions for well-being take place, pointing to the enduring importance of socialities, and of narrative, in relation to therapy and healing.

Surprisingly, however, whereas scholars in media studies have devoted a great deal of attention to the role of media platforms and tools in social movements, relatively little energy has been devoted to critically considering the ways that media platforms and tools are being employed increasingly in public health and in therapeutic interventions (exceptions include Hendry et al., 2017; McCosker, 2018). The role of media production in intervention has been a topic of increased interest in sister fields such as public and clinical health in recent years, however (Buchbinder et al., 2005; Byrne et al., 2018; Matthews and Sunderland, 2017; Rich et al., 2000, 2002; Rich and Patashnick, 2002). Media studies scholars do question pop psychology and point out the dangers of substituting the therapeutic for the political (Illouz, 2008; Peck,

2015). Therapy surely is put to work for neoliberalism, as these and other scholars note. Yet this way of viewing therapy may have left us ill equipped to think about the lived experiences of poverty in relation to trauma, suffering, and pain, which in turn limits our abilities to consider media in relation to the traumas of mobility. I argue that a mobile socialities approach that foregrounds the importance of narrative can provide a corrective to the current shift in practice toward internet and cognitive behavioral interventions and away from narrative, and can also contribute a different perspective on therapy that is grounded in critical race and feminist lived experiences. Pinchevski makes a relevant point here:

> While supposedly superfluous clinically, the exclusion of talk has significant moral consequences. For it is through talk that one is able to address one's experience to another and thereby take a stand with respect to violence and suffering, one's own as well as another's. Engaging with trauma discursively takes it out from clinical considerations and into social and moral concerns.
>
> (Pinchevski, 2019: 138)

Interestingly, scholars in the fields of public health and social work increasingly view civic engagement as not an afterthought to health, but as *constitutive* of it, as Ballard and Ozer (2016) have argued. This suggests a need to rethink the role of narrative interventions in relation to media and trauma.

In media studies, scholars have been concerned with retraumatization and with poverty's saleable shock value, as poverty is exoticized and packaged for sale in reality television, drama, social media memes, or even in journalism in a form that can devolve into "poverty porn" (Brooker et al., 2015; Ouellette, 2017). This exoticization of poverty and of trauma occurs because media "trade in otherness," as Roger Silverstone (2007) argued, "in the spectacular and the visible," exaggerating otherness or eliminating it. What we need, Silverstone argued, is a media system governed by what he termed an "ethic of hospitality" which guarantees that those with less power are given the opportunity to be seen and heard on their own terms. Such an ethic also requires a more extensive program of media literacy, for media regulation alone cannot help citizen-viewer/readers to appreciate the non-transparency of the media and the moral implications of its non-transparency. We need to consider the limitations of the cosmopolitan imagination, as John Durham Peters (2015) has argued, and as we do this, we are compelled to take a moral stance and to act in response (Peters, 2001). Part of this response may rest in the possibilities for compassion that can emerge out of new imaginaries (Hoijer, 2004).

U.S. sociologists have lambasted therapeutic culture, Arlene Stein (2011) argues, as they have assumed that "when therapeutic discourse reigns, political participation suffers, the citizen is displaced by the consumer/manager, and introspection, rather than communal obligation, becomes an end unto itself." Stein further observes that a different critique of therapy took root in the U.K.: not that it unleashed a focus on the self, but that it subjected selfhood to professional expertise and power. Following Nicholas Rose's (1990) powerful critique in *Governing the Soul* and Eva Illouz's (2008) critique of popular psychology, Stein argues, sociologists assumed that "power had come to permeate subjectivity" and had "diminished the self by subjecting it to rationalization and control" (Stein, 2011: 188). Yet, Stein points out, the reality on the ground seemed to be quite different:

> Rather than defuse political aspirations, the popularization of therapeutic discourse frequently accomplished the opposite, at times energizing social movements, opening up discursive spaces, and empowering individuals who had previously been

> marginalized. . . . It helped facilitate the rise, among other things, of movements for sex education, and those against rape, child sexual abuse, and other forms of violence that had before been silenced, opened up space for the acknowledgement of shame and suffering, and helped make "victimhood" a source of moral authority.
>
> (p. 189)

Activists confronted sexual shame, Stein argues, and viewed the transformation of selfhood as a political act: one that enabled victims to see themselves as survivors who could publicly testify in their own defense and change public opinion and policy in the process. Stories of migration that are mediated and shared, like those of Ishmael and Rahma, have similar potential to change public opinion and to open up new ways for thinking about policy and political processes.

Articulating a narrative of "one who has suffered, survived, and surpassed" (Stein, 2011: 192 quoting Plummer, 1995: 76) is important, as it enables individuals to address stigma, develop and articulate resistance, and nurture resilience that can be called upon to work against injustices faced by oneself and one's community. Stein then makes this key argument:

> Rather than leading to de-politicization, narrative communities have emerged, linked therapy and politics, the inner and the outer, individualistic and communal aspirations. Therapeutic politics encourages individuals, and the groups they are part of, to display painful feelings in public rather than hold them back, and it can prod them to utilize those feelings as a way of mobilizing politically.
>
> (Stein, 2011: 192)

Politics, and our interactions with governance, will always be in part related to the relentless quest for personal meaning, and young people – all of us, really – must learn to "feel our way into politics," as Zizi Papacharissi (2015) has argued. Papacharissi notes that we become sutured into affective publics as we recognize the shared emotional responses to our common situations. Narratives that can evoke such emotions are full of possibility for reshaping what Henry Jenkins and his colleagues have called the "civic imagination." (Jenkins et al., 2016)

Eighteen percent of U.S. adolescents live in poverty, with multiracial and Latinx youth comprising a rapidly growing segment of the U.S. population (Haffejee and East, 2016). Black communities like the ones in which Ishmael and Rahma live have been long denied opportunities to thrive due to the long-term disinvestment in the U.S. social safety net that has rendered health care, affordable housing, education, and employment inaccessible. We need to be mindful of the limits of such media interventions while recognizing that voice is key to civic agency in media studies. But exploring the role of such anchoring narratives over time may help us to better understand how young people who have experienced trauma are able to draw upon resilience and utilize their own experiences of trauma as a platform for the collective recognition of injustices that then result in collective civic action.

Thus, while we must recognize that narratives are never sufficient on their own, we also need to acknowledge that anchoring narratives may be a catalyst for civic engagement, and may be important for the fostering of the civic imaginary among young people living through the traumas of mobility and immobility. When youth choose to articulate anchoring narratives in the context of programs of media literacy, then, I argue, following Paul Mihailidis (2018), media competency development is brought together with aims of political involvement. Such efforts must be rooted in an ethic of caring, Mihailidis argues, which necessitates interventions that seek to address and integrate experiences of trauma into the narratives of one's past, present, and future.

The mobile socialities framework foregrounds the traumas of mobility and migration, and highlights the need for media scholars to engage in conversations with public health practitioners in relation to the use of media-making in activities of protection, prevention, well-being, and civic engagement. We in media studies can learn and engage with the ethical duties that come into play when interacting with vulnerable populations. And those in the fields of public health, working with those in media studies, can further study the perceived costs, as well as the possible benefits, of the making public of these stories as they consider the risks associated with publicness (Lee, 2020; Marwick et al., 2017; Marwick and boyd, 2014, 2018).

In this chapter, I have observed that programs of digital storytelling can be especially effective in inviting the creation of anchoring narratives. These anchoring narratives are addressed to our existential needs for constancy and persistence, which are especially important in a time in which digital life is largely experienced as a series of ever-new outrages, and daily life is defined by continuous upheaval and uncertainty. Social workers and public health officials have introduced this practice of creating anchoring narratives as a form of therapeutic healing in relation to trauma, and I have argued that those in media can contribute to further developing and testing the concept of anchoring narratives, particularly when they are mediated. By highlighting the media product itself, with its ability to engender recollections both of a time long past and of a time in which the product itself was created, I have demonstrated the ways in which an anchoring narrative maintains its therapeutic potential for future times that are not yet known to its creator. The production of an effective anchoring narrative may be an enduring part of the tool kit for resilience. As is clear given Ishmael's story, this is a tool that may be helpful but not sufficient for the building of resilience among at-risk young people. Future research is needed to explore how the media product itself may function in relation to prospective therapeutic endeavors in resilience building for its creator.

Conclusion: theorizing anchoring narratives

> Narrative is radical, creating us at the very moment it is being created.
>
> *(Toni Morrison, Nobel Prize Lecture, 1993)*

As Margaret Somers (1994: 606) noted, we "come to be who we are (however ephemeral, multiple, and changing) by being located or locating ourselves (usually unconsciously) in social narratives." This chapter has highlighted the role of narrative in stories of mobility, and has introduced the concept of anchoring narratives, in order to make three important contributions to the mobile socialities framework.

First, the foregrounding of narrative recognizes that those who are most vulnerable need social spaces to bring their own stories into a narrative framework. The therapeutic, or the ability to heal, is understood as central to being human and to claiming agency in our increasingly mobile lives and societies. The concept of anchoring narratives therefore promises to bring to the field of mobile socialities an approach that unites theories of narrative from both cultural sociology and cultural anthropology (Ochs and Capps, 1996; Somers, 1994; Swidler, 2008) with insights on sociality as understood in relation to the human desire for relationship and connectivity that is increasingly mediated by platforms (Van Dijck, 2013), and in relation to insights on the fetishism of mobility (Polson, 2019), as it plays out in the lives of those who experience forced mobility and displacement.

Second, the concept of an anchoring narrative underscores the importance of not only encouraging the telling of migration stories (important as that is), but also the creating of spaces in which those stories can become meaningful for others. This is what Couldry (2010) means by "voice": people need both an opportunity for the expression of a distinctive perspective on the world, and they – we – need for that perspective to be acknowledged. Technologically mediated and media content creation interventions for amplifying voice have been engaged, and increasingly are being tested, as strategies for the prevention of negative outcomes among vulnerable youth populations. Drawing on critical race, Indigenous epistemologies, and critical youth research, I have argued that media-making strategies of prevention are particularly important to research in relation to goals of enhanced individual agency and civic engagement among communities that have been marginalized, including those of refugees, immigrants, and asylum seekers. More research is needed to better understand the roles that creating and expressing narratives can play in relation to the goals of people in these communities.

Third, examples like those relayed in this chapter remind us that the ability to put together a narrative is not a given. It is a skill that can be learned. We are largely unaware of how existing narratives and narrative traditions regulate how we narrate the past, understand the present and imagine and orient ourselves to the future, as Meretoja (2017) observed. As we become aware of the stories that are culturally available to us and that lead us to narrate our experiences in certain ways, we can imagine other narratives. It therefore can be therapeutic, and empowering, to be able to formulate a narrative that brings together experiences of the past with aspirations for the future, particularly when that narrative finds an audience and contributes to positive change. Examining the processes of working with migrants to tell their stories helps to raise important questions such as: which social groups need to hear these narratives? How can these narratives be mobilized for the individual as well as for the social good? Who listens, and what is the responsibility to do something with what we have heard?

There are therefore three consequences of developing a theory of a particular kind of narrative, an anchoring narrative, for the interdisciplinary fields related to media studies.

First, the focus on narrative in the mobile socialities framework invites scholars to center the experiences of migration in conversations about citizenship and political engagement. It takes seriously the need to challenge the traditional starting point of human stasis in relation to the state. Following Nail's (2015) admonition that we understand political theory as a politics of movement, this chapter has proposed that we view migration as the rule rather than the exception, and that we take seriously the traumas involved and the social as well as the physical and economic needs that emerge as a result. The mobile socialities framework looks at what these experiences of migration and of trauma mean for those living with various consequences of that movement, particularly in light of the need for spaces that demand that those most affected be heard.

Second, the concept of an anchoring narrative builds on the idea that shared trauma can be a foundation for collective action. We connect with one another through our vulnerabilities. Then, as we come to see ourselves and others we care about as having agency both over our own stories and over how those stories can be leveraged to seek change, we embrace a narrative of ourselves as politically agentive. An anchoring narrative about one's past exercise of political agency, when united with other narratives of trauma and of demands for response, can thus become a platform for collective action. I have therefore argued that we in media studies would do well to recognize the need for an acknowledgement of traumas associated with mobility as a common starting point to civic engagement, particularly among young people who have experienced traumas such as those related to forced mobility and resettlement. We also need to acknowledge that narratives of migration are too often inaccessible, particularly by those involved in policy debates; they are also unlikely to be easily translated into policy discussions. There is a need, therefore, for intermediaries who can help those stories to find the right audience at the right time. This involves challenging the broadcast, one-to-many model of communication and introducing a different model of communication that is networked and mobile – and that therefore can consider the smallest socialities that can serve as audiences to the narratives of those most in need of being heard. Such an approach affirms that it is our collective and individual narratives – rather than our technologies – that constitute the connective tissue of our relationships, and of our societies. People trained in understanding processes of mediation are well-placed to contribute to reconceptualizing and understanding how narratives of trauma come to relate to collective action.

Third, foregrounding the particular kind of anchoring narrative that is therapeutic for those who have experienced traumas related to migration allows us to conceptualize the social conditions that gave rise to an industry that has monetized narrative interventionist approaches and has thus constrained the view of what narrative is and does. Interventionist approaches emerged historically in relation to progressive movements, and these movements have approached communication in particular, applied ways. Future studies can explore these relationships to provide better contexts for evaluating those that are emerging today.

A focus on anchoring narratives therefore opens an agenda for future scholars to trace these relationships of movement, trauma, and media-rich narrative interventions. Some techniques of narrative intervention have fallen out of favor over time, whereas others have persisted and given shape to the ways that people experiment with mediated narratives today. This history is worth tracing, too, in order to better understand the interrelations of communication technologies with human relationships and social change.

This chapter has called attention to the fact that migrants are enacting political change both through the physical actualizing of their mobility and as they give shape to what I have termed anchoring narratives: narratives that relate the experiences of migration to both their personal and their collective aspirations for a better future. As we are now entering a time in history when unprecedented migration is anticipated due to climate change, understanding the role of narrative and media in relation to both mobility and to shared governance in shared spaces is more important than ever. Scholarship that is opening new spaces for the consideration of media, trauma, and politics is important terrain for future research. We have an obligation, as Ricoeur has argued, to record history that follows "the plot of suffering rather than that of power and glory" (1991: 74). Such work will, we can hope, open spaces for imagining, and narrating, new possibilities.

References

Akom, A. A., Cammarota, J. and Ginwright, S. (2008) "Youthtopias: Towards a new paradigm of critical youth studies", *Youth Media Reporter*, 2(4): 1–30.

Anyon, Y., Bender, K., Kennedy, H. and Dechants, J. (2018) "A systematic review of youth participatory action research (YPAR) in the United States: Methodologies, youth outcomes, and future directions", *Health Education & Behavior*, 45(6): 865–878. https://doi.org/10.1177/1090198118769357

Bailey, J., Steeves, V., Burkell, J., Regan Shade, L., Ruparelia, R. and Regan, P. (2019) "Getting at equality: Research methods informed by the lessons of intersectionality", *International Journal of Qualitative Methods*, 18. https://doi.org/10.1177/1609406919846753

Ballard, P. J. and Ozer, E. (2016) "The implications of youth activism for health and well-being", in J. Conner and M. Rosen (eds.) *Contemporary Youth Activism: Advancing Social Justice in the United States*. Westport, CT: Praeger.

Barman-Adhikari, A., DeChants, J. P., Brydon, D. M., Portillo, A. and Bender, K. (2019) "On the fringes: How youth experiencing homelessness conceptualize social and economic inequality – a photovoice study", *Journal of Community Psychology*, 47(4): 924–942. https://doi.org/10.1002/jcop.22164

Barthes, R. (1981) *Camera Lucida: Reflections on Photography*. New York: Hill and Wang.

Beltrán, R. and Begun, S. (2014) "'It is medicine': Narratives of healing from the Aotearoa digital storytelling as indigenous media project (ADSIMP)", *Psychology and Developing Societies*, 26(2): 155–179. https://doi.org/10.1177/0971333614549137

Beltrán, R., Olsen, P., Ramey, A., Klawetter, S. and Walters, K. (2014) "Digital tapestry: Weaving stories of empowerment with Indigenous youth", in P. McCardle and V. Berninger (eds.) *Narrowing the Achievement Gap for Native American Students: Paying the Educational Debt*. London: Routledge.

Benford, R. and Snow, D. (2000) "Framing processes and social movements: An overview and assessment", *Annual Review of Sociology*, 26: 611–639.

Brooker, P., Vines, J., Sutton, S., Barnett, J., Feltwell, T. and Lawson, S. (2015) "Debating poverty porn on Twitter: Social media as a place for everyday socio-political talk", *Proceedings of the 33rd Annual ACM Conference on Human Factors in Computing Systems*, 3177–3186. https://doi.org/10.1145/2702123.2702291

Bruner, J. (1991) "The narrative construction of reality", *Critical Inquiry*, 18(1): 1–21. https://doi.org/10.1086/448619

Bruner, J. (2002) *Making Stories: Law, Literature, Life*. Cambridge: Harvard University Press.

Buchbinder, M. H., Detzer, M. J., Welsch, R. L., Christiano, A. S., Patashnick, J. L. and Rich, M. (2005) "Assessing adolescents with insulin-dependent diabetes mellitus: A multiple perspective pilot study using visual illness narratives and interviews", *Journal of Adolescent Health*, 36(1): 71.e9–71.e13. https://doi.org/10.1016/j.jadohealth.2004.02.019

Burgess, J. (2006) "Hearing ordinary voices: Cultural studies, vernacular creativity and digital storytelling", *Continuum*, 20(2): 201–214. https://doi.org/10.1080/10304310600641737

Byrne, E., Elliott, E., Saltus, R. and Angharad, J. (2018) "The creative turn in evidence for public health: Community and arts-based methodologies", *Journal of Public Health*, 40(suppl_1): i24–i30. https://doi.org/10.1093/pubmed/fdx151

Castleden, H., Garvin, T. and First Nation, H. (2008) "Modifying photovoice for community-based participatory Indigenous research", *Social Science & Medicine*, 66(6): 1393–1405. https://doi.org/10.1016/j.socscimed.2007.11.030

Chávez, V. and Soep, E. (2005) "Youth radio and the pedagogy of collegiality", *Harvard Educational Review; Cambridge*, 75(4): 409–434, 488.

Clark, L. S. and Marchi, R. (2017) *Young People and the Future of News: Social Media and the Rise of Connective Journalism*. Cambridge: Cambridge University Press.

Cohen, S. A. and Gössling, S. (2015) "A darker side of hypermobility", *Environment and Planning A: Economy and Space*, 47(8): 166–1679. https://doi.org/10.1177/0308518X15597124

Collins, P. H. (1986) "Learning from the outsider within: The sociological significance of Black feminist thought", *Social Problems*, 33(6): s14–s32. https://doi.org/10.2307/800672

Couldry, N. (2010) *Why Voice Matters: Culture and Politics After Neoliberalism*. London: Sage.

Crenshaw, K. (1990) "Mapping the margins: Intersectionality, identity politics, and violence against women of color", *Stanford Law Review*, 43(6): 1241–1300.

Cresswell, T. (2010) "Towards a politics of mobility", *Environment and Planning D: Society and Space*, 28(1): 17–31. https://doi.org/10.1068/d11407

Dreher, T., McCallum, K. and Waller, L. (2016) "Indigenous voices and mediatized policy-making in the digital age", *Information, Communication & Society*, 19(1): 23–39. https://doi.org/10.1080/1369118X.2015.1093534

Fortunati, L., Pertierra, R. and Vincent, J. (2013) *Migration, Diaspora and Information Technology in Global Societies*. London: Routledge.

Fortunati, L. and Taipale, S. (2017) "A different glimpse into mobilities: On the interrelations between daily spatial mobility and social mobility", *The Information Society*, 33(5): 261–270. https://doi.org/10.1080/01972243.2017.1354110

Ghaziani, A. and Baldassarri, D. (2011) "Cultural anchors and the organization of differences: A Multimethod analysis of LGBT marches on Washington", *American Sociological Review*, 76(2): 179–206. https://doi.org/10.1177/0003122411401252

Ginwright, S. and James, T. (2002) "From assets to agents of change: Social justice, organizing, and youth development", *New Directions For Youth Development: Theory, Practice, Research*, 96: 27–46.

Granovetter, M. (1983) "The strength of weak ties: A network theory revisited", *Sociological Theory*, 1: 201. https://doi.org/10.2307/202051

Haas, A. (2015) "Toward a decolonial digital and visual American Indian rhetorics pedagogy", in L. King, R. Gubele and J. R. Anderson (eds.) *Survivance, Sovereignty, and Story: Teaching American Indian Rhetorics*. Logan: Utah State University Press, 188–208.

Haffejee, B. (2015) "African refugee youths' stories of surviving trauma and transition in U.S. public schools", *Journal of Muslim Mental Health*, 9(1). https://doi.org/10.3998/jmmh.10381607.0009.101

Haffejee, B. and East, J. F. (2016) "African women refugee resettlement: A womanist analysis", *Affilia*, 31(2): 232–242. https://doi.org/10.1177/0886109915595840

Hartmann, M. (2018) "Mobilism in translation: Putting a new research paradigm to the test", in *Geomedia Studies: Spaces and Mobilities in Mediatized Worlds*. Routledge. Available at: https://books.google.com/books?hl=en&lr=&id=RNErDwAAQBAJ&oi=fnd&pg=PT196&dq=info:BPUa3rAGTnkJ:scholar.google.com&ots=5CUjg9ohge&sig=IIPjySca8xQznz7H2LnYFfPM3UE#v=onepage&q&f=false

Hendry, N., Robards, B. and Stanford, S. (2017) "Beyond social media panics for "at risk" youth in mental health practice", in S. Stanford, E. Sharland and N. R. Heller (eds.) *Beyond the Risk Paradigm in Mental Health Policy and Practice*. New York: Palgrave Macmillan.

Hill, C. et al. (2018) "Navigating the development and dissemination of internet cognitive behavioral therapy (ICBT) for anxiety disorders in children and young people: A consensus statement with recommendations from the #iCBT Lorenz workshop group", *Internet Interventions*, 12: 1–10.

Hoijer, B. (2004) "The discourse of global compassion: The audience and media reporting of human suffering", *Media Culture & Society*, 26(4): 513–531.

Hollis, C. et al. (2017) "Annual research review: Diital health interventions for children and young people with mental health problems – A systematic and meta-review", *Journal of Child Psychology and Psychiatry*, 58(4): 474–503.

Horst, H. and Miller, D. (2006) *The Cell Phone: An Anthropology of Communication*. Oxford: Berg.

Illouz, E. (2008) *Saving the Modern Soul: Therapy, Emotions, and the Culture of Self-Help*. Berkeley, CA: University of California Press.

Ito, M., Odgers, C. and Schueller, S. (2020) *Social Media and Well-being: What We Know and Where We Could Go*. Irvine, CA: Connected Learning Alliance.

Jenkins, H., Shresthova, S., Gamber-Thompson, L., Kligler-Vilenchik, N. and Zimmerman, A. (2016) *By Any Media Necessary: The New Youth Activism*. New York: New York University Press.

Kelly, L. L. (2018) "A snapchat story: How black girls develop strategies for critical resistance in school", *Learning, Media and Technology*, 43(4): 374–389. https://doi.org/10.1080/17439884.2018.1498352

Lambert, J. (2002) *Digital Storytelling: Capturing Lives, Creating Community*. 1st edition. Life on the Water. London: Routledge.

Lee, A. (2020) *Youth, Media Activism, and Communication Counterpower: A Comparative Study*. Cambridge: Harvard University Press.

Madianou, M. and Miller, D. (2013) *Migration and New Media: Transnational Families and Polymedia*. London: Routledge.

Maira, S. and Soep, E. (2005) *Youthscapes: The Popular, the National, the Global*. Philadelphia: University of Pennsylvania Press.

Mallapragada, M. (2017) "Immigrant activism: Narratives of the 'H-4 life' by Indian women on YouTube", *Communication, Culture & Critique*, 10(1): 76–92. https://doi.org/10.1111/cccr.12147

Marwick, A. E. and boyd, d. (2014) "Networked privacy: How teenagers negotiate context in social media", *New Media & Society*, 16(7): 1051–1067. https://doi.org/10.1177/1461444814543995

Marwick, A. E. and boyd, d. (2018) "Understanding privacy at the margins", *International Journal of Communication*, 12: 1157–1165.

Marwick, A. E., Fontaine, C. and boyd, d. (2017) " 'Nobody sees it, nobody gets mad': Social media, privacy, and personal responsibility among low-SES youth", *Social Media + Society*, 3(2): 2056305117710455. https://doi.org/10.1177/2056305117710455

Matthews, N. and Sunderland, N. (2017) *Digital Storytelling in Health and Social Policy: Listening to Marginalised Voices*. London/New York: Taylor & Francis.

McCosker, A. (2018) "Engaging mental health online: Insights from beyond Blue's forum influencers", *New Media & Society*, 20(12):, 4748–4764.

McIlwain, C. D. (2004) *When Death Goes Pop: Death, Media and the Remaking of Community*. Peter Lang Inc., Bern: International Academic Publishers.

Meretoja, H. (2017) *The Ethics of Storytelling: Narrative Hermeneutics, History, and the Possible*. Oxford: Oxford University Press.

Mihailidis, P. (2018) *Civic Media Literacies*. 1st edition. London: Routledge.

Morley, D. (2017) *Communications and Mobility: The Migrant, the Mobile Phone, and the Container Box*. Hoboken, NJ: John Wiley,.

Nail, T. (2015) *The Figure of the Migrant*. Redwood City, CA: Stanford University Press.

Njeru, J. W., Patten, C. A., Hanza, M. M. K., Brockman, T. A., Ridgeway, J. L., Weis, J. A., Clark, M. M., Goodson, M., Osman, A., Porraz-Capetillo, G., Hared, A., Myers, A., Sia, I. G. and Wieland, M. L. (2015) "Stories for change: Development of a diabetes digital storytelling intervention for refugees and immigrants to Minnesota using qualitative methods", *BMC Public Health*, 15(1): 1311. https://doi.org/10.1186/s12889-015-2628-y

Nussbaum, M. (1992) *Love's Knowledge: Essays on Philosophy and Literature*. Oxford: Oxford University Press.

Ochs, E. and Capps, L. (1996) "Narrating the self", *Annual Review of Anthropology*, 25(1): 19–43. https://doi.org/10.1146/annurev.anthro.25.1.19

Ouellette, L. (2017) "How the other half lives: The will to document from poverty to precarity", in J. Deery and A. Press (eds.) *Media and Class*. Routledge. Available at: www.academia.edu/35113397/How_the_Other_Half_Lives_The_Will_to_Document_from_Poverty_to_Precarity

Papacharissi, Z. (2015) *Affective Publics: Sentiment, Technology, and Politics*. Oxford: Oxford University Press.

Peck, J. (2015) *Age of Oprah: Cultural Icon for the Neoliberal Era*. London: Routledge.

Pertierra, R. (2013) "Diasporas, the new media and the global homeland", in L. Fortunati, R. Pertierra and J. Vincent (eds.) *Migration, Diaspora and Information Technology in Global Societies*. London: Routledge.

Peters, J. D. (2001) "Witnessing", *Media, Culture and Society*, 23(6): 707–723.

Peters, J. D. (2015) *The Marvelous Clouds: Toward a Philosophy of Elemental Media*. Chicago: University of Chicago Press.

Pinchevski, A. (2011) "Archive, media, trauma", in M. Neiger, O. Meyers and E. Zandberg (eds.) *On Media Memory*. New York: Palgrave Macmillan, 253–264. https://doi.org/10.1057/9780230307070_19

Pinchevski, A. (2019) *Transmitted Wounds: Media and the Mediation of Trauma*. Oxford: Oxford University Press.

Plummer, K. (1995) *Telling Sexual Stories*. New York/London: Routledge.

Polson, E. (2019) "Information SuperCalle: The social internet of Havana's Wi-Fi streets," in R. Wilken, G. Goggin and H. A. Horst (eds.) *Location Technologies in International Context*. London: Routledge.

Ramirez, J. (2020) "Class Presentation, Hip Hip as Medicine", University of Denver.

Ricoeur, P. (1991) "Narrative identity", *Philosophy Today*, 35(1): 73–81.

Ricœur, P. (1988) "L'identité narrative", *Esprit (1940-)*, (140/141 (7/8)): 295–304. Available at: http://www.jstor.org/stable/24278849 (accessed January 12, 2021).

Rich, M., Lamola, S., Gordon, J. and Chalfen, R. (2000) "Video intervention/prevention assessment: A patient-centered methodology for understanding the adolescent illness experience", *Journal of Adolescent Health*, 27(3): 155–165. https://doi.org/10.1016/S1054-139X(00)00114-2

Rich, M. and Patashnick, J. (2002) "Narrative research with audiovisual data: Video intervention/prevention assessment (VIA) and NVivo", *International Journal of Social Research Methodology*, 5(3): 245–261. https://doi.org/10.1080/13645570210166373

Rich, M., Patashnick, J. and Chalfen, R. (2002) "Visual illness narratives of asthma: Explanatory models and health-related behavior", *American Journal of Health Behavior*, 26(6): 442–453. https://doi.org/10.5993/AJHB.26.6.5

Rose, N. (1990) *Governing the Soul: The Shaping of the Private Self*. London/New York: Taylor & Francis/Routledge.

Silverstone, R. (2007) *Media and Morality: On the Rise of the Mediapolis*. London: Polity Press.

Somers, M. R. (1994) "The narrative constitution of identity: A relational and network approach", *Theory and Society*, 23(5): 605–649. JSTOR.

Stein, A. (2011) "The forum: Therapeutic politics-an oxymoron?", *Sociological Forum*, 26(1): 187–193. https://doi.org/10.1111/j.1573-7861.2010.01231.x

Strack, R. W., Magill, C. and McDonagh, K. (2004) "Engaging youth through photovoice", *Health Promotion Practice*, 5(1): 49–58. https://doi.org/10.1177/1524839903258015

Swidler, A. (2001) *Talk of Love: How Culture Matters*. University of Chicago Press. Available at: www.press.uchicago.edu/ucp/books/book/chicago/T/bo3637950.html

Swidler, A. (2008) "Comment on Stephen Vaisey's 'Socrates, Skinner, and Aristotle: Three ways of thinking about culture in action'," *Sociological Forum*, 23(3): 614–618.

Van Dijck, J. (2007) *Mediated Memories in the Digital Age*. Stanford, CA: Stanford University Press.

Van Dijck, J. (2013) *The Culture of Connectivity: A Critical History of Social Media*. Oxford: Oxford University Press.

Vizenor, G. R. (1994) *Manifest Manners: Postindian Warriors of Survivance*. Hanover: Wesleyan University Press.

6

DATING APP LOGIC AND GEO-ENABLED MOBILE SOCIALITIES

Deborah Chambers

Introduction

By exploiting smartphones' global positioning system (GPS) location capabilities, dating applications (apps) offer dramatically new opportunities to meet nearby others for real-time experiences according to defined proximities. This proximity-based location system implies a mode of sociality conducted 'on the move'. Targeting a mainly young demographic group, with 85% of users aged between 18 and 34 (Smith, 2016), geolocational connectivity has transformed how a generation of singles meet potential partners. However, with users' spatial proximity as a primary identifier, little is known about the role played by this mode of geosocial communication in shaping mobile socialities. The soaring popularity of dating platforms and decline in non-digital ways of meeting people means that exclusion from a geo-enabled dating culture can have adverse effects. It may curtail individuals' opportunities to gain access to prospective partners, create relationships and networks, and develop long-term ties (Duguay, 2017: 363). An exploration of how dating apps mobilise geolocational intimacy is therefore vital for an understanding of a now a major dimension of young people's mobile socialities.

By asking how geolocational dating is algorithmically steered, this chapter explains how dating app practices signal a new kind of mobile sociality. With an accent on dating app platform infrastructures and on women and LGBTQ+ users[1], it considers whether dating apps empower mobility practices and, if so, in what ways. Mobile dating apps comprise a synthesis of technical, bodily and symbolic attributes. First, they encompass temporal features by imparting a sense of speed and immediacy in selecting matches. Second, the GPS location capabilities of dating apps involve spatial and temporal tracing, mapping and tracking of nearby strangers to coordinate physical co-presence. Third, observing and connecting with strangers requires dating app users to navigate transformed public-private boundaries. These attributes involve critical decisions about self-presentation and the competences to gauge strangers' identities and motives in co-situated virtual and physical spaces. The concept of 'hybrid space' (de Souza e Silva, 2006) enables us to approach these digital, temporal and geographical elements of the dating app ecology as hybridised features of online-offline connectedness. Within a process of physical and digital hybridisation (de Souza e Silva and Frith, 2012), these techno-social dating app qualities entail situated, bodily and emotional qualities, confirming that material and virtual space are no longer separate. As such, they suggest a dramatic reorganisation of social connectedness described here as geo-enabled mobile sociality.

Framed by scholarship within communication studies, critical algorithm studies and science and technology studies, the chapter engages with germane research to explore geolocational mobile dating as a mode of physical and digital hybridisation. A conceptual and interpretive approach is employed within a meta-analysis by drawing on Walter Benjamin's concept of the 'flâneur' as a heuristic tool to consider how dating app use is shaped by 'discursive and algorithmic regulatory conventions' (David and Cambre, 2016: 2). The concept of the flâneur foregrounds key opportunities and tensions involved in the patterning of dating app practices. Supported by this concept, algorithmic and discursive contradictions emmeshed in dating apps' geo-enabled mobile sociality are revealed: the navigation of mobilities and immobilities, digital structures and user agency, social norms and resistances.

The first section develops the concept of the 'geosocial flâneur'. The privileged status of the modern, nineteenth-century 'flâneur' as a mobile, cultured and autonomous male figure reveals the hierarchical nature of mobilities and immobilities. The dating app enables users to explore new hybridised intimate encounters but initial figurations of a masculinised 'flâneur' draw attention to heteronormative values that, by regulating sexual and gendered identities, immobilise women and LGBTQ+ people in public spaces. This prompts us to enquire whether a potentially queered geosocial flâneur can emerge as a late modern figure and enables us to assess how far dating app technology empowers users to transgress heteronormative conventions. Do dating apps empower sexual minorities to challenge these conventions?

The following four sections consider the intersecting digital and discursive tensions entangled in dating apps' geo-enabled mobile sociality. Users' potential to contest heteronormativity and foster a queered geosocial flâneury is circumvented by tensions triggered by dating apps' coded social norms. These coded social norms involve ethical complications. Centred on the gender identity binary, heteronormativity refers to the principle that heterosexuality is a standard, 'normal' mode of sexual orientation. The process of 'queering' challenges these hetero-dominant norms. The first of these four sections explains that dating apps' algorithms are value laden. User profiles, identities and mobile dating conventions are algorithmically coded according to hetero-dominant norms. The second considers how dating apps' geolocational structures reproduce social norms by promoting and regulating the interplay between tracking and being tracked. It reveals the moral ambiguities and risks involved in being stalked and gauging trust in 'the stranger'. The third part examines the challenges of dealing with collapsed geo-mobile social contexts to uncover the moral dilemmas associated with the disclosure and exposure of one's sexual identity which, in turn, curbs user agency. How users negotiate and resist these algorithmically imposed constraints are considered. The fourth part explains how user agency is impeded by an algorithmic reversion to traditional gender and sexual norms and identities, exemplified by Bumble's heteronormative digital logic. The chapter concludes that although dating app culture promises to empower sexual minorities by queering geolocational space, dating apps' software and interfaces are framed by heteronormative values that impede these opportunities.

The geosocial flâneur

While all present-day personal ties are now, in various ways, mediated (Chambers, 2013), dating apps do more than simply 'mediate' intimacy. The rise of geolocation-based app dating is a distinctive episode in mobile intimacies that opens up new modes of social coordination to realise encounters with others nearby (Sutko and de Souza e Silva, 2011). Supporting real-time dating embedded in offline geographical places, Grindr, launched in 2009, uses location-based

GPS capabilities for gay, bisexual or curious men (Blackwell et al., 2015). Other popular dating apps followed, such as Tinder in 2012. The real-time geolocational affordances of these apps communicate geographical positioning in several ways: by calculating the distance between the user and the profile, offered by Tinder and Grindr; by delivering details about the users' location at a particular time to identify the vicinity, offered by Bumble; or by showing a map where the user and potential match have crossed, as offered by happn.

These geo-enabled platforms support varied aspirations. Users seek face-to-face meetings, friendship, occupational networks or some form of visual and mediated fantasy. They create a dating profile by identifying their name, age, occupation and education level, and by allowing their smartphone's location tracking function to grant access to others' profiles. Users then set filters for the profiles and proximity of potential matches. Together, these steps determine the search criteria to locate prospective partners within a given geographical range. On apps such as Grindr, Tinder and Bumble, users steer the dating source by administering a left swipe to discard a potential match and a right swipe to accept. Two people who form a match can message each other. To offer matches tailored according to compatibility, this 'see-and-screen' system enables users to leaf through a database of profiles, using filters to select partners (Tong et al., 2016). Bounded by the software, users' intentions determine how they present their bodies, identities and desired outcomes, and how they negotiate geographical space to initiate encounters. As such, geolocation technologies generate new spatially and temporally organised modes of intimacy mapping (Veel and Thylstrup, 2018: 43).

These late modern dating tools are regarded as user-friendly but morally dubious. Early moral panics revolved around on the spread of hook-up cultures and sexually transmitted diseases (Duguay, 2017). In its initial promotional videos, Tinder fuelled this image by projecting itself as an adjunct to the all-night partying and drinking hook-up culture. To offset the negative publicity following their launch and counteract the social stigma surrounding app culture, dating app companies reconfigured their marketing strategies. For example, Tinder re-branded itself to convey the dynamic lifestyles of young, attractive and affluent upper-class urbanites inhabiting loft apartments, attending concerts and rooftop soirées, and holidaying in Europe (Duguay, 2017). The success of its revamped marketing imagery brought Tinder into mainstream culture, giving dating app culture a new, albeit qualified, respectability (Smith, 2016).

Buoyed by the fantasy of cosmopolitan individualism and globetrotting freedom, dating app imagery captures a sense of cultured physical mobility and urbane normality. This marketing vision invokes the figure of the flâneur: a privileged and cultured citizen who wanders, incognito, through city streets to capture a rich sensory experience that becomes an attitude to life. The term 'flâneur', used as a metaphor since the nineteenth century, refers to artists and intellectuals who critically surveyed people's behaviour while strolling cityscapes. In his influential early twentieth-century work on the 'passages' (arcades) of nineteenth-century Paris, Walter Benjamin (1999) developed the term as a conceptual device to explore a distinct contemplative relationship between people and public spaces. For Benjamin, the flâneur symbolised a transitional phase of urban modernity marked by escalating consumer capitalism. Revelling in the joys of observing, as a 'a passionate spectator, moving around amongst movement' (Kociatkiewicz and Kostera, 2019), the flâneur's immersive experience is described by Balzac as 'the gastronomy of the eye' (Balzac 1847 in Ferguson, 1994: 35). This urban spectatorship illuminates a mode of travel and phantasmagoric exploration of cities (Nuvolati, 2014) that resonates with today's imaginary of the free-floating and self-aware cosmopolitan observer. Supported by app culture, today's geosocial flâneur delights in the richness of the cityscape as a spectacle.

Importantly, Benjamin approached the flâneur as an equivocal figure who views the familiar city as a phantasmagoric public space through a defamiliarising veil. This unknown 'man of the crowd' embodies a dynamic presence that empowers the viewer to reflect, as if waking from a dream, on the uniqueness and strangeness of a familiar cityscape. However, this modern archetype oscillates from a radical to a conformist figure. He could acknowledge his commodification, yet is so enchanted by the city's maze of arcades, commodities and commercial entertainment that he loses the ability to transcend his intoxicated gaze. Ultimately, the flâneur is 'the viewer who takes pleasure in abandoning himself to the artificial world of high capitalist civilization' (Lauster, 2007: 140).

Through the power of geosocial technology, the mobile dater is positioned as both stranger and voyeur like a window shopper who is bestowed the power to navigate the city's throng of phantoms, taking pleasure in the details and attractions of urban life. This observer-participant dialectic of the geosocial flâneur captures the dating app users' dream-like fascination with voyeurism, anonymity, speed and the immediacy of urban life. But when unlocked by a 'match', the spectator's anonymity is suspended. Together, matched individuals must then negotiate the pleasures and risks of meeting nearby strangers. The metaphor of the geosocial flâneur points, then, to the dynamic complexities and challenges of late modern life, marked by highly developed global networks and a post-traditional, mobile cosmopolitanism. Related terms introduced by scholars include 'cyberflâneur', 'digital flâneur' and 'post-flâneur' to explain other kinds of digital navigations and experiences (see, for example, Argin et al., 2019; Hartmann, 2004; Skees, 2010). In the case of dating apps, this geolocational technology is characterised by navigating a digitally layered geographical space and a geographically layered digital space. We can refer, then, to the geosocial flâneur's encounter as geo-enabled mobile sociality. This encounter also extends beyond the metropolis. Absorbed into 'cosmopolitanised' geo-enabled experiences, rural app users are invited to participate in this geosocial flânerie.

However, as a historically masculine concept, the 'flâneur' exposes the uneven distribution of sexual freedoms. Unaccompanied women and gay, bisexual or curious men located in public space are stigmatised and immobilised as social and sexual minorities. In a society that puts a high value on mobility, structural immobility corresponds with inequality and access to poor resources. 'Mobility' is an agent of freedom and transformation that foregrounds gendered and sexual power structures. The depiction of women and gay men as 'out of place' in public space is part of a wider patriarchal and hetero-dominant process of regulating, containing and immobilising social minorities. However, while immobility is viewed as the binary opposite of mobility, it is more productive to consider mobility-immobility as a relational continuum (Pellegrino, 2011). This allows 'mobility' to be approached as relationally interconnected (Hanson, 2010). It reminds us that gendered and sexual immobilities are structurally and discursively constituted. In this respect, the 'queering' of the flâneur offers a productive way of considering how far this new geo-enabled app culture offers opportunities to reclassify heteronormative digital-physical settings as safe, women-centred and queer spaces.

The success of dating app culture depends on the disruption of those heteronormative intimacy discourses that traditionally constrain and marginalise intimacies 'that bear no necessary relation to domestic space, to kinship, to the couple, to property or to the nation' (Berlant and Warner, 1998). Symbolising a democratised sexual resourcefulness, dating apps pledge to overturn sexual minorities' immobile status. They claim to confer a safe virtual space for women and the LGBTQ+ community, to free them to engage in the dynamic mobility of navigating matches in real space and time. Significantly then, from Grindr (which caters to gay, bisexual and curious men) to Bumble (branded as a 'female-friendly' app), these platforms position themselves as agents of gendered and sexual transformation. By summoning sexual minorities to assume the role of a new kind of flâneur, this geo-enabled technology presents itself as profoundly emancipatory – as a liberating tool that opens up new, geo-enabled mobilised socialities. However, its emancipatory potential is constrained by its algorithmically coded values. The repurposing and reinterpretation of virtual and physical-geographical space is programmed according to techno-social principles that support circumscribed performances and positionings which pose new challenges for individuals' self-presentations, self-disclosure, and identities.

Algorithmically coded identities and mobilities

Although presented to users as agentic, the matchmaking dynamics programmed into dating app software activate narrow identity categories. Dating apps seem to generate fluid sexual codes, but the user interface sets limits on communicative protocols. They summon performance and positioning within highly standardised self-presentation conventions involving selfies, photographs, short text, and voice recording. Pre-coded identity categories, including gender and sex preferences, set conditions and relations that gauge partner desirability. Via classifications and scrolling lists, narrow identity categories are offered that reduce users' capacity to convey fluid, non-binary identities. Design settings guiding the disclosure of personal information define individuals in ways that can be disempowering. They often fail to convey the fluidity of raced, gendered or sexual articulations of self (Biven and Hoque, 2018; Cheney-Lippold, 2017; Noble, 2013; Rosamond, 2018; Sweeney, 2016).

As such, dating apps reproduce locational, gender and racialised inequalities via software-sorted geographies based on normative judgements that 'continuously and invisibly classify and demarcate rights, privileges, inclusions, exclusions, and mobilities' (Veel and Thylstrup, 2018: 45). This is redolent of a wider history of modern bureaucracy and techno-governance that categorises individuals and their identities in files and index cards (see Foucault, 1991; Krajewski, 2011). Remediating earlier index card procedures, users curate their own file entry as an intrinsic feature of the dating app genre, infrastructures and procedures. Users mediate the categorisation of granulated identities in bounded and abridged profile categories. Yet those who resist dating apps' prescribed criteria that categorise the self can find themselves omitted from grids that filter users according to set norms. This reduces the user's likelihood of being networked and browsed.

Given the goal of meeting face-to-face, users present identities to appear appealing yet credible (Ellison et al., 2012). However, concerns about the potential risks to personal safety involved in meeting prospective suitors face-to-face raises questions about how digital daters verify their identity. In geosocial networking where a face-to-face encounter is anticipated, the online/offline identity dualism must be reconfigured. To deter deception, users must provide some form of visual authentication of this 'self', but within parameters that allow meaningful participation. Like related apps, Tinder's mechanism for establishing authenticity is to appropriate a user's Facebook or Instagram social media profile as proof of a coherent biography. This allows users to interlink their username and profile and display selected photos. Referred to as a 'warranting' technique (Walther and Parks, 2002), the aim is to generate trust and confirm common connections. Ultimately, the social media profile makes authenticity claims on behalf of users as the 'primacy actor in verifying that people are 'real' (Duguay, 2017: 356).

This alignment of Tinder self-presentations with impressions from the user's social media account, to safeguard against uncertainty, involves referral to a preceding digital regime that vouches for the 'realness' of the user. This process of preparing a physical encounter entails a double digital framing of the self, thereby precluding alternative, more fluid ways of constructing self-identities. However, many users unwilling to share their personal images with unidentified audiences find ways to resist reliance on Facebook as a testimonial. Identity can be concealed through bogus profiles, blurred photos and nicknames, yet users who depart from mainstream norms risk discrimination and safety threats (Duguay, 2017).

Dating platforms like Grindr, Tinder and Bumble are designed to operate through a glance and swipe action. This enables instant access to potential matches. Based on sorting, filtering, ranking and assessing data, location-defined audiences make snap decisions based on first impressions (Gillespie, 2014; Kitchin, 2017; Klinger and Svensson, 2018). Young people are already immersed in mobile cultures where the touch-sensitive features of swiping, thumbing and pinching mobile phones have been domesticated. The effortless 'swipe logic' of the user interface offers a yes/no choice. This binary swipe logic relies on a deceptive compression of information to ease interface movement (David and Cambre, 2016). Like flipping through cards or magazine pages, swiping camouflages the complexities and the 'maybe' of earlier dating apps. The temporal velocity of fast-paced browsing and matching offer users a sense of validation and agency, allowing users to play the system like a game.

As the main criterion in negotiating potential matches, fleeting glances at photos invokes the image of a geosocial flâneur. This swiping spectator and connoisseur of the matchmaking game revels in the sensualities of watching and gathering impressions from the dating pool. But rather than strolling incognito through city streets, the digital flâneur's free-floating exploration and movement through the geo-digital landscape is enabled by the velocity of image swiping. Underlining the relationship between touch, speed and mobility as a symbolic manifestation of individual expression, this swiping summons screened relations of intimacy that indicate 'aspects of speed, ethereality, fragmentation, and volatility' (David and Cambre, 2016: 1). Tinder claims that users typically log on 9–11 times a day (Smith, 2016). On average, men spend 7.2 minutes swiping during a session while women spend 8.5 minutes (LeFebvre, 2018), with men comprising 62% of users and women 38% (McGrath, 2015).

Swiping allows users to initiate multiple relationship interests instantaneously. The sense of velocity, immediacy and spontaneity conjures agency. But swiping also prompts impulsive decisions leading to mistakes. By curtailing the forethought needed to make decisions, the speed designed into the user interface undermines choice. If a person swipes left, they cannot 'undo' the deletion. To reverse and revisit the discarded profile, they must sign up to Tinder's paid option, Tinder Plus. And by paying for Passport, users can alter their geolocalisation to connect with people more widely. David and Cambre argue that although the swipe gesture offers the potential for meaningful connections, it performs as a discursive hindrance to intimacy by validating the casual and transient nature of dating.

However, users find various ways to 'game' the algorithmic logic of app culture by resisting and subverting the algorithms' protocols and matching schemes (David and Cambre, 2016; Light et al., 2016). For instance, Tinder's algorithms restrict the number of available profiles to 20 repeated right-positive swipes. When users spend their allotted collection, Tinder notifies the user of no contactable others in the vicinity. This forces users to wait or pay for the upgrade. But users sidestep this barrier. Changing the setting to search for members of alternate sex enables new profiles to appear. Notwithstanding, the user interface fuels the impression of curatorial agency. The pedagogical processes of editing and self-curating coach users to self-brand and respond to algorithmic profile recommendations according to bounded conventions. These conventions involve a 'pseudo self-authenticity' that commodifies bodies and identities. Although the geosocial flâneur is aware of this self-commodification, the app's maze of potential matches causes this digital voyeur to lose the detachment required to transcend the illusion. Ultimately, the voyeur takes pleasure in revelling in the marketing world of match-making.

The geosocial flâneur no longer needs to move from the armchair to observe the spectacle. The acceleration generated by geolocational dating is not about travelling faster but about speed of transmission (David and Cambre, 2016: 8). Yet the stylish, mesmerising interface often gets passed around friends during nights out and used 'on the move'. The screen image 'brackets the physical presence of the participants, either by omitting or by simulating corporeal immediacy' (Heim, 2001: 74 quoted in Roth, 2014: 2115). The swipe logic capitalises on the impression of speed to create a sense of navigational agency. But the geolocational logic of navigating proximity between strangers creates a division between the time of virtual interactivity and the time of embodied presence. This complicates 'the split between the private self and public self' (Rojek, 2001: 11). When someone nearby is matched with the user, the potential to meet strangers changes the symbolic dynamic of time and space. The voyeur must step out of the screen and connect – face to face, and bodily – with proximate others.

The performance of selecting and matching comprises, then, two distinctive experiences: first, a screen gaze, and second, a physical encounter. Geolocation assimilates the situated, embodied, and emotional attributes of movement within a 'hybrid space' (de Souza e Silva, 2006). But this parallel set of practices – virtual gazing followed by spatial meeting – is a sequential process: a process of digital and physical hybridisation (de Souza e Silva and Frith, 2012). Licoppe and Inada describe this geolocational dating process as a situated 'unfolding' that creates 'location-aware connected bodies on the move' (Licoppe and Inada, 2016: 264). The following sections show how dating apps' unfolding process of spatio-temporal events, shaped by the hybridised affordances of online connectedness, poses risks and moral challenges as key features of geo-enabled mobile sociality.

The moral precarities of geolocating strangers

The ambiguity of geolocating and then encountering strangers involves uncertainties and risks. While the flâneur's detached reconnoitring invokes the figure of a stranger perusing crowds of strangers, this voyeur is reconfigured as a digital presence to be unveiled by the stranger as the coveted 'other'. The potential to meet the 'stranger' conjures fears of the unknown. It involves juggling two contradictory experiences: managing levels of trust in the stranger and avoiding the risk of being stalked. Offering new ways of relating to strangers, this dual geolocational role poses new moral dilemmas. Users' information-seeking actions become a mode of surveillance. Veel and Thylstrup (2018) draw on Simmel's approach to the stranger as a figure of urban modernity who is both symbolised and determined by 'a state of detachment from every given point in space' (Simmel, 1971 [1908]: 143). Redolent of the flâneur, Simmel's stranger is simultaneously 'attached' and 'detached' by inhabiting our space while not part of our social group. On what basis do we decide whether to engage with this stranger?

How dating apps convey users' location vary. Some update the location when the users open the app. Others track the users' location, undetected. This geo-tracking affordance offers opportunities to experiment with the alignment of anonymity and physical proximity. Veel and Thylstrup (2018) distinguish between the use of physical proximity as a technique for matching users (to generate some connection between the user and potential partners) and as a warranting technique (to find ways to manage the discrepancies, provoked by coded virtual profiles, between performance and authenticity). These 'matching' and 'warranting' techniques enable users to decide whether to trust the virtual profile. The term 'warrant' returns us to the challenges of authenticity: how users make judgements about digital profiles to negotiate this field of uncertainty. The value of a warrant, to determine the profile's reliability, increases as the information becomes more difficult to navigate (Walther and Parks, 2002). A person's network can be verified by checking discrepancies in self-presentation. Yet users may wish to conceal their sexual preferences or relationship status from friendship networks. Matching and warranting are, then, intertwined. Sometimes they work in alignment and often they contradict one another (Veel and Thylstrup, 2018). This raises questions about how far users are prepared to follow or resist the app's algorithmically coded social norms. Users who resist these norms may be misinterpreted, causing potential personal risks.

Dating apps are differentiated by two further systems, one whereby the app's geo-logging functions solely in real time (exemplified by Tinder and Grindr) and one that also collates information over time to map a location history (exemplified by happn). The latter records the number of times the user's location has overlapped with someone, within 250 metres. A person's route travelled through the city is shown on a map, divided into chains of historical moments when users crossed paths. This 'always on' mapping of proximity foregrounds the intersection of experiential and representational visualisation, enabling users to track and navigate their random encounters. The geolocational logic that selects then pinpoints the strangers who passed near the user are ideological and discursive. Allowing users to perceive their activities as serial movements through space, the app adopts the role of voyeur, summoning the notion of 'blind lovers' (Veel and Thylstrup, 2018). These virtual encounters transform perceptions of the geophysical environment. Like the flâneur, the user becomes both remote spectator and embedded pedestrian moving through space. The app's capacity to adopt this dual position of 'overview and embedment' is a major appeal, delivering a sense of control in an otherwise precarious situation. It offers users the potential to stalk strangers yet also generates a sense of protection from stalkers by activating 'a distanced gaze, while also immersing themselves in the messy realities of dating' (Veel and Thylstrup, 2018: 48)

Shaped by this mode of geo-enabled mobile sociality, the hybridised spatio-temporal feature of connecting strangers involves an emotional vacillation between feelings of familiarity and stalking. While the alignment of matching and warranting procedures can go wrong, with the stranger perceived as a dangerous outsider, the stranger is presented not as 'dangerous Other' but as an intimate, coveted figure (Veel and Thylstrup, 2018). Framed by flirtatious desire, tracking, stalking and lurking are no longer marked as sites of oppression or shame. Rather, stalking is normalised, woven into the very fabric of this new form of geo-enabled mobile sociality. Managed by imperceptible social codes that are foregrounded only when transgressed, app culture unlocks a new but troubling terrain of mediated mobile sociality. Positioned as the stalker-stranger, individuals are liberated from the confines of already known contacts, conjuring the impulse of the voyeuristic flâneur.

Geo-mobile context collapse

This reconfiguration of the dating app user as flirtatious stalker has moral implications for users, involving 'context collapse'. Context collapse refers to a digital context in which a person who divulges information or discloses their sexual identity for one set of audiences can be viewed by others lurking in the same online space (Marwick and boyd, 2011). In the case of Grindr – a dating app for men seeking encounters with men – meeting nearby others can cause disclosure problems. As Blackwell et al. (2015: 1122) explain, 'This conflation of geography and multiple intentions has the potential to cause tension, as people may not wish to be immediately identifiable to anybody nearby who downloads Grindr'. They may not want others to know they are seeking sex on a mobile app. Although regarded as a 'sexualised' app, Grindr's online space supports multiple and diverse intentions. Users might be seeking conversations, instant sexual encounters or dates. Yet they receive few visible cues to verify or counter expectations.

Geography has historically played a key role in the intersection between gay identity and physical space. Risks of being outed or found cruising underlines the significance of creating recognisable gay locations that offer a degree of assurance: bars, cruising areas or whole city districts where being visible and identifiable are culturally meaningful (Roth, 2014). Grindr dissolves the contextual cues normally relied on to detect strangers' norms, values, and intentions. The complications of detecting the exact location of the other user, or what they seek from Grindr, affects how users interpret others' intentions and behaviour, particularly with discrepancies between profiles and chat. Given the public regulation of intimacy, contact must be carefully negotiated to avoid sexual stigma. This is particularly relevant for gay men who may be assigned, by a match, to a heteronormative context that marks them negatively as outsiders. The wish to be visible yet to preserve some degree of privacy against self-disclosure leads many Grindr users to adopt identity-masking tactics. This affects how users craft their self-presentations, triggering further uncertainties. Grindr's aggregation or 'co-situating' of diverse sets of individuals highlights, then, the variegated sexualisation of physical and virtual spaces (Blackwell et al., 2015: 1130).

Issues of context collapse posed by Grindr's layering of digital and geographical space compel gay users to participate in complex granular profile searching and filtering practices. In a study of Scruff, a gay geosocial networking service targeted primarily at 'bears', Roth (2014) argues that rather than preserving anonymity, members are acutely attentive to and highly adept at curating profiles. Despite the reputation of dating apps and the problems of context collapse resulting from scant information in user profiles, Scruff and Grindrs' grid interfaces 'offer unique abilities to organise and efficiently navigate the "crowd" of other users' (Roth, 2014: 2127).

Dating app use is much higher among urban populations, but these apps support rural as well as urban mobile socialities. While the flâneur historically referred to cityscapes, today's queered, geosocial flâneur cosmopolitanises rural contexts. Unless a space is already defined as gay, most public spaces are normalised as heterosexual social spaces. Accessing apps like Grindr and Scruff is therefore highly valued by those living in rural or remote locations with no access to gay offline settings. As one user put it: 'What I like about Grindr is that it makes every space a potentially gay space' (Blackwell et al., 2015: 1126). Dating apps' affordances reach into rural settings highlighting the challenges for young people isolated and immobilised in remote locations served by poor transport networks. The technology enables lively geo-enabled mobile socialities across the rural-urban divide.

Heteronormatively coded logic

Galvanised by the #MeToo movement, public debate about 'safe spaces' offline and online underscore the level of harassment on dating apps where hostile hypersexualised messages and unwelcome graphic pictures are commonplace. Relating to dating platforms' identity-based infrastructural practices, such harassment is unevenly targeted at women, minority ethnic groups and gender non-conforming identities (Noble and Tynes, 2016; Srnicek, 2017). Exceptional among dating apps, Bumble was launched in 2014 to pursue gender equality as a deliberate marketing strategy. Labelled the 'Feminist Tinder', 'Bumble entered the apps market in the midst of a climate of user discord' (Biven and Hoque, 2018: 442). The aim was to combat sexual harassment by designing an app in which 'the woman always makes the first move' (Gross, 2019). Research confirms that dating apps are predisposed to support toxic masculinity and that women are regular recipients of male harassment within a climate that marks women as sexually promiscuous if they 'play the field' (Hess and Flores, 2016). This problem extends to queer, non-binary and transgender users, highlighting inadequate reporting mechanisms to confront transphobia (Biven and Hoque, 2018).

Conceding women's structural and discursive immobility, Bumble's philosophy presents gender as the sole designation of oppression to be tackled. However, Biven and Hoque argue that while Bumble offers control and safety to straight cisgender women, this is contingent on labelling straight male bodies as 'an aggressive form of masculinity'. The platform's attempts to manage aggressive responses from certain men – assumed to stem from their fear of rejection – is to challenge conventional perceptions that women who initiate dates are desperate (Anderson, 2005). Bumble's promotional tag line to women is: 'Make the first move: Start meeting new people in your area!' (https://bumble.com/en/).

To contest their traditionally immobile status, Bumble aims to democratise the dating app's hybridised digital/physical space by presenting a 'safe' digital setting for women to navigate a match on their own terms (see MacLeod and McArthur, 2019). It invites women to subvert conventional romance scripts of male initiation and female passivity. By projecting a feminised mobile agency, women are summoned to perform as geosocial flâneurs. However, the suggestion that digital space is safer for women than physical space is contentious, and the idea that making the first move counteracts the discursive and physical injustices of dating deals only with the digital initiation ritual. The sense of control that women may feel in 'making the first move' is soon lost. As one of Biven and Hoque's interviewee states, '[A]fter sending the first message as the female, everything went right back into the male's court. He asked for my number. He asked me out. (He also stood me up 20 minutes before we were supposed to meet)' (Biven and Hoque, 2018: 450).

Another feature of Bumble's equality principle, concerning safety, is the platform's reliance on Facebook log-in information as the double digital process of identity authentication. Yet Facebook's custom gender, enabling users to enter their preferred phrasing in a text box, is ignored. By offering only binary gender options (male and female), Bumble follows a cisnormative logic that digitally regulates gender and sexual identities (Biven and Hoque, 2018: 451). This heteronormative logic excludes those who do not conform to a binary gender identity. Bumble's feminist stance, of allowing women to send the first message, is pointless for lesbian users who find their searches automatically changed to the 'Best Friends Forever' (BFF) feature, accessible to all users. This can place queer women at risk of detection to heterosexual Bumble users who expected a BFF match, and could encourage transphobic messaging and reporting. By managing predictions of aggression within a heterosexual context, only the feminine and masculine binary can be digitally presented within Bumble's infrastructure. This conflates gender, sex and sexuality (Biven and Hoque, 2018).

Conclusion

This chapter explains that the conflation of the digital and the geographical supports a distinctive kind of geo-enabled mobile sociality as a unique episode within mobile intimacies. The geolocational feature of dating apps complicate spatial forms and meanings by overlaying physical spaces with dynamic data and overlaying digital spaces with geo-physical data. Described as a process of physical and digital hybridisation (de Souza e Silva and Frith, 2012), this overlaying process subverts traditional online/offline binaries by embedding materiality within the digital realm. Within these heterogenous and hybrid spaces, the physical and the digital no longer have ascendency over one another (Roth, 2014: 2126). This reconfiguration of physical spaces enables the formation of co-constituted networks that characterise new, geo-enabled mobile socialities.

The geolocational agency offered by dating apps evokes a sense of control over the spatial-virtual 'self' and 'other'. A doubly digitised process of verifying the user steers the transfer of an algorithmically constituted self from virtual presence to offline encounter to create a multiply projected digital 'self'. Together with geolocational functions, this process enables the embedding of bodies in digital and physical space, both as digital traces and as 'unfolding' corporeal entities (Licoppe and Inada, 2016). The geo-enabled flâneur's fascination with the anonymity, speed and impulsiveness of urban life is rewarded by a pleasurable sense of performative agency derived from algorithmically engineered acts of self-curation, observation, reconnoitring, stalking and cruising. The co-constitution of bodies as online/offline digital-physical entities shifts the meaning, status and currency of the human body. Counteracting descriptions of a 'split' between online-offline worlds, it confirms Mark Deuze's notion of 'media life': that we live in media and not alongside them (Deuze, 2012).

However, the algorithmic configuration of geographical proximity and embodiment are not neutral. They are steered by value-laden codes. Geo-enabled dating apps are 'as ideological and discursive as the maps of yesteryear' (Veel and Thylstrup, 2018: 48). These tools form part of a wider geolocational ecology to include apps and digital services such as Couchsurfing, Meetup, Facebook Marketplace and even Uber, which have adopted similar technological infrastructures. Such communication technologies gather people together within a 'capitalist sociality' to exchange value. Enmeshed in dating app logic are neoliberal values of late capitalism that urge a speeding up of user judgement and spur an unending fantasy that there might be something or someone better just around the corner. Moreover, the geosocial intersection of identities, bodies and places conjures ideologically coded agential subjectivities that empower some users and marginalise others. The values and assumptions embedded in dating app conventions point to four conflicting dynamics that generate inequalities in the patterning of geo-enabled mobile socialities.

First, the notion of the geosocial flâneur as a queered figure offers a productive way of uncovering the tensions and inconsistencies involved in mobile daters' geolocational negotiation of inhabited space. Digital manifestations of instantaneous matches offer users a sense of control over the precariousness of initiating encounters and forming attachments. Users are depicted as sophisticated, discriminating citizens with the cultural flair and competences to play the urban and rural scene. This can also generate a powerful sense of freedom from the constraints of a traditional moral community. Dating app culture promises to sexualise 'space' and empower sexual minorities by reclassifying and reconfiguring heteronormative spaces as digital-physical queer spaces. By activating the potential to queer dating app as hybridised and co-situated spaces, the agency offered by this virtual-physical process infers a democratisation of mobile sociality.

This brings us to the second point. Geolocational logic may offer new powers of mapping, tracking and lurking that liberate individuals from the confines of known contacts by reconfiguring the relationship between the 'self' and 'stranger'. But to manage the precarious correspondence between voyeuristic detachment and intimacy, this logic sanctions stalking as a central impulse to control the initiation of intimate encounters. Dating app users find themselves reassigned as the stranger-stalker, as a pleasurable yet risky endeavour. The transfiguration of the user from voyeur-flâneur to 'stalker' is a moral predicament, albeit deflected by shrewd marketing and algorithmic strategies. As such, the geosocial flâneur, as a late modern figure, works to foreground ethical issues concerning the sexual and gendered precarities of negotiating algorithmically coded identities and physical encounters with strangers. Exposing a troubling feature of geo-enabled mobile sociality, the social risks involved in meeting strangers are structured according to patriarchal, hetero-dominant norms. Consequently, ensuing vulnerabilities are unevenly distributed.

A third, related issue is that the technology's algorithmic coding generates new privacy challenges that may have negative consequences. The geolocational potential to transform this hybridised space into a site of queer sociability holds the promise of a spatially differentiated sexuality that breaches hetero-dominant norms. But, as Berlant and Warner (1998) observe, the queering of fixed ideas about the 'public' and 'citizenship' within digitally mediated intimate relations traverse between online and offline contexts to reframe the public and private. Hjorth states: 'This traversing sees physically public worlds entangled by electronic privacy and an electronic public that is geographically private' (Hjorth, 2018: 63). As geolocational technologies render profiles visible to multiple audiences simultaneously, users must contend with the challenges of negotiating this entanglement of privacy, self-disclosure and risks of context collapse. To avoid stigma and negative encounters, the threat of context collapse affects self-presentation strategies. In response, opaque techniques are adopted, making profile verification difficult. These challenges result in discrepancies between authenticity and performance, throwing up additional complications in understanding each other's norms and intentions. The risks of context collapse reveal, then, constraints imposed on the queering potential of the geosocial flâneur. Imposed by hetero-dominant protocols, these constraints compel users to confront and find ways to work around heteronormative identities and conventions.

Correspondingly, a fourth issue is that software-sorted geographies based on normative judgements flatten the complexities of gendered, raced and sexual self-manifestations. Bumble's algorithmic regulation of self-representations steered by gender binary coded intimacy scripts generates gender and sexual inequalities by excluding certain identities that disempower certain users. These tensions between algorithmic structures and user agency, and between social norms and resistances, place further constraints on the queering potential of the geosocial flâneur. As such, the value of the concept of the flâneur lies in its dual function of bearing the promise of a queered geo-enabled sociality and exposing the heteronormative techno-social constraints that block this promise.

A final point is that the effects of geolocational dating app culture on mobile socialities are continuously evolving. The tensions between algorithmically engineered identities and physically engaged social interactions are routinely resisted or subverted by dating app users, albeit within prescribed marketing limits. These resistances can then act on the algorithmic logic that they challenge (Bucher, 2017). Over time, designers may respond by realigning the algorithmic codes that support identity classifications and allied filtering and ranking procedures. This may create new conditions for geolocational intimacy and support more agile and unfettered geo-enabled mobile socialities in the future.

Note

1 This chapter uses the abbreviation 'LGBTQ+' (lesbian, gay, bisexual, transgender, queer/questioning or other personal identity label) to describe a range of sexual and gender identities (Parent et al., 2013).

References

Anderson, T. L. (2005) "Relationships among internet attitudes, internet use, romantic beliefs, and perceptions of online romantic relationships", *Cyberpsychology & Behaviour*, 8(6): 521–531.

Argin, G., Pak, B. and Turkoglu, H. (2019) "Post-flânerie: How do mobile devices shape the experience of the city?", *SHS Web of Conferences*, 64: 01005. https://doi.org/10.1051/shsconf/20196401005.

Balzac, H. de. (1847) *Physiologie du Mariage*. Paris: Charpentier.

Benjamin, W. (1999) *The Arcades Project*. Edited by R. Tiedemann and translated by H. Eiland and K. McLaughlin. Cambridge, MA: Belknap Press of Harvard University Press.

Berlant, L. and Warner, M. (1998) "Sex in public", *Critical Inquiry*, 24(2): 547–566.

Biven, R. and Hoque, S. (2018) "Programming sex, gender, and sexuality: Infrastructural failures in the 'feminist' dating app bumble", *Canadian Journal of Communication*, 43(3): 441–459.

Blackwell, C., Birnholtz, J. and Abbott, C. (2015) "Seeing and being seen: Co-situation and impression formation using Grindr, a location-aware gay dating app", *New Media & Society*, 17(7): 1117–1136.

Bucher, T. (2017) "The algorithmic imaginary: Exploring the ordinary affects of Facebook algorithms", *Information, Communication & Society*, 20(1): 30–44.

Chambers, D. (2013) *Social Media and Personal Relationships: Online Intimacies and Networked Friendship*. London: Palgrave Macmillan.

Cheney-Lippold, J. (2017) *We are Data: Algorithms and the Making of Our Digital Selves*. New York: New York University Press.

David, G. and Cambre, C. (2016) "Screened intimacies: Tinder and the swipe logic", *Social Media+ Society*, 2(2). https://doi.org/10.1177/2056305116641976

de Souza e Silva, A. (2006) "From cyber to hybrid: Mobile technologies as interfaces of hybrid spaces", *Space & Culture*, 9(3): 261–278.

de Souza e Silva, A. and Frith, J. (2012) *Mobile Interfaces in Public Spaces: Locational Privacy, Control, and Urban Sociability*. New York: Routledge.

Deuze, M. (2012) *Media Life*. Cambridge: Polity Press.

Duguay, S. (2017) "Dressing up Tinderella: Interrogating authenticity claims on the mobile dating app Tinder", *Information, Communication and Society*, 20(3): 351–367.

Ellison, N. B., Hancock, J. T. and Toma, C. L. (2012) "Profile as promise: A framework for conceptualizing veracity in online dating self-presentations", *New Media and Society*, 14(1): 45–62.

Ferguson, P. P. (1994) "The Flâneur on and off the streets of Paris", in K. Tester (ed.) *The Flâneur*. London: Routledge, 22–42.

Foucault, M. (1991) *Discipline and Punish*. London: Penguin.

Gillespie, T. (2014) "The relevance of algorithms", in T. Gillespie, P. J. Boczkowski and K. A. Foot (eds.) *Media Technologies: Essays on Communication, Materiality, and Society*. Cambridge, MA: MIT Press, 167–194.

Gross, E. L. (2019) "Bumble launched a new initiative to support a cause whenever a woman makes the first move", *Forbes*. Available at: www.forbes.com/sites/elanagross/2019/05/10/bumble-moves-making-impact/#4694cb685257 (accessed September 12, 2020).

Hanson, S. (2010) "Gender and mobility: New approaches for informing sustainability", *Gender, Place and Culture*, 17(1): 5–23.

Hartmann, M. (2004) *Technologies and Utopias: The Cyberflâneur and the Experience of 'Being Online'*. Baden-Baden: Nomos Verlagsgesellschaft.

Heim, M. (2001) "The erotic ontology of cyberspace", in D. Trend (ed.) *Reading Digital Culture*. Malden, MA: Wiley-Blackwell.

Hess, A. and Flores, C. (2016) "'Simply more than swiping left': A critical analysis of toxic masculine performances on Tinder nightmares", *New Media & Society*, 20(3): 1085–1102.

Hjorth, L. (2018) "Snapshots of afterlife: The cultural intimacies of posthuman camera phone practices", in A. M. Dobson, B. Robards and N. Carah (eds.) *Digital Intimate Publics and Social Media*. London: Palgrave Macmillan, 59–74.

Kitchin R. (2017) "Thinking critically about and researching algorithms", *Information, Communication & Society*, 20(1): 14–29.

Klinger, U. and Svensson, J. (2018) "The end of media logics? On algorithms and agency", *New Media and Society*, 20(12): 4653–4670.

Kociatkiewicz, J. and Kostera, M. (2019) "Textual flâneurie: Writing management with Walter Benjamin", *Ephemera: Theory and Politics in Organization*, 19(1): 163–178.

Krajewski, M. (2011) *Paper Machines: About Cards and Catalogues 1548–1929*. Cambridge, MA: MIT Press.

Lauster, M. (2007) "Walter Benjamin's myth of the 'Flâneur'", *The Modern Language Review*, 102(1): 139–156.

LeFebvre, L. E. (2018) "Swiping me off my feet: Explicating relationship initiation on Tinder", *Journal of Social and Personal Relationships*, 35(9): 1205–1229.

Licoppe, C. and Inada, Y. (2016) "Mobility and sociality in proximity-sensitive digital urban ecologies: 'Timid encounters' and 'seam-sensitive walks'", *Mobilities*, 11(2): 264–283.

Light, B., Burgess, J. and Duguay, S. (2016) "The walkthrough method: An approach to the study of apps", *New Media and Society*, 20(3): 881–900.

MacLeod, C. and McArthur, V. (2019) "The construction of gender in dating apps: An interface analysis of Tinder and Bumble", *Feminist Media Studies*, 19(6): 822–840.

Marwick, A. and boyd, d. (2011) "I tweet honestly, I tweet passionately: Twitter users, context collapse, and the imagined audience", *New Media and Society*, 13(1): 96–113.

McGrath, F. (2015) "What to know about Tinder in 5 graphs", *Globalwbindex*. Available at: https://blog.globalwebindex.com/trends/what-to-know-about-tinder/ (accessed September 12, 2020).

Noble, S. U. (2013) "Google search: Hyper-visibility as a means of rendering Black women and girls invisible", *InVisible Culture*, (19). Available at: http://ivc.lib.rochester.edu/google-search-hyper-visibility-as-a-means-of-rendering-black-women-and-girls-invisible/

Noble, S. U. and Tynes, B. M. (eds.) (2016) *The Intersectional Internet: Race, Sex, Class, and Culture Online*. New York: Peter Lang.

Nuvolati, G. (2014) "The Flâneur: A way of walking, exploring and interpreting the city", in T. Shortell and E. Brown (eds.) *Walking in the European City*. Farnham: Ashgate, 21–40.

Parent, M. C., De Blaere, C. and Moradi, B. (2013) "Approaches to research on intersectionality: Perspectives on gender, LGBT, and racial/ethnic identities", *Sex Roles*, 68(11–12): 639–645.

Pellegrino, G. (2011) *The Politics of Proximity: Mobility and Immobility in Practice*. Farnham: Ashgate.

Rojek, J. (2001) *Celebrity*. London: Reaktion Books.

Rosamond, E. (2018) "To sort, to match and to share: Addressivity in online dating platforms", *Journal of Aesthetics and Culture*, 10(2): https://doi.org/10.1080/20004214.2017.1400864.

Roth, Y. (2014) "Locating the 'scruff guy': Theorising body and space in gay geosocial media", *International Journal of Communication*, 8: 2113–2133.

Simmel, G. (1971 [1908]) "The stranger", in D. N. Levine (ed.) *Georg Simmel*. Chicago, IL: University of Chicago Press, 143–150.

Skees, M. (2010) "Digital Flanerie: Illustrative seeing in the digital age", *Critical Horizons*, 11(2): 266–287.

Smith, A. (2016) "15% of American adults have used online dating sites or mobile dating apps", Pew Research Center. Available at: www.pewresearch.org/internet/2016/02/11/15-percent-of-american-adults-have-used-online-dating-sites-or-mobile-dating-apps/ (accessed September 12, 2020).

Srnicek, N. (2017) *Platform Capitalism*. Cambridge: Polity Press.

Sutko, D. M and de Souza e Silva, A. (2011) "Location aware mobile media and urban sociability", *New Media & Society*, 13(5): 807–823.

Sweeney, M. E. (2016) "The Ms. Dewey 'experience': Technoculture, gender, and race", in Jessie Daniels, Karen Gregory, and Tressie McMillan Cottom (eds.) *Digital Sociologies*. Bristol: Policy Press, 401–420.

Tong, S. T., Hancock, J. T. and Slatcher, R. B. (2016) "Online dating system design and relational decision making: Choice, algorithms, and control", *Personal Relationships*, 23(4): 645–662.

Veel, K. and Thylstrup, N. B. (2018) "Geolocating the stranger: The mapping of uncertainty as a configuration of matching and warranting techniques in dating apps", *Journal of Aesthetics and Culture*, 10(3): 43–52.

Walther, J. B. and Parks, M. R. (2002) "Cues filtered out, cues filtered in", in M. L. Knapp, J. A. Daly and G. R. Miller (eds.) *Handbook of Interpersonal Communication*. Thousand Oaks: Sage, 529–563.

7

REPRESENTING MOBILE SOCIALITIES IN THE SINO-JAPANESE CONTEXT

A keyword approach

Jamie Coates

Introduction

This chapter is an exploration of how a mobile socialities approach helps us question the ways practices and imaginaries are 'scaled'. It draws from ethnographic insights among young mobile Chinese people in Japan, but speaks to my broader research interest in how interpersonal and quotidian phenomena take on new meanings when imagined at differing scales. I ask why it is that micro-scale practices, such as the presentation of self (Goffman, 1959), the cultivation of friendships, and viral media are often re-framed as signifying something at a greater scale or magnitude. These wider scales can be political and moral, from debates around 'community' to 'international relations', but also include personal questions of purpose, belonging and destiny (Coates, 2019). Broadly speaking, I ask 'How is it that the seemingly personal can come to signify so much?'. These lofty questions grow out of a long-term desire to understand the fraught Sino-Japanese context from an anthropological perspective. Frustrated with the focus on formal political institutions and actors within discussions of international relations in East Asia, I found inspiration in mobilities approaches that emphasised detailed accounts of what is moving, when it is moving and why. This interest allowed me to include a variety of phenomena 'on the ground' between China and Japan, but it has also necessitated considering imaginaries at transnational scales, the 'methodological nationalism' (Wimmer and Schiller, 2002) of commentators in my field, and the ethno-nationalism of those I work with.

I have conducted two periods of fieldwork (2009–2011 and 2014–2016) among young Chinese people in Japan, with a particular focus on social networks connected to Ikebukuro in northwest Tokyo. Ikebukuro is the second busiest train station in Tokyo (after Shinjuku) and has become a major hub for Chinese migrant sociality in Tokyo (Yamashita, 2010). Most of the people I have researched are Chinese educationally-channeled migrants under 45 years old who played, worked or lived in Ikebukuro and were either studying or came to Japan on student visas before transitioning to work. As Gracia Liu-Farrer has shown (Liu-Farrer, 2011, 2020), educationally channeled visas in Japan served as a proxy labour migration scheme in Japan from the 1990s up until the recent introduction of migrant worker visas in 2019. Today, many of these students have found work or started businesses in Japan since first studying there, attracting relatives to Japan through kin-based 'chain migration' (Brettell and Hollifield, 2000). Through my participant-observation, I regularly conversed with over 300 under-45 mobile urbanites in Tokyo who lived in intensely mediated ways. This cohort who live transnationally, both in their travels and media practice, have also inspired me to think about different ways we might approach the question of 'Sino-Japanese relations'.

A significant proportion of studies on the Sino-Japanese context focus on inter-governmental relations and formal economic ties (Morris-Suzuki and Soh, 2017). The antagonistic ways in which Chinese and Japanese institutions and discourses frame one another also mean that it is difficult to escape a certain degree of 'methodological nationalism' (Wimmer and Schiller, 2002) when discussing the Sino-Japanese context because the assemblages researchers trace are posited as 'bounded nations' by those we work with. Yet, a focus on mobility and sociality can foster new ways around this ongoing issue. In order to do so, however, we need to ensure we have a suitable multi-scalar approach (Eisenhart, 2017), and readers need to be aware of some of the ongoing political issues surrounding formal Sino-Japanese relations. There is a fraught history of war between China and Japan from the late nineteenth century onwards, with ongoing disagreements about the territorial spoils of these wars (Hara et al., 2015) and the ways in which Japan's wartime aggression should be remembered (Gustafsson, 2014). These wider issues feed into contemporary governmental disagreements between the two countries, and are referenced as one of the original causes of significant animosity between large portions of the Chinese and Japanese population (Genron, 2017; Stokes, 2015). Events such as the 2012 anti-Japan riots in China also demonstrate how Sino-Japanese relations are a tinderbox of heated imaginaries in the media (Schneider, 2018; Zhou and Wang, 2017). At the same time, the increase in flows of people, popular culture and products between China and Japan suggests that mobilities in the Sino-Japanese context afford a range of imaginative opportunities. For example, Japanese popular culture is highly popular in the Chinese-speaking world (Otmazgin, 2008), and many young Chinese people desire to travel to Japan (Mostafanezhad et al., 2018; Seaton and Yamamura, 2018), study at Japanese universities (Liu-Farrer, 2011; Fong, 2011), or even change their nationality to Japanese as part of their personal project to become cosmopolitan (Coates, 2018). As one of my interlocutors once called it, the desire to come play in Japan's 'consumption paradise' (*xiaofei tiantang*) is irresistible for some, and much of the research on mobilities in Japan suggests that it is an important site for the exploration of new identities among young Chinese people, whether in terms of gender, class, ethnicity and/or sexuality (Baudinette, 2016; Lai, 2015; Liu-Farrer, 2020; Tajima, 2005).

I have found both mobilities and socialities approaches helpful in thinking about how my fieldwork research might speak to this multi-scalar milieu. Yet, this volume is the first time I have been invited to think about how the two approaches might relate to each other. Nicholas Long and Henrietta Moore's discussion of 'sociality' (2013: 2) posits it as 'a dynamic and interactive relational matrix through which human beings come to know the world they live in and to find their purpose and meaning within it'. It is the way in which we understand our relationships with both the world and other people, including a variety of actors, both human and non-human. Most significantly, it is a methodology for trying to think past classical social scientific references to 'society' and 'the social' as a distinct domain of activity. A mobility approach also targets 'society' and 'the social', but focuses on these two terms' historically static connotations (Sheller and Urry, 2006). I see mobility as 'a complex assemblage of movement, social imaginaries and experience . . . infused with both attributed and self-ascribed meanings' (Elliot et al., 2017: 2). Rather than understanding it as movement in itself, as some scholars might, I see mobility as the ways we translate different kinds of movement into shared imaginaries. Forms of 'mobility', such as migration, transport and tourism, are assemblages of movement, legal frameworks, infrastructures and imagination. 'Mobility' as a term can never be movement in itself, either, because the word conjures various imaginaries, from social mobility to notions of freedom. This fact remains before we interrogate its potential meanings in other languages.

Combining mobility and sociality through the phrase 'mobile socialities' is a means of asking how connectedness and movement inspire differing experiences and imaginaries. While many within this volume (including myself) focus on media within their explorations of mobile socialities, I include the movement of things as a form of mobility, from the interpretation of circulations of objects as economic exchange, to the relabelling of these items and their associated practices as gifts. My broad and inclusive definition of mobile socialities takes inspiration from Long and Moore's (2013) emphasis on sociality as a concept metaphor (Moore, 2004) that encourages scholars to develop a processual *theory* of sociality rather than defining sociality or seeing 'the social' as a product of other phenomena. I thus interpret the phrase 'mobile socialities' as an invitation to explore how we might develop a dynamic theory of the ways mobility and sociality shape one another. I do not imagine that such a theory could be achieved within a single paper, however, and use this chapter to provide some ethnographic provocations from a context that is often neglected in Anglo-European mobilities research.

Within this chapter, I combine a focus on mobile socialities with a 'keyword approach' (Williams, 1985). In his *Keywords: Vocabulary of Culture and Society*, Raymond Williams posits an approach that explores the significance of commonly occurring terms in English-language discourse. This approach has more recently been used to identify keywords of mobility (Salazar and Jayaram, 2016), as well as inspiring research into key figures of mobility (Hartmann, Chapter 17 in this volume; Salazar, 2017). Within Chinese studies, a keyword approach has similarly been used to trace how governmental discourse intersects with everyday practices through the examination of pivotal Chinese terms (Kipnis, 2006). I explore how the mobilities and socialities tied to two keywords in the Sino-Japanese context (*daigou* and *bakugai*) are imagined and enacted in differing contexts. I interpret these keywords as representing two interconnected assemblages of mobile sociality with implications for how Chinese and Japanese people imagine their relationships with each other at differing scales. I borrow the point of Carr and Lempert (2016) that scale, in its epistemological and ontological sense, never precedes representation; or in other words, the words and images we use to describe phenomena such as mobile socialities are the only way they can be imagined at differing scales. Anna Tsing raises a similar point in her recent work on the global matsutake mushroom trade (Tsing, 2013). By tracing the movements of a highly valued mushroom industry, Tsing shows how a variety of sorting practices are used to translate commodities, gifts, and localities into globally valued networks. The language we use to describe mobile socialities is a similar process of sorting that allows everyday actions to take on political significance at differing scales. Consequently, I use two reoccurring keywords within my fieldwork observations, *daigou* and *bakugai*, to examine how mobile socialities are imagined at different scales and in different contexts. Jumping from personal practices to wider discourse, I use this ethnographic material to provoke a series of questions about the relationship between scale, imagination and a mobile socialities approach, arguing that attention to the way assemblages are invoked – as keywords, key figures and other forms of representation – should be an important part of a mobile socialities approach.

Ethnographic encounters with *daigou* and *bakugai*

In the autumn of 2017, I went to one of Tokyo's luxury shopping areas, popularly known as Ginza, to meet one of my friends, Bao,[1] who had been a participant in my study from 2014–2016. He had asked to meet me nearby as he had rented a parking space and wanted to take me for a drive in his new car. Before meeting Bao, I decided to have a quick walk around the area to gauge how things had changed in the six months since I left Tokyo and moved to Kyoto to be with my partner. It was roughly 10am, and the wide street was already filling up with bustling pedestrians eager to shop in one of Tokyo's famous commercial districts. Ginza is home to many of Japan's flagship stores for luxury brands, with equally impressive stores showcasing less expensive brands, such as MUJI and UNIQLO, which have become global fashion labels in themselves. As I walked, announcements from doorways repeated various deals in Japanese, Chinese, English and Korean, and reminders on how to conduct oneself in the store were repeated on speakers at the entryways to many of the stores. Cutting through the Mitsukoshi department store on the way to where I would meet Bao, I saw a young woman holding up her smartphone to take a photograph of a stand with various cosmetics. She then held her phone up to her lips to leave a short voice message in Chinese: 'Are these the kind you are looking for?' Soon after passing this woman, I saw a small sign in Chinese and English stating '2 items per customer' and looked back to see the woman filling up a basket with a large number of a specific eyeliner. Further throughout the store you could see shoppers inspecting expensive handbags and fashion items, as well as talking about where they might go to next. Coming out the back of the store, I was greeted by a large group of Chinese-speaking tourists getting off a coach and another group wandering up to their chartered bus, struggling to carry all the items they had purchased nearby.

Scenes such as these were a common sight across tourist shopping hotspots around the world prior to the COVID-19 pandemic. And indeed, the decline of Chinese tourist spending during COVID-19 has caused concern in many economies (Onishi, 2020). The enthusiastic shopping of Chinese tourists in the five years prior to 2020 also attracted a wide range of media attention, from articles about the concerns of shopping behaviour of Chinese tourists in Paris (Callan, 2016) to geopolitical tensions spurred from Chinese tourism to Thailand (Mostafanezhad et al., 2018). In this sense, the scene I described previously was typical of many high streets in cities and tourist hotspots around the globe. The Japanese instantiation of this boom in Chinese tourist spending is particularly interesting, however, because it spurred debates about the relations between China and Japan, as well as the morality of Chinese mobilities and practices. At the same time, these issues intersected with several other mobilities and socialities in the Japanese context making it an ideal example of the increasingly entangled nature of global mobilities today and their imaginative consequences.

By late 2019, there were over one million registered Chinese nationals living in Japan, making them the largest non-Japanese minority in the country (Ministry of Justice, Japan, 2019). A significant majority of this population are relatively recent migrants, having grown from some 45,000 in 1985 to current figures today. However these figures only represent citizens of the People's Republic of China and do not account for the increasing diversity of Japan (Liu-Farrer, 2020). Even among those of Chinese-speaking heritage, there are a growing number of permanent residents, naturalised citizens and people from Taiwan, as well as overseas Chinese from other countries such as Canada and Malaysia. Alongside these migration patterns, favourable currency exchange rates and changes in visa regulations have supported a boom in trade and tourism between China and Japan. For example, mainland Chinese visitors to Japan grew from 942,439 in 2007 to over 8.3 million in 2018 (MLIT, 2019), becoming one of Japan's fastest growing industries in the 2010s.

These human mobilities intersect with both formal and informal trade between China and Japan, too. China is Japan's largest trading partner as of 2020, and Japan is China's third largest trading partner (OEC, 2020). The popularity of Japanese products has been particularly pronounced since Chinese consumer faith in Chinese products declined. For example, since one of China's major e-commerce companies, Alibaba, introduced Tmall Global in 2014, there has been consistent growth in the volume of ecommerce imports, by roughly 30 per cent a year (CBNData, 2017). Tmall Global is an online platform that allows Chinese shoppers to access international brands. The most popular commodities accessed through these services, according to CBNData's analysis, are beauty products, food and mother/baby products (2017). Among the countries that Tmall shoppers import from, Japan ranks first. It has been suggested that there is a growing relationship between Japan's image as a site of modernity (Iwabuchi, 2004), perceived trust in the quality of its commodities, and Chinese consumer's efforts to circumvent the perceived risks of Chinese commodities in recent years. This perception has been used to explain why Japanese commodities are often subject to purchasing 'crazes' such as a boom in washlet toilet purchases in 2014.

The scene I just described is not simply a representation of Chinese tourist shopping, but connects to issues of migration, study abroad and social media economies. In particular, Bao's story shows how mobile socialities are constituted by multiple mobilities and relationships. Bao had lived in Japan for almost eight years, originally enrolling as a language student before apprenticing in a hair salon. After making some extra money through various side projects, such as selling men's fashion items via social media online, he managed to save enough money to stop his work in the hair salon and start a portfolio of small ventures. One of these ventures was a business where he would run private chauffeured day tours around Tokyo and its surroundings, taking small wealthy groups between attractions and shopping hotspots. When I finally met Bao on that day, he pulled up in a shiny new Toyota hybrid minivan with leather seats and custom decals which he had bought explicitly for his business. It was an expensive car, but he proudly explained how he would pay it off taking groups for roughly 400 USD a day and receiving small commissions from luxury stores. He sourced his clients through WeChat, both in collaboration with some official tourism companies in China and word of mouth through reposts in his friends' private groups. Bao was not formally registered with the Japanese Tourism Board, though. The work was 'a little bit black', in Bao's words, which is a turn of phrase commonly used in Chinese to mean that it is not formally legal. Nonetheless, he explained, it was honest work and a good business opportunity.

Both the practices Bao had used to save up the capital for this venture, and the practices he was facilitating among Chinese tourists, had been given specific labels within wider discourse in the Sino-Japanese context. Bao's side ventures where he sold men's fashion items and hair products online were an example of a practice commonly called *daigou* in Chinese. *Daigou* means to buy something on behalf of someone else (Martin, 2017; Zhang, 2017). Although it has existed as an economic practice for a long time, it has taken on new form and meaning via the network sociality (Wittel, 2001) afforded by Chinese social media. Typically, it involves a buyer-seller who purchases items in places outside of mainland China, such as Japan or Australia, and then posts them to private buyers sourced through Chinese social media. Starting in the middle of the first decade of the 2000s, *daigou* became increasingly popular as young Chinese consumers became more brand aware (Zhang, 2017). However, as Lin Zhang notes, it became even more prominent after 2010 when the global recession and new social media platforms put Chinese buyers in an advantageous position. Today, buy-sell opportunities are often coordinated through WeChat, while resellers tend to pass on goods via platforms such as Taobao.com. *Daigou* has varying levels of trade, but mostly operates as an 'informal economy' (Hart, 1985) because of its legally ambiguous status and its diversity of practices. As a practice, *daigou*'s diversity and its reliance on private social media contacts makes it difficult to pin down exactly what 'form' the *daigou* economy takes. During the time of my fieldwork in Japan, *daigou* typically involved networks of buyers and sellers circumventing import-export taxes by mailing items labelled as 'gifts' from Japan to China. Making use of this taxation loophole, many of the young Chinese people whom I worked with were able to buy and sell expensive luxury items bought in Japan at a lower cost than they would be in China. They would often talk about sending 'a gift' to a friend back in China as a euphemism for a *daigou* transaction. Sometimes this was discussed ironically, but their buyers were often framed as friends and so it would be unwise to interpret these actions using Anglo-European distinctions between instrumental and sentimental activities. Indeed, Marcel Mauss' original discussion of 'the gift' posits it as an often-strategic political act (1954).

As suggested in the earlier portrayal of Bao in Ginza, *daigou* often overlaps with other practices, such as the enthusiastic purchasing of large quantities of items, which had been popularly dubbed *bakugai* in Japanese. The term *bakugai* literally means 'explosive buying', and the practice emerged as a response to the upsurge of consumption among tourists since 2014. The origin of the term is contested, but as Kei Nakajima notes, it became popular in Japanese talk shows by the end of 2014, and references to *bakugai* were typically accompanied by footage of Chinese tourists buying all sorts of items, or pictures of the empty shelves they left behind them (Nakajima, 2015). By 2015, it was voted one of two top 'new terms' in an annual competition run by Jiyu Kokuminsha publishing company and has since taken on a variety of popular meanings. Its origins connote a certain violent quality to China's consumption boom and is usually related to images of the 'uncivilised' manner of visitors from China. It was considered offensive by some Chinese people I spoke to in Japan, while others saw it as rude but enjoyed making jokes with the term. Today, *bakugai* is used in reference to all kinds of shopping sprees in Japanese, and while associated with Chinese tourists, is no longer limited to describing Chinese people and their practices.

Stories such as Bao's, alongside the Sino-Japanese mobilities context more broadly, stand as key examples of why we need a mobile socialities approach that examines the experience of how media intersect with multiple mobilities. Bao was not simply a student, or a migrant, and his livelihood depended on a tangle of digital socialities and mobilities. His story was but one among many in the Sino-Japanese mobile socialities context. The Sino-Japanese context is entangled with everyday circulations of people, things and practices. These various mobilities, from overseas study to overseas trade, have created a context whereby, on an everyday level, Japan and China increasingly resemble a transnational social field. From a mobilities perspective, it is the movement of these various phenomena that constitutes China and Japan as places (Ingold, 2007; Massey, 1994), affording different modes of (im)mobility as imagination (Salazar and Schiller, 2016). Transnational socialities and media have long been associated with imagination as a practice that shapes regional and global relationships (Appadurai, 1996).

The two keywords of *daigou* and *bakugai* represent overlapping assemblages of human mobilities, media mobilities and their emergent socio-economic socialities. They also embody the importance of keywords in how mobile socialities – as assemblages of imagination, actions and 'stuff' – are framed as distinct practices in discourse. Within this chapter, I use the term *daigou* and *bakugai* to trace a cluster of mobile socialities constituted by the social relationships and imaginaries that emerge from migrants and tourists buying and exchanging items in the Sino-Japanese context. I do so to interrogate how it is that seemingly quotidian practices can be reimagined as, on the one hand, serving as a personal commercial activity, and on the other hand, taking on moral and political connotations that are posited at a national scale. As I will show in what follows, the two terms have played important roles in framing how Chinese people describe their own mobile socialities, as well as the ways the mobile socialities of Chinese people are described by others. In particular, the ways these terms have been deployed in debates around the morality of shopping, tourism and informal social media trade suggests the powerful imaginative effects of these quotidian practices.

Daigou as young Chinese mobile socialities

Among the young mobile Chinese people whom I worked with from 2014–2016, *daigou* became a popular way to avoid the dirty, dangerous and demeaning jobs that many migrants did in Japan. The level of commitment to *daigou* varied, though. Some of those I spoke to had bought something specific once or twice and then sold it on to try and make some quick cash, whereas others had elaborately organized networks of customers and contacts that they used to support themselves entirely. There were even a few individuals who had organized multilevel marketing (MLM) schemes to supply wholesalers in the mainland. One of my oldest informants, Dayuan,[2] whom I first met in 2010 during my first period of fieldwork, had positioned himself as a mentor for a few aspiring *daigou* sellers, and coordinated networks of sellers in ways that bordered on an MLM scheme. Having lived in Japan for 15 years, he was one of the earliest people to start selling Japanese electronic goods back in his hometown in northern China, and he was an experienced and connected businessman, albeit not particularly wealthy when we first met. However, between 2011 and 2014, he had managed to trade enough that he had developed an elaborate portfolio of businesses. By 2015, he had a range of real estate investments in Tokyo, including billboards, and had used his existing contacts to set up wider *daigou* operations: hiring others to mail commodities to a warehouse his father owned in Liaoning province, labelling them as gifts, and limiting the size of each shipment so as to forego import tax. He promoted the practice of *daigou* through educational posts on WeChat that told people how to do it, and encouraged them to collaborate with him. For example, an image circulated among multiple private groups on WeChat setting out plans for sharing information about what was in demand in mainland China. It was part of a series of animations that took users through the steps for joining specific private groups specifically for *daigou*, and how to better coordinate *daigou* activities overall. Parodying a poster from the cultural revolution, the post ironically paraphrased a famous slogan '*Daigou* for the masses! Add a brick to the building of socialism, struggle until the end for the great pot of socialism'. Humorous imagery and slogans such as these embody the conviviality (see Duru, Chapter 15 in this volume) of even seemingly instrumental activities and connote a sense of how 'everyone' was doing, or could do, a little bit of *daigou*.

Dayuan's position within these *daigou* arrangements had become so prominent that many young people started joking about him as if he were the leader of a religious sect or kung fu school. On drinking nights with Dayuan, some would jokingly call him '*Shifu*', a term usually reserved for master artisans and religious teachers who take disciples, such as in martial arts networks. In addition, Dayuan had a small network of young people who would ask him for advice with their small business ventures and he would jokingly call them his 'disciples' (*tudi*). While this business network may have seemed purely commercial, Dayuan's efforts to mentor less experienced people was widely appreciated, and the network itself became a source of friendships and romances. Finding a new source for a coveted commodity, such as limited-edition boots on one occasion, was also framed as a 'gift' within the network and generally considered something a friend should do if they are able. Beyond *daigou* trade, Dayuan organized many parties and social outings, and took care of people in need. For example, on one occasion, a young woman went through a period of severe mental distress and Dayuan helped her find two friends to move in with while she recovered.

The craze for *daigou* from 2015 onwards meant that posts and promotions filled the various private WeChat groups I joined during my fieldwork, as well as the cascading wall of individual posts known as 'moments' in WeChat's interface. Typically, people would repeat the same posts across moments and private groups, or tailor lifestyle posts in moments that promoted products they were eager to sell. The affective and gendered labour poured into some of these posts borrowed from the aesthetics of micro-celebrities known in Chinese as *wanghong*, which is short for '*wangluo hongren*' ('people who have gone viral on the internet') (Han, 2020; Liao, 2019), and the wider world of influencer culture (Abidin, 2015, 2016, 2017). In particular, their posts wavered between more glamorous shots emphasizing how specific products fit into their lifestyle, and a degree of 'calibrated amateurism' as Crystal Abidin has theorized, whereby actors craft a 'contrived authenticity that portrays the raw aesthetic of an amateur' (2017: 1) despite the significant labour and expertise that often goes into these kinds of performances.

For example, one of Dayuan's 'disciples' was a woman named Chenjin[3] who cultivated a glamorous and highly sexualized online presence to promote the beauty products she sold both in Tokyo and online in China via *daigou*. In particular, she had travelled to Korea to secure a vendor for a much sought-after product called Latoja, and had opened a small beauty salon in her apartment specialising in weight loss treatments. Latoja is a cream that when applied to the desired area through vigorous massage, is purported to stimulate fat loss in that desired area. Chenjin does not make much money through her treatments. Rather, her online sales of Latoja are her major source of income, while her salon provides a way for her to promote herself as a successful businesswoman. Chenjin would often promote her *daigou* business through short videos of before and after treatments on her friends, personal posts about changes in her life, and more explicit advertisements that connected slender women's bodies to the product itself, such as one where she photographed her friends 'working out' on various weight machines with a bottle of Latoja placed next to them. In this instance, the calibrated amateurism depended on the performance not only of personal beauty, but also care for friends and the display of influence she has by being able to call upon her friends for favours to promote her products.

Extant scholarship on *daigou* has shown the ways young Chinese women have utilized the practice to navigate local disadvantages while overseas. Fran Martin (2017), in her longitudinal ethnographic study of young female students, for example, shows how exploitative hiring practices among local Chinese employers in Australia are supplemented and circumvented through *daigou*. Martin's participants' stories are very similar to that of Lu lu,[4] where disappointment in local employment encouraged flexible commercial tactics such as *daigou*. Lu lu was a twenty-something woman from Fujian who came to Japan seven years ago on a dependent visa with her mother. In 2016, she was studying at a Japanese music and dance school, and was trying to live as independently as possible from her mother. At first, Lu lu worked in a convenience store to make some extra cash while studying, but frustrated with the low wages (800 yen an hour) and late hours, Lu lu started selling various Japanese luxury items to buyers in China via her social media contacts in 2015. She would mostly use WeChat, and had a few friends who would help her resell beauty products and fashion items among their contacts. Borrowing money from her mother, Lu lu starting larger and larger operations. From cosmetics to fish oil and branded clothing, her income fluctuated with the tide of demand, making 200,000 yen (1800 Euro) most months and on a few occasions making as much as twice that. Through this income, she managed to move out of home and share a flat with several of her friends. With dreams of being a professional singer one day, she only works one day a week in a convenience store – 'just in case' – and spends her spare time performing in a Chinese-owned bar in Ikebukuro.

As Lin Zhang and Martin both note, *daigou* epitomizes how feminine labour is increasingly important since the rise of social media and 'prosumer capitalism'(Martin, 2017; Zhang, 2017). This kind of feminine network capital represents the gendered emotional labour utilized to gather alternative weak ties in otherwise unequal relations. It is also an example of the labour needed to maintain mobile socialities. In line with this extant research, many of my interlocutors, such as Lu lu and Chenjin, were highly motivated and organized, gauging demand through constant online activity that involved a significant amount of gendered labour. Having secured reliable buyers through friends of friends, they would constantly work to keep them interested through posts about new products, their own health, or fashion gossip they heard while in Japan.

Yet, while extant scholarship has focused on women who practice *daigou*, I met many men who would engage in *daigou* in more casual and ad hoc ways. For example, Kai[5] had been in Japan for two years on a student visa when I interviewed him. He first studied in a Japanese language school, but had enrolled in a contemporary music production school around the time I interviewed him. While Kai had the financial support of his parents, he wanted to earn more, largely because of his appetite for new musical instruments. The last time I visited his apartment in 2017, for example, he had a large pile of Japan-only limited-edition Adidas sneakers that he was slowly selling to friends and their contacts at twice the original price. The gendered differences in these approaches seemed to rely mostly on how online gender performances facilitated particular kinds of networks. Lulu mostly sold to female friends and older women known to her, whereas Kai and the other men I spoke to mostly found their contacts through other men. These differences arguably represented two sides of a similar phenomenon. Gendered socialities in Chinese-speaking networks have a strong tendency towards same-sex friendships, or homosociality. Consequently, reliance on digitally mediated gendered homosociality in response to inadequacies in local networks in Japan, appeared to be a relatively easy extension of this pre-existing tendency. That being said, they were often experienced in different ways. Kai's activities were used to supplement his income, and he utilized his online performances to promote a quirky, cosmopolitan image of his daily life in Japan. He enjoyed these practices, and once explained how his *daigou* would help facilitate interest in his own dream, which was to become a successful musician. In contrast, while she never spoke to me directly about it, Chen Jin's would often post parallel, more personal commentaries on platforms with fewer of her Chinese friends (such as Facebook). These posts betrayed a strong sense of emotional fatigue from the constant gender performances she engaged in online.[6]

In summary, this ethnographic fieldwork suggests how mobilities are entangled with different forms of sociality. On the one hand, we have the socialities of *daigou* as a community of practice (Wenger, 1998), where friendships, the mentorship of Dayuan and collaborative commercial activities foster tangible imaginaries of success and togetherness in Japan. On the other hand, these activities connect to the transnational socialities and mobilities of actual *daigou* trade, which often depend on selling to, or through, friends and acquaintances back in mainland China.

Re-scaling Sino-Japanese mobile socialities

At the same time as many of the young Chinese people in my field site were engaging in *daigou* and providing informal shopping tours to tourists, increasing concern was voiced in Japanese broadcast media about the effects of prolific Chinese consumption in Japan. In broadcast media talk shows, for example, these economic practices were often represented as reflecting the moral qualities of Chinese nationals more broadly, or a sign of Japan's economic future. *Bakugai* became a key signifier around which many disparate conversations about Sino-Japanese relations and Chinese people came to circulate. And although the debates around *bakugai* were often enthusiastic or curious, the framing of the term and its wider discourse afforded many conservative and xenophobic pundits to muster panic among Japanese audiences. A brief Google keywords trends analysis shows that *bakugai* emerged from 2012 onwards, reaching two peaks in early 2016 and early 2020 (Google Trends Labs, 2020). These trends parallel two developments, first as a series of new tourism visa allowances were made between China and Japan from 2012 onwards; and secondly in response to consumer activities during the early stages of the COVID-19 pandemic.

During the Chinese New Year festival in 2015, an unprecedented number of Chinese shoppers travelled to Japan, clearing the shelves of many major stores in shopping districts such as Ginza. For the following year, an ongoing debate continued in the Japanese media about China's role in Japan's economic future, with the parallel practice of *daigou* entering Japanese discussions of Chinese economic practices more generally. Japanese news items about *bakugai* during this period ranged from the informative to the inflammatory, but the mostly tended to follow similar themes. For example, an Asahi-TV Ann News Channel segment in 2015 informatively discussed how the increase in Chinese *bakugai* has necessitated hiring more Chinese staff within luxury stores, and they portrayed the failure to hire enough Chinese-speaking staff as the reason why many illegal tourism operations had become necessary. In contrast, the current affairs talk show *Ki ni naru!* (Attention!) held a similar segment with a panel of discussants. Within this show the tone was more suspicious, with sighs of disbelief at images of Chinese customers carrying piles of suitcases and a litany of questions such as 'why are they coming here?', 'what is wrong with shopping in Japan?', and 'will there be anything left for Japanese people?' Similarly, in 2019 when it was revealed that Chinese investors had bought significant amounts of land in Hokkaido, a series of items about Japan's land being subject to *bakugai* caused panicked discussions on Twitter.

Chinese migrant responses to the invention of the term *bakugai*, as well as the wider practice of *daigou*, have expressed a sense of injustice about how the Japanese media has represented the whole issue, arguing that to conduct *daigou* is morally good. One interlocutor in my social media network posted the following statement on WeChat after the news reported a group of Chinese shoppers had cleared out a pharmacy of its nappies, milk formula and cosmetics 'like a swarm of locusts'

> *Bakugai* has become a derogatory term within the Japanese media. I cannot understand why customers should be likened to locusts. Meanwhile, Chinese people's massive consumption is providing comfort to failing Japanese industries, carrying the Japanese economy. This is clearly a good thing, they should hold gratitude in their hearts.

Similar responses have emphasised the important contribution *daigou* and *bakugai* have brought to the economy. With a common trope likening these practices to a 'gift' or *hongbao* to the Japanese economy. *Hongbao* are traditionally red envelopes with money inside them that are given over the Chinese New Year period or other festive occasions where gift-giving is necessitated (such as weddings). Previous scholarship has argued that *hongbao* exemplify the way gifts are framed within Han Chinese socialities, where material exchange has less distinction between instrumental and sentimental activity than is popularly thought in Anglo-European social theory (Osburg, 2013; Strickland, 2010; Yang, 1994). As noted earlier, the distinction between gift and commodity is more blurred within many Chinese discussions of economic exchange than classical theorists such as Mauss or Marx argue. While *daigou* is a practice that is part of capitalist consumption culture, it also depends on other modes of exchange in the tactics used to disseminate it, and the items distributed through *daigou* are often used as gifts. Moreover, as Tsing notes in her work on matsutake trade: 'Despite all the apparatus of private property, markets, commodity fetishism, and more, taking the gift out of the commodity is never easy' (2013: 21). Yet what fascinates me here is not so much the distinction between commodities and gifts as it is the possibility of framing a commercial practice as a 'gift' to a national economy. The invocation of the term *hongbao* suggests the differing imaginaries surrounding exchanges of gifts or commodities across cultures, and the possibility to personally reimagine who or what might be a giver or recipient, even at the scale of 'the Japanese economy'. The reason why these sorts of questions are worth considering is that the imaginative leaps necessary to make such statements are an important part of the ways we foster both understanding and misunderstanding between people. As Marcel Mauss originally noted, gifts are important political tools for bonding people together through a moral sense of obligation (1954). But what happens when perceptions of whether something is a gift differs? When one party sees their commercial engagement as something morally 'good' while the other party frames it as uncouth and aggressive, it is easy to imagine why tensions may arise. Given the historical tensions in the Sino-Japanese context, where it was only a few decades ago when travel from China to Japan was unthinkable for many, the imaginative leaps that underpin the tensions in this one small example help us understand the importance of imagination and representation in processes that may otherwise be simply relegated as 'economic'. By recognizing that the status of a particular practice (*daigou*) or its representation (*bakugai*) is in the eye of the beholder, we are reminded that socialities are also a matter of perception. As Reading notes in Chapter 16 of this volume, one person's sociality may be another's anti-sociality. Paying attention to how mobile socialities are represented and rescaled provides one methodology for not taking the perception of these socialities for granted.

Indeed, the moral and political imaginaries of *daigou* and *bakugai* continue to play prominently in how young mobile Chinese people imagine their relationship to China *and* Japan. Just before the COVID-19 lockdown was implemented in Wuhan in late January 2020, many of my former interlocutors were posting pictures of themselves buying up large quantities of N95 face masks to post back to China. Emotions ran high in the posts, with many emojis of the Chinese flag and short statements of hope for the people of Wuhan, such as one post which coupled several pictures of them posting a box of masks alongside the statement 'hoping to send a humble bit of support to China, hoping the people of Wuhan can overcome this quickly'. Another emotive video shared among many was a short video of a Japanese teller working in a pharmacy in Tokyo saying 'You can do it, China. Good luck!'. Some leapt at this opportunity to conduct some *daigou*; indeed, Chenjin posted a picture of boxes of face masks alongside the phrase 'The old businesses must be cursing us new *daigou* retailers, and those in 9–5 jobs have already decided to change their vocation!' A similarly jovial post of a car boot full of masks stated 'if going on a long journey one cannot go empty handed!'

Unfortunately, as the severity of the global COVID-19 situation unfolded, the mass buying of facemasks took on a new meaning within Japanese media. In particular, *bakugai* debates took on a new sense of urgency with images of empty shelves in pharmacies and supermarkets posted on Twitter, printed in magazines and broadcast on television shows. Confronted by these images anti-Chinese sentiment seemed to be building during the time of writing this chapter under COVID-19. Chinese shoppers buying large quantities of masks were dubbed 'bio-terrorists' in a series of Tweets and a hashtag of #ChineseDontComeToJapan! (*#chuugokujinnihonnikuruna*) encouraged a spate of anti-Chinese posts. Within similar posts, references to *bakugai*, empty shelves and the inherently 'unclean' nature of 'the Chinese race' circulated widely. Thankfully, Japanese anti-hate speech activists also intervened to challenge the more egregious posts, but the COVID-19 pandemic, much like elsewhere has seen a dramatic rise in anti-Chinese racism in Japan. Regardless of the content of the posts, these events demonstrate the capacity for a seemingly quotidian mobile sociality to take on political significance in the Sino-Japanese context. Furthermore, given the volatile nature of these mobile socialities at the time of writing, it is likely that further attention to how mobile socialities are politicized will be necessary in the near future.

Discussion: imagining mobile socialities

In this chapter, I combine a keyword approach with ethnographic insights to trace two entangled assemblages of mobile sociality in the Sino-Japanese context. I do so to explore how these terms are enacted and imagined by young mobile Chinese people in Japan, both among their personal networks and in response to wider media discourse. I started by introducing the context of *daigou* and *bakugai*, while also providing an example of the significance these practices had for one of my friends and participants. On a personal scale, *daigou* as a mobile sociality represented an opportunity for success and social mobility, while also alluding to a wider scale of Sino-Japanese mobilities. In the section following this ethnographic vignette, I explained the importance of *daigou* as a form of mobile sociality among networks of young Chinese people in Tokyo. It is a form of sociality generated from the transnational mobility of commodities between China and Japan. But it also involved young mobile Chinese people, who as mobile subjects themselves also moved between China and Japan. In doing so, I explore how they imagine its relationship to their own life projects and provide examples of how they re-present *daigou* to each other and to potential customers, both online and offline. These examples are not merely individualistic or instrumental, however, as the collaborative enthusiasm that emerged from creating networks of *daigou*, particularly through the mentorship of Dayuan, fostered collaborative opportunities and friendships. In the final section, I discussed how the wider discursive representation of *daigou* and its related term *bakugai* rescales the mobile socialities discussed in the earlier sections. In particular, I provide some examples of how young mobile Chinese people in Tokyo have responded to the negative portrayal of their activities, as these examples are particularly evocative for how we think about the imaginative rescaling of mobile socialities.

The way in which their mobile socialities were represented and imagined was both practically and politically important to many of my participants. The presentation of self, the presentation of one's social position and the relationships *daigou* afforded, reflected their socialities in its most common sense of the term. However, the economic ties they forged with each other and their buyers, alongside their response to the politicization of *daigou* and *bakugai*, were also forms of sociality. As Long and Moore (2013: 4) argue, sociality is a 'dynamic relational matrix' within which we come to understand our place in the world. Imaginaries, from the political to the personal, are an important part of sociality 'within which subjects [and objects] are constantly interacting in ways that are co-productive, and continually plastic and malleable' (Long and Moore, 2013: 4). I combine an ethnographic focus on mobile socialities with attention to the keywords used to describe how these mobile socialities are framed in discourse, such as through the term *daigou*. This approach has allowed me to trace the continually plastic and rescalable imagination of a set of young Chinese mobile socialities in Japan. In this final discussion, I would like to take inspiration from a few suggestive turns of phrase within the ethnographic example I provide in this chapter to speculate on what else might be learned from the case of *daigou* and *bakugai*. These reflections include the relationship between imagination and mobility, and the need to further theorise imagination in order to ensure that a mobile socialities approach is scalable.

In an effort think beyond earlier scholarship that depended on methodological nationalism (Wimmer and Schiller, 2002) or epochal claims about globalization (Giddens, 1990), fieldwork-based mobilities and socialities research has tended to focus on inter-personal scales. The 'turn' to socialities within anthropology, geography, sociology and their cognate fields saw increased emphasis on a 'flat ontology' in how we theorise our encounters with the world (Ash, 2020). In efforts to move away from analyses where 'the social' is used to explain almost anything (Latour, 2005), sociality became a term that signified the researcher's desire to move away from the idea that societies or 'the social' stand as an umbrella under which other activities take place (Long and Moore, 2013). Within work on socialities and mobilities, for example, this call for a 'flat ontology' in global research provoked shifts away from nested scales, whereby the local fit into the national which fit into the international, and so on. Yet, in daily life and fieldwork, the assemblage of mobilities and socialities that we theorize as 'flat' are often described and imagined in representational forms that are scaled. One of these forms is the nation, which as Florian Schneider (2018) has noted, is particularly resilient as a bounded imaginary despite our best critiques. My interlocutor's objection to *bakugai* discourse, and their proposal that Chinese economic activities in Japan are a gift *to Japan*, suggests the power of bounded national imaginaries, even when discussing transnational mobile socialities. Furthermore, their identification of an assemblage of mobile socialities as *daigou* or *bakugai*, and offence at these mobile socialities' (mis)representation, suggests that the desire to define and bind complex phenomena so that it can be imagined as a 'thing' applies to socialities and mobilities as much as a territory or an object. 'People use language to scale the world around them' (Carr and Lempert, 2016: 3), and the attribution of a name to a set of mobile socialities allows them to be re-imagined at multiple scales.

Imagining destinations or places left behind has been a central concern of ethnographic studies of migration and tourism for at least the past 30 years (Polson, 2016; Salazar, 2011). And wider global processes, as Arjun Appadurai argued, have ensured that

> The imagination has become an organized field of social practices, a form of work (in the sense of both labour and culturally organized practice), and a form of negotiation between sites of agency (individuals) and globally defined fields of possibility.
> (1996: 31)

(Im)mobility is at the center of both processes of imagination and our shared imaginaries (Adey et al., 2014; Elliot et al., 2017; Salazar, 2020). Historically within media research, such as Marshall Mcluhan's original declaration that 'the medium is the message' (McLuhan and Lapham, 1994) the power of printed materials (Anderson, 1991), television broadcasts (Silverstone, 1994) and online activism to rescale perceived collectivities (Postill, 2014) has been one of the field's strongest legacies. Despite criticisms of the potential for technological determinism in these analyses, most agree that these various media are what make the imagined communities of contemporary nations possible (Anderson, 1991). In these studies, tracing how technologies afforded new ways for content to move is used to explain how imaginaries can spread and become reimagined as representing a new scale of sociality (also see Morley in Chapter 1 of this volume). More recently within the anthropology of social media, the phrase 'scalable sociality' has become a popular way to examine how media platforms such as Facebook, WhatsApp and Twitter, allow agents to frame the seemingly personal as public, or vice versa (Miller et al., 2016). As Miller et al. argue, 'social media has colonised the space of group sociality between the private and the public. In so doing it has created scales, including the size of the group and the degree of privacy' (2016: x). What these studies all suggest is that theories of mobility and sociality require close attention to imagination and scale, as well. Tracing how mobile socialities are represented through keywords is a useful methodological starting point, as are the many other approaches provided in this volume (cf. Hartmann, Chapter 17).

The exchange and circulation of 'things' has long been understood as representing both relationships and modes of imagination. Marx's essay on 'commodity fetishism', for example, argued how imaginative leaps were necessary to separate an object from the social relationships that produced it, which allowed for the profitable revaluation of the object and the erasure of its morally dubious production (Marx, 1976). However, the role of imagination in exchange is not limited to capitalism or commerce, as Marcel Mauss reminds us in *The Gift* (Mauss, 1954). Perhaps the earliest form of exchange, gifts maintain social relationship and create opportunities for new ones, communicating moral obligations between people. The gift, through its mobility between persons and groups, creates a force that binds giver and receiver (or *hau* in Mauss' original analysis of Maori gift-giving), with moral quandaries arising when the gift is repaid too soon, or never repaid. Gifts can stand for the virtue and prestige of the powerful, or simply relationships of care. Regardless of their dynamics, however, gifts show how the practice of circulating items both produces and re-presents socialities, with wide-ranging imaginative effects. One might even say that gifts are the prototypical mobile sociality. Classical ethnographies, such as Malinowski's *Argonauts of the Western Pacific* (Malinowski, 1922) spent considerable time tracing how gifts circulated between geographically separated groups produced social bonds over considerable distances. The journeys required to send these gifts required various mobilities, from sea travel to trekking, with socialities of obligation and prestige produced via these mobilities. My participants' commodity-based mobile socialities and their repeated reference to gifts are strangely reminiscent of these earlier social theoretical fixations with 'things' and the imaginative affordances of their circulation.

The mobile socialities of *daigou* and *bakugai*, with slippages between commodities and gifts, as well as personal, transnational and national scales, provides a vivid example of the imaginative power and scalability of mobile socialities. The declaration of a commodity as a 'gift' to avoid a *daigou* transaction being taxed, the connections that developed from working together to start *daigou* networks, and the claim that *daigou* and *bakugai* are gifts to the Japanese economy all demonstrate the flexibility found in how mobile socialities are represented and rescaled. As a mobile sociality that centres on the circulation of 'things', whether as gifts or commodities, *daigou* also provides ripe fruit for considering how the movement of 'things' and the representation of a practice where things are circulated (*daigou* and *bakugai*) inspires differing imaginaries (Sneath et al., 2009). Given the significance attributed to N95 face masks in the COVID-19 era and the tense debates surrounding the informal trade of items such as masks in the Sino-Japanese context, it is clear that 'things' that may not explicitly convey media content, nonetheless convey considerable meaning. As David Morley notes in Chapter 1 of this volume, drawing on materials developed for a workshop hosted in Cologne in 2017,[7] 'while mobility has been defined as movement ascribed with meaning, one might in similar fashion define media as meaning ascribed with movement' (Coates et al., 2017). Cases such as *daigou* add credence to this proposition, as they show how the trade or 'gifting' of objects can be reframed as acts of communication. Within this context we need to extend what we consider as 'media' (Morley, 2017) and focus more on 'things', how they move, how their movements are interpreted and what the effects of their movements might be. A focus on mobile socialities affords closer attention to things and people, their associated mobilities and practices, and the socialities, scales and imaginaries that may emerge from them.

Notes

1 This name is a pseudonym.
2 This name is a pseudonym.
3 This name is a pseudonym.
4 This name is a pseudonym.
5 This is the participant's real name, he featured in an ethnographic film I made and is happy to have his public name included.

6 As these were personal posts on a closed platform, I have not included any of the specific content of the posts.

7 This quotation originally comes from the CFP for a workshop written by Jamie Coates, Roger Norum and Anna Lisa Ramella. The workshop was hosted on 14–16 September 2017 at The University of Cologne, Germany and was a collaboration between the Anthropology and Mobility Network (ANTHROMOB) and the Media Anthropology Network (MEDIANTH) for the European Association of Social Anthropologists (EASA).
The CFP states:

> This international workshop seeks to theorize the relationship between media and mobility. While mobility has been defined as movement ascribed with meaning, one might in similar fashion define media as meaning ascribed with movement. Interrogating the linkages between media and mobility can enable more thorough understandings of how various power structures produce, transform and reproduce social, material and discursive orders. People, devices, and data are increasingly on the move – movements that may transgress borders and boundaries, but which are also integral to the constitution and regulation of the barriers themselves. The movement of people triggers new imaginaries of territories and social spaces, which circulate through media, questioning and forging new ties between people, signs and things. More broadly, the mobilisation of tangible and intangible things demands a reconceptualization of what a 'thing' is, what constitutes the human, and what defines human collectivity. In such circumstances, reimagining circulations through the lens of media and mobility becomes an important step towards understanding current socio-cultural and political changes. While this lens has been applied broadly within anthropological research, its theoretical consequences merit further investigation and discussion.

With a focus on a comparative approach, this workshop invites papers that rethink the theoretical underpinnings of media and mobility studies in anthropology. In particular, we hope to encourage papers based on multidisciplinary and mixed-methods research between social anthropology and other disciplines, including sociology, geography, communication studies, and the digital humanities.
Further details about the workshop are available at: https://mediamobility.wordpress.com/.

References

Abidin, C. (2015) "Communicative intimacies: Influencers and perceived interconnectedness", *Ada*, 8: 1–16.

Abidin, C. (2016) "Aren't these just young, rich women doing vain things online? Influencer selfies as subversive frivolity", *Social Media + Society*, 2(2): 1–17.

Abidin, C. (2017) "#familygoals: Family influencers, calibrated amateurism, and justifying young digital labor", *Social Media + Society*, 3(2): 1–17.

Adey, P., Bissell, D., Hannam, K., Merriman, P. and Sheller, M. (2014) *The Routledge Handbook of Mobilities*. London: Routledge.

Anderson, B. (1991) *Imagined Communities: Reflections on the Origin and Spread of Nationalism*. New York: Verso.

Appadurai, A. (1996) *Modernity at Large*. Minneapolis: University of Minnesota Press.

Ash, J. (2020) *Flat Ontology and Geography. Dialogues in Human Geography*. New York: Sage.

Baudinette, T. (2016) "Ethnosexual frontiers in queer Tokyo: The production of racialised desire in Japan", *Japan Forum*, 28(4): 465–485.

Brettell, C. and Hollifield, J. (2000) *Migration Theory: Talking across Disciplines*. Hove: Psychology Press.

Callan, A. (2016) "Pardon your French: Chinese tourists and a clash of cultures in Paris", *Aljazeera*. Available at: www.aljazeera.com/blogs/2016/07/pardon-french-chinese-tourists-clash-cultures-paris-160727093734703.html (accessed July 23, 2020).

Carr, E. S. and Lempert, M. (2016) *Scale: Discourse and Dimensions of Social Life*. Berkeley, CA: University of California Press.

CBNData (2017) "Jìnkǒu xiāofèi dòngchá bàogào" [Import consumption report CBNData]. Available at: http://cbndata.com/report/92 (accessed June 16, 2020).

Coates, J. (2018) "Persona, politics, and Chinese masculinity in Japan", in D. Hird and G. Song (eds.) *The Cosmopolitan Dream: Transnational Chinese Masculinities in a Global Age*. Hong Kong: Hong Kong University Press, 127–148.

Coates, J. (2019) "The cruel optimism of mobility: Aspiration, belonging, and the 'good life' among transnational Chinese migrants in Tokyo", *positions. Asia Critique*, 27(3): 469–497.

Coates, J., Norum, R. and Ramella, A. (2017) "Media and mobility workshop call for papers", *Anthropologies of Media and Mobility: Theorizing Movement and Circulations across Entangled Fields*, The University of Cologne, Germany, September 14–16. Available at: https://mediamobility.wordpress.com/

Eisenhart, M. (2017) "A matter of scale: Multi-scale ethnographic research on education in the United States", *Ethnography and Education*, 12(2): 134–147.

Elliot, A., Norum, R. and Salazar, N. (2017) *Methodologies of Mobility: Ethnography and Experiment*. Oxford: Berghahn Books.

Fong, V. (2011) *Paradise Redefined: Transnational Chinese Students and the Quest for Flexible Citizenship in the Developed World*. Stanford, CA: Stanford University Press.

Genron (2017) "The 10th Japan-China public opinion poll analysis report on the comparative data". Available at: www.genron-npo.net/en/pp/archives/5153.html (accessed February 22, 2018).

Giddens, A. (1990) *The Consequences of Modernity*. Stanford, CA: Stanford University Press.

Goffman, E. (1959) *The Presentation of Self in Everyday Life*. New York: Double day.

Google Trends Labs (2020) "About Google trends". Available at: www.google.com/intl/en/trends/about.html#7 (accessed July 27, 2020).

Gustafsson, K. (2014) "Memory politics and ontological security in Sino-Japanese relations", *Asian Studies Review*, 38(1): 71–86.

Han, X. (2020) "Historicising Wanghong economy: Connecting platforms through Wanghong and Wanghong incubators", *Celebrity Studies*, 0(0): 1–9.

Hara, K., Liao, T. and Wiegand, K. (2015) *The China-Japan Border Dispute: Islands of Contention in Multidisciplinary Perspective*. London: Ashgate.

Hart, K. (1985) The informal economy", *Cambridge Anthropology*, 10(2). Oxford: Berghahn Books, 54–58.

Ingold, T. (2007) *Lines: A Brief History*. London: Routledge.

Iwabuchi, K. (2004) *Feeling Asian Modernities: Transnational Consumption of Japanese TV Dramas*. Hong Kong: Hong Kong University Press.

Kipnis, A. (2006) "Suzhi: A keyword approach", *The China Quarterly*, 186: 295–313.

Lai, H. (2015) "Engagement and reflexivity: Approaches to Chinese – Japanese political relations by Chinese students in Japan", *Journal of Current Chinese Affairs*, 44(3): 183–212.

Latour, B. (2005) *Reassembling the Social: An Introduction to Actor-Network-Theory*. Oxford: Oxford University Press.

Liao, S. (2019) "Wang Hong fashion culture and the postfeminist time in China", *Fashion Theory*, 0(0): 1–23.

Liu-Farrer, G. (2011) *Labour Migration from China to Japan International Students, Transnational Migrants*. London: Routledge.

Liu-Farrer, G. (2020) *Immigrant Japan: Mobility and Belonging in an Ethno-Nationalist Society*. New York: Cornell University Press.

Long, N. and Moore, H. (2013) *Sociality: New Directions*. Oxford: Berghahn Books.

Malinowski, B. (1922) *Argonauts of the Western Pacific*. London: Routledge.

Martin, F. (2017) "Rethinking network capital: Hospitality work and parallel trading among Chinese students in Melbourne", *Mobilities*, 12(6): 890–907.

Marx, K. (1976) *Capital: Volume 1*. London: Penguin, New Left Review.

Massey, D. (1994) *Space, Place and Gender*. Minneapolis, MN: University of Minnesota.

Mauss, M. (1954) *The Gift: Forms and Functions of Exchange in Archaic Societies*. Translated by Ian Cunnison. New York: W.W. Norton.

McLuhan, M. and Lapham, L. (1994) *Understanding Media: The Extensions of Man*. Cambridge, MA: MIT Press.

Miller, D. et al. (2016) *How the World Changed Social Media*. London: UCL Press.

Ministry of Justice [MOJ] (2019) "Zairyu gaikokujin tōkei (kyū tōroku gaikoku hito tōkei) tōkei-hyō" [Statistical summary of foreign residents in Japan (including old foreign registration statistics)]. Available at: http://www.moj.go.jp/housei/toukei/toukei_ichiran_touroku.html (accessed June 16, 2020).

MLIT [Ministry of Land, Infrastructure, Transport and Tourism] (2019) "Hōnichi gaikoku hito ryokō-sha-sū shukkoku nihonjin-sū" [Foreign visitors in Japan: figures for tourists and number of Japanese leaving]. Available at: https://www.mlit.go.jp/kankocho/siryou/toukei/in_out.html (accessed November 15, 2019).

Moore, H. (2004) "Global anxieties concept-metaphors and pre-theoretical commitments in Anthropology", *Anthropological Theory*, 4(1): 71–88.
Morley, D. (2017) *Communications and Mobility: The Migrant, the Mobile Phone, and the Container Box*. Hoboken, NJ: Wiley-Blackwell.
Morris-Suzuki, T. and Soh, E. J. (2017) *New Worlds from Below: Informal Life Politics and Grassroots Action in Twenty-First-Century Northeast Asia: Volume 9*. Canberra: ANU Press.
Mostafanezhad, M., Coates, J. and Coates, J. (2018) "Journeys from the east: The popular geopolitics of film motivated Chinese tourism", *International Journal of Tourism Anthropology*, 6(3): 219–236.
Nakajima, K. (2015) *Bakugai-go: Karera wa doko ni mukau no ka. Explosive Shopping: What are their Goals?* Tokyo: President Publishing.
OEC (2020) *Country Profile: Japan*. The Observatory of Economic Complexity. Available at: https://oec.world/en/profile/country/jpn (accessed January 8, 2021).
Onishi, N. (2020) "Coronavirus empties European cities of Chinese tourists", *The New York Times*, February 17. Available at: www.nytimes.com/2020/02/17/world/europe/coronavirus-tourism-europe.html (accessed July 24, 2020).
Osburg, J. (2013) *Anxious Wealth: Money and Morality Among China's New Rich*. Stanford, CA: Stanford University Press.
Otmazgin, N. (2008) "Contesting soft power: Japanese popular culture in east and southeast Asia", *International Relations of the Asia-Pacific*, 8(1): 73–101.
Polson, E. (2016) *Privileged Mobilities: Professional Migration, Geo-social Media, and a New Global Middle Class*. New York: Peter Lang.
Postill, J. (2014) "Democracy in an age of viral reality: A media epidemiography of Spain's indignados movement", *Ethnography*, 15(1): 51–69.
Salazar, N. (2011) "The power of imagination in transnational mobilities", *Identities: Global Studies in Culture and Power*, 18: 576–598.
Salazar, N. (2017) "Key figures of mobility: An introduction", *Social Anthropology*, 25(1): 5–12.
Salazar, N. (2020) *On Imagination and Imaginaries, Mobility and Immobility: Seeing the Forest for the Trees: Culture & Psychology*. London: Sage. Available at: https://journals.sagepub.com/doi/10.1177/1354067X20936927 (accessed July 26, 2020).
Salazar, N. and Jayaram, K. (2016) *Keywords of Mobility: Critical Engagements*. Oxford: Berghahn Books.
Salazar, N. and Schiller, N. (2016) *Regimes of Mobility: Imaginaries and Relationalities of Power*. London: Routledge.
Schneider, F. (2018) *China's Digital Nationalism*. Oxford: Oxford University Press.
Seaton, P. and Yamamura, T. (2018) *Japanese Popular Culture and Contents Tourism*. London: Routledge.
Sheller, M. and Urry, J. (2006) "The new mobilities paradigm", *Environment and Planning A*, 38(2): 207–226.
Silverstone, R. (1994) *Television and Everyday Life*. London: Routledge.
Sneath, D., Holbraad, M. and Axel Pedersen, M. (2009) "Technologies of the imagination: An introduction", *Ethnos*, 74(1): 5–30.
Stokes, B. (2015) *How Asia-Pacific Publics See Each Other and Their National Leaders*. Pew Research Centre Global Attitudes and Trends. Available at: https://www.pewresearch.org/global/2015/09/02/how-asia-pacific-publics-see-each-other-and-their-national-leaders/ (accessed January 8 2021).
Strickland, M. (2010) "Aid and affect in the friendships of young Chinese men", *Journal of the Royal Anthropological Institute*, 16: 102–118.
Tajima, J. (2005) "Daitoshi Ni Okeru Chūgoku Kei Ijūsha Chōsa" [A survey of Chinese migrants in the metropolis], in Junko Tajima (ed.) *Chūgoku Kei Ijūsha Kara Mita Nihon Shakai No Sho Mondai* [Japanese Social Problems from the Perspective of Chinese Migrants]. Tokyo: Research Foundation for a Safe Society, 7–37.
Tsing, A. (2013) "Sorting out commodities: How capitalist value is made through gifts", *HAU: Journal of Ethnographic Theory*, 3(1): 21–43.
Wenger, E. (1998) *Communities of Practice: Learning, Meaning, and Identity*. Cambridge: Cambridge University Press.
Williams, R. (1985) *Keywords: A Vocabulary of Culture and Society*. Oxford: Oxford University Press.
Wimmer, A., and Schiller, N. (2002) "Methodological nationalism and beyond: Nation – state building, migration and the social sciences", *Global Networks*, 2(4): 301–334.
Wittel, A. (2001) "Toward a network sociality", *Theory, Culture & Society*, 18(6): 51–76.

Yamashita, K. (2010) *Ikebukuro Chainataun: Tonai Saidai No Shinkakyogai No Jitsuzou Ni Semaru* [Ikebukuro Chinatown: The Emergence of Tokyo's Biggest New Overseas Chinese Street]. Tokyo: Yousen Publishing.

Yang, M. (1994) *Gifts, Favours and Banquets: The Art of Social Relationships in China.* New York: Cornell University Press.

Zhang, L. (2017) "Fashioning the feminine self in 'prosumer capitalism': Women's work and the transnational reselling of Western luxury online", *Journal of Consumer Culture*, 17(2): 184–204.

Zhou, M. and Wang, H. (2017) "Anti-Japanese sentiment among Chinese university students: The influence of contemporary nationalist propaganda", *Journal of Current Chinese Affairs*, 46(1): 167–185.

8

DIGITIZING DESIRES

Immobile mobility and social media in southeast Turkey

Elisabetta Costa

Introduction

This chapter is an ethnographic exploration of the mediated practices of young homebound women who escape into the place of social media to create and maintain new forms of social relationships. In the last decade, research on digital technologies has shed light on the multiple ways in which digital platforms and mobile media are embedded in people's everyday lives, showing that boundaries between online and offline environments are blurred, traversed, and continuously crossed. Extensive ethnographic literature suggests that the internet and social media are strictly interwoven into different domains of people's lives within multiple contexts and fields (among others, see Miller et al., 2016; Miller and Slater, 2000; Hine, 2015, 2017). This is not a universal social fact determined by the properties of the platforms. Social media can also be used as places in their own right that are neatly divided from the offline world. In Mardin, a medium-sized town in southeast Turkey, young Muslim homebound women from conservative families commonly use social media to create and maintain relationships that mostly pertain to the online world. They engage in personal communications and interactions with people they never met face to face. Online and offline remain two different discrete places.

These findings recall the conceptualization of the internet formulated by early scholars of cyberspace, who emphasized the capacity of digital technologies to create a new separate environment where people could construct virtual identities that are detached from their physical ones (among others, see Graham and Khosravi, 2002; Holmes, 1997; Jones, 1997; Jordan, 1999). Cyberspace was viewed as a site of experience where people could leave their physical world's life behind. Unlike those studies, though, this chapter is grounded in traditional long-term anthropological fieldwork that views the use of the internet as a set of embodied practices embedded in specific social, cultural, and physical environments. The online place that I describe in this chapter is brought into being by specific cultural practices. I observed and compared the full spectrum of women's interactions and socialities online and offline. By doing so, I located my analytical focus away from media-centric perspectives that associate the old internet with cyberspace, and social media with the routines of everyday life, and I discovered that also social media can be experienced as a place divided from the offline world, a place where people can transgress social norms, enjoy new freedoms, and engage in diverse forms of socialities.

In Mardin, social media are often used to satisfy wishes and desires that remain unfulfilled offline under the constraints imposed by social norms ruling interactions between gender, ethnic and age groups. Facebook and WhatsApp facilitated the flourishing of new types of gender-mixed friendships, flirting, and romances that escaped existing social constraints. Social media are spaces of flirtation, interaction, and fun (see Costa, 2016 for an overview of the uses of social media in Mardin). As eloquently stated by Güliz,[1] a 25-year-old Kurdish woman, "In Mardin, we [young women] use Facebook to break barriers". Although in many cases, the relationships created online moved offline, and the other way around, in the case of homebound women living with restrictive parents or husbands, relationships developed online remained online. On social media, these women could secretly meet new people, befriend strangers, engage in online romances, and explore new communication opportunities without disappointing family members, neighbors, and friends. In Mardin, homebound women with limited opportunities for interactions with people outside their family used Facebook to interact or flirt with strangers, look at their public profile out of everybody's gaze, and secretly chat with them without meeting them face to face.

Drawing on the experiences of women who lived under constraints on their freedom of movements and relationships, this chapter shows that social media can be viewed as a discrete place that provides opportunities to fulfill desires that cannot be realized in the physical world. I define this movement from offline to online as "immobile mobility" (see also Wallis, 2011, 2013; Ureta, 2004).[2] Building on and reversing Roland Barthes' concept of "transported immobility" (1979: 143), which describes the paradox of travelers that experience the illusion of immobility while moving, "immobile mobility" captures the experience of getting away while sitting motionless at home. This experience is interwoven with the practices of social media usage in the offline context of forced immobility. In the backdrop of a patriarchal society characterized by older men's control of young women's movements and relationships, social media became a getaway world where women realize their unfulfilled desires of romances, friendships, and socialities. The online place of social media is very real, but at the same time is different and divided from the actual physical world (see also Boellstorff, 2015, 2020; Waltorp, 2020).

By examining social media as a place with its own specific forms, the chapter sheds light on the spatial elements that shape online social interactions and relations. In the digital place of social media, young women privately liberate themselves from constraints that are experienced as oppressive, while publicly conforming to the expectations that the society has on them. In the social media topography, they can satisfy wishes and desires out of the gaze of older relatives and adults, while reproducing offline the public normative understandings of social relationships. They appropriate and domesticate (Silverstone et al., 1992) digital technologies to satisfy desires – emotions that represent a longing, a yearning, or a wishing for something (Joseph, 2005), in such a way that relationships with family members and neighborhoods are not compromised, and relations of male domination are reproduced. In the last section of this chapter, I build on the concept of "immobile mobility" to capture how female agency unfolds in the practices of social media usage. I conclude with a discussion of how a "mobile sociality" approach allows us to shed light on the specific forms of women's abilities for actions within the context of unequal gender relations.

Methods and site

The chapter draws from the results of a long-term ethnographic research that examined uses and consequences of social media on people's everyday life in the town of Mardin, in south-east Turkey, a region inhabited by a large majority Kurdish population. Fieldwork lasted 15 months and was carried out in 2013 and 2014 as part of a larger comparative project based at University College London Department of Anthropology.[3] In Mardin, I lived on my own and hung out with locals all day. I sat in cafés and had dinners with friends; I took part in weddings, family gatherings, and holiday trips. I friended my research participants on Facebook, Twitter, and WhatsApp, and I followed what they were doing online. At that time, Facebook was the most used social media and, for this reason, dominates the ethnographic material on which this chapter is based. I also carried out 100 in-depth interviews and submitted two different sets of questionnaires.

Mardin is a fast-growing urban center with a population of around 87,000, which includes a majority of Muslim Sunni Arabs and Kurds, a small minority of Orthodox Syriacs and Catholic Armenians, an increasing number of Syrian refugees, and a few Turks. Located within the larger Kurdish region of Turkey, Mardin has a big Arabic-speaking minority, which makes it a unique place in the country. In the last 20 years, the city underwent significant social transformations, brought by processes of urbanization and economic development. Surrounded by the poorer Kurdish cities of Kızıltepe and Nusaybin, Mardin stands out as the most "modern" and wealthiest town in the province, also thanks to the increasing number of residing state employees. Fieldwork took place in the new part of the city, called *Yenişehir,* one of the three main city's areas. This is the well-off part of the town. Here, single nuclear family households with few children are more common than extended families, the traditional family unit in southeast Turkey. Young adults tend to be more educated and digitally literate than in other provinces and regions of the southeast. Women often go to school and universities, and work outside the home. They often have access to public spaces, such as schools, university, streets, and modern cafés (Costa, 2016, 2019). Mardin, especially the New City *(Yenişehir)*, is inhabited by a diverse population that includes more progressive and educated inhabitants, along with more conservative and religious ones. This diversity reflects both the economic transformations that affected the southeast of Turkey in the first decade of the 2000s, the rising of a better-off class of state employees, and the different religious and political ideologies crossing the region. Internet penetration in the less developed region of southeast Turkey and Mardin is high. At time of fieldwork, in Mardin, everybody under the age of 35 used the internet, and almost every house in the New City had a computer. Almost all 102 respondents to my questionnaire owned a mobile phone, and 80 percent owned a smartphone. Almost 80 percent of the respondents had either a wi-fi connection in their houses or a USB dongle, and only 20 percent did not have any internet at home (Costa, 2016). The most significant inequality in internet access was generational. All teen-agers and young adults in Mardin were massive internet and social media users. Only the elderly rarely accessed computers and smartphones.

Despite the undergoing social changes, Mardin continues to be dominated by patriarchal relations and gender segregation. In most cases, women are expected to stay at home or in the neighborhood after school or work, and their movements are highly controlled by male family members. Female reputation is built on modesty, piety, lack of suspicious contacts with men, and strict adherence to public social norms. Yet, there are significant differences in women's behaviors. Whereas young women from more progressive families walk around the city, sit in cafés, go to universities, work outside the home, and engage in conversations with male neighbors, colleagues, and friends, young women from more conservative and less educated backgrounds have limited access to places outside their homes. They are more often excluded from higher education and work activity, and their social life is mostly limited to the extended family. Most of those I spoke to perceive the limitation of movement and the exclusion from social relations beyond the household as an impediment to their happiness and well-being. They rarely question these constraining social norms, though. They rather try to circumvent the restrictions with the tools available, and social media is one of them. Facebook, more than any other social media platform, offered them the opportunity to secretly start and maintain new interactions and relationships with strangers, and at the same time keep these relationships hidden from the public view and not damage their reputation. Facebook provided women with limited opportunities of movement, the possibility to live inside a place that perfectly fulfills their desires.

Suad Joseph (2005) defined the desires of Arab women living in the neighborhood of Borj Hammoud in Beirut as "relational desires", to include erotic and non-erotic desires that are learned and practiced in the context of patriarchal relationships. Women's desires originate within a system of patriarchal connectivity that privileges the initiatives of males and elders (Joseph, 1999: 12). In Beirut, women's desires are invested both as a property of a single person and of her relationships (Joseph, 2005). Questions such as "What do women want?" should be addressed within the context of relational selfhood (Joseph, 2005). Women's self is embedded in gendered and hierarchical relations defined by their subordinate role within the kinship system (Joseph, 2005). In Mardin, very much like in Joseph's field-site in Beirut, women's desires are "relational" and the formation of desire is organized around domination. Women rarely openly question their father's or husband's wishes – they rather tend to indulge in them – and they do not openly express desires that can make them unhappy. Like in many other areas of the Middle East, also in Mardin, family, kin structures, and communities are necessary for a successful social existence and are often more important than individual freedoms. People feel responsible for others in their family. For example, Gül, a 19-year-old woman, explained to me that she would never have pre-marital romances with a man because her dad had heart problems and might become ill if he found out about it. And above all, she wanted to make him happy. I later found out that she was engaging in secret mobile phone conversations with a male neighbor. When I talked to her again, she told me that she was extremely careful to keep this interaction completely confidential, and she would be ready to end it if it would ever become public. Pleasing her father and family was more important than pleasing herself. In Mardin, expressing responsibilities towards other members of the family is more culturally appropriate than presenting an autonomous and independent self.

Digitizing desires in the social media topography

Gülüm is a Kurdish married housewife in her early thirties who lives in Mardin with her unloved husband and their two children. During the day, she stays at home, takes care of the house, makes food, and looks after the children. She lives her life within the house in front of her smartphone and laptop. She spends several hours every day on Facebook, either chatting and gossiping privately with her female friends, or engaging in a romantic relationship with a married man living in another province in southeast Turkey. Gülüm and the Facebook man have never met face to face, but spend several hours every week texting and video calling each other. He approached her on Facebook by sending a private message a few months earlier, and they then started texting each other. Gülüm keeps the romantic relationship secret, and only a trusted female cousin and I knew about it. Her Facebook profile is kept rigorously anonymous, the profile picture portrays a flower, her name is fake, and the list of Facebook friends is hidden from public view. Gülüm talks to the man on the phone when she is alone in the house and the kids are in school, and frequently sends him text messages or photos at different times of the day. Gülüm described this online relationship through the register of liberal romantic love, characterized by her passionate and honest feeling (Schielke, 2009). Gülüm's husband is violent, not caring, and distant with her and the children, and very likely involved in romantic relationships with other women. Her marriage was arranged by their families with the consent of the husband, but unlike other arranged marriages in town, it did not turn into a happy one. Adultery obviously clashes with Gülüm's notion of a respectful and honorable woman, and she preferred not to talk extensively about it with me. Yet, at the same time, a mediated romance that does not involve physical co-presence and sexual intercourse was more compatible with her moral values of Islam and honor than a relationship in physical co-presence. Gülüm mentioned that she did not consider the option to meet the "Facebook man" face to face, and even less did she take into account divorce. The online romantic flirting made her life more livable and fulfilled while cultivating her duties as wife and mother. Gülüm has been suffering from the consequences of an arranged marriage with a man she did not love, and who did not prove to be a reliable and caring husband and father. Trapped in an unhappy relationship and life, she enjoyed the freedoms enabled by the architecture of Facebook, without risking the compromising of her reputation in town, nor her relationship with her family. She engaged in a digital extra-marital relationship without leaving her house. Gülüm is one of the many young homebound women who use social media to flirt and have romantic relationships with men. Many others used social media to experience new forms of socialities detached from their offline life, either friendships or pre-marital and extra-marital romances.

Güliz, an unmarried 25-year-old Kurdish woman who lives with her parents argued:

> I use Facebook for two main reasons. To be in touch with my foreign friends, most of my Facebook friends are foreigners; and to talk to men because I am not allowed to do it in the offline world. It is not just me. This is the reason why people use social media here in Mardin.

Seda, a 17-year-old woman from an Arab speaking family affiliated to a Muslim brotherhood said:

> I have only one Facebook account, and I don't have any real friend from my offline life there. I use Facebook to look at online friends' profiles because I cannot leave home after school. But I want to know other people and discover more about their life.

Social media enabled the creation and maintenance of social relationships that depart from established modes of relations embedded in kinship, patriarchal family, Muslim values, and ordinary gender roles. Although sometimes relationships created online moved offline, in the case of women relegated at home, relationships developed online remained online. I highlight this separation between the online and offline world to demonstrate that social media can be treated as places in their own right. They do not simply mediate relationships between people, but create an environment in which these relationships can be started, lived, experienced, and ended. Further to this, digital places are not immutable and given, but are crafted, molded, built, and shaped by their users. Users have agency over the degree of privacy and the size of groups they want to interact with. Facebook is not only a "scalable sociality" (Miller et al., 2016), an ecology of spaces with different degrees of publicness and privacy that are inhabited by smaller and bigger groups of people. In Mardin, Facebook is a configuration of very private and very public environments that facilitate secret encounters, flirting, dating, and chatting with a potentially unlimited number of people, and where users actively engage in the making of online places (Costa, 2016). In what follows, I illustrate the social media practices enacted by those women who have a very limited social life outside the extended families, and could not openly be on social media. These practices included communications and interactions with strangers, flirting, pre-marital and extra-marital romances, and the crafting of public and private online environments.

First, many women accessed social media secretly on their phones. Younger women living with their parents could have the complicity of sisters and mothers. They could go online while sitting in the same room with their sisters who would check that their father or brother did not enter the room. When they do not own a phone, they use older sisters' or cousins' phones. Teen-agers often use the phones of school friends. When they have no wi-fi at home, in those days or weeks in which they could not afford internet data, they connect to other wi-fi connections in the building. I encountered a few families where young adult daughters knew exactly what wi-fi was available at different times of the day, and they moved accordingly around the house. It was also possible to ask the internet password of a trusted neighbor.

Second, identities in the public-facing Facebook are anonymous. There are no photos portraying real faces and bodies as profile pictures. Women preferred images of flowers, colorful nature, children, or verses from the Koran. There are no real names either, but rather acronyms of their favorite stars like actors, singers, or musicians, or invented names. Privacy settings were changed in such a way that only anonymous profile pictures and fake names were visible to the entire public or friends of friends. I have previously shown that in Mardin people mastered privacy settings pretty well to keep different social spheres apart (Costa, 2018). Women were more skillful than men, because the consequences of the unintentional disclosure of private information would have been more severe for them. The public-facing walls were usually visible only to friends and most of the time did not include photos or any other references revealing the real identity. The walls are rather filled with memes, religious images, news, and photos without recognizable faces. These posts are usually followed by "likes", but very few or no comments at all. The wall usually aims at presenting a female self that conforms to moral values of Islam, female piety, and modesty, or Kurdish guerrilla aesthetic, and do not disclose any personal information. It is not a place for direct interactions and communications with others, but sometimes it is used for indirect communications. For these women, the Facebook wall is a (semi)public identity card that included very little detail and gave them access to other public places, such as Facebook groups, Facebook games, and other personal Facebook walls, and to very private chats.

Third, it was common to spend long hours looking at the public profiles of relatives, friends from school, or acquaintances from the neighborhood, and above all to gaze into the lives of strangers living in other regions in Turkey or abroad. Facebook was used to know more about the lives of people who were unreachable offline. I collected several accounts of female teen-agers and young adults that described Facebook as a getaway to strangers living far away. In some cases, the observation of strangers' Facebook walls was followed by a friend request and the start of private conversations. Teen-agers learned English on Facebook more than at school. Social media was the only place where they could practice this language with foreigners.

Fourth, playing Facebook online games was another common way to meet strangers and interact with them, while remaining anonymous. Okey, Criminal Case, Tavla, and Poker were particularly suitable for these goals.

Fifth, public presence with an anonymous Facebook profile elicited flirtatious approaches from men in the private chat. It was common for men to start flirting with women never met face to face. They would write "hello, how are you?" (*Merhaba nasılsın?*), and the woman could then decide whether to answer or not (Costa, 2016). In the case of women with less restriction of movements, online flirting could turn into face-to-face encounters, secret offline dating, and occasionally even more stable relationships or marriage. I met a significant number of young women who met their partner online and then discreetly started dating him face to face. For example, Ceyda, a woman in her late thirties, met her husband on MSN more than ten years earlier. I bumped into her while I was submitting questionnaires in the little park in the center of the town, and she described to me her happy love story: she was approached by her current husband on MSN, they chatted for months, saw each other face to face, got engaged, and eventually married and had children together. But in the case of the women discussed in this chapter, private flirting remained limited to Facebook and could turn into online long-term correspondence and romances that were completely detached from embodied relationships of family and kin groups. Women could engage in online relationships for weeks or months, chatting several hours every day and night, without never meeting their online partners face to face. Most of the time, teen-agers and very young women were aware that online romances would not evolve into more serious relationships, and they were accepting of this. Indeed, in more conservative families, arranged marriages were the norm and the most common way to escape it was to elope with the man, something that not every woman and man was willing to do. Only a few unmarried women told me that they dreamed to turn these romances into a formal engagement and possibly a marriage, but they were too young for a wedding, and there were too many obstacles to make this happen. I did not meet these women again to know exactly what occurred, but the online romances very likely ended at a certain point.

Tom Boellstorff (2020) developed the notion of "digital topography" to differentiate digital media from digital place. He argued that the online virtual world of Second Life is a digital place characterized by the presence of specific virtual geographies with their logics and structures, what he calls "digital topographies". Second Life does not simply mediate between different places or people, but rather shapes human experiences as places can do. In the context of my study, the notion of "digital topography" allows us to capture the ways in which Facebook's affordances bring into being a configuration of online private and public environments. Boellstorff's concept of "digital topography" emerged from the study of a platform that, unlike Facebook, was designed to be a virtual world. But in my case study, Facebook is also experienced as a place detached from other social contexts of the physical realm. Here, the concept of "digital topography" highlights the ways in which the spatial forms of Facebook are intertwined with – and actively shape – practices and notions of private and public.

In Mardin, like in many other Muslim-majority countries, boundaries between the private and public have always been carefully patrolled. Women have been traditionally associated with the private place of the home, and in different ways and degrees excluded from public spaces. This exclusion is made possible also by the lack of anonymity that characterizes offline physical places. Indeed, in big Muslim cities, where a certain degree of anonymity is more easily achievable, women tend to be more visible in streets, cafés, shopping malls, and restaurants, whereas in small Muslim villages, where women are immediately recognized as specific individuals or members of specific families, their presence in public tends to be more limited. In Mardin, access to public places, such as streets, cafés, and restaurants was possible for women – but only under specific circumstances, for example in the presence of other family members. By contrast, the public Facebook safeguards complete anonymity. And unlike other internet websites, it generates a feeling of co-presence with people who have not been met face to face. Facebook generates the experience of being in a public place. Mirca Madianou calls "ambient co-presence" (2016) what is generated by peripheral awareness of the actions of others that is made possible by social media affordances. She refers, for example, to the news feed of Facebook, a personal news aggregator that shows what people do, think, and post (Ellison and boyd, 2013; Vitak, 2014). Being on Facebook enables the awareness of several others to whom these women are connected, but that they do not see face to face in their daily life.

I call the spatial configuration made of the public places of "co-presence", the semi-public places of groups, and very private chats, "social media topography". This specific topography enabled the actualization of a desired state of things: the experience of mixed-gendered friendships, pre-marital and extra-marital romances, and the accomplishment of society's wishes. The social relationships embedded in the "social media topography" are the realization of the desires of women and their male relatives at the same time. They are the realization of "relational desires" (Joseph, 2005). In the following section, I illustrate how the concept of "immobile mobility" captures the agency of homebound women and sheds light on the main dilemma of practice theories, which is the explanation of how social reproduction becomes and always includes – to some extent – social transformations.

Women's agency as immobile mobility

Neval is a 17-year-old Kurdish student from Mardin who goes on Facebook to meet strangers, know about their lives, and engage in conversations with them. I met Neval at school. She was one of the 45 high-school students whom I interviewed during fieldwork. She was excited to talk to me, and she happily opened up with me about her life and the role of social media in it. She did not like Mardin. She described it as a backward city dominated by traditional values and old-fashioned people, and she found her life boring. After school, she sat at home all afternoon and evening. Her parents did not want her to go out, sit in cafés, or see friends. She did not have sisters. She occasionally went to her female cousins' home, or received their visits at her house. For the rest of the day, she sat long hours in the bedroom with a small smartphone in her hands connected to the neighbor's wi-fi. In the afternoon, her father and two older brothers were out at work, and her mother was busy cleaning, cooking, looking after her younger brother, or chatting in the sitting room with relatives. Neval used Facebook to friend strangers and be in touch with them. She had an anonymous profile with no friends from the physical world and 250 friends from all around Turkey, Europe, and the world. Her parents did not allow her to be on social media, and she didn't want anybody to know that she was there. The best way to keep her "real" identity concealed and her Facebook account hidden from the family, was to not friend anyone from Mardin. Neval described to me how much she enjoyed looking at her Facebook friends' updates, visual posts, and photos. She also used to have private chats with men from Turkey, Turkish-speaking men from different countries in Europe, and sometimes English speakers. Neval, like other high-school students, was taking advantage of Facebook to practice her English. Learning this language was considered to be beneficial to young people's future job opportunities, but English language education at the school was not enough.

Neval's accounts of her Facebook use did not focus on online romances but on the encounters with strangers online.[4] She was fascinated by the wonderful opportunity to meet unknown others and feel part of a public that she could not access in the physical world. Online strangers remained strictly separate from the networks of social relations she had offline. Interactions with strangers online allowed her to secretly enact a femininity that was different from the socially accepted one, which did not entail that women engage in forms of socialities outside their family network. In Mardin, like in other Muslim contexts, women are expected to occupy the private domain of the household, and their social relations to be mediated by the family. The entire social world is locally classified by the two categories of kin (*akraba*) and non-kin (*yabancı*), with one family's reputation being strictly related to the limitation of women's access to social contacts outside the family network. As pointed out by McDonald (2019), the proliferation of social media in rural China has transformed local concepts of stranger and have created new opportunities for judgment of stranger relationships by others. In a similar manner, in Mardin, the new forms socialities made possible by digital platforms were strictly concealed by women. Relationships with strangers remained unproblematic – as far as nobody knew about them and they remained restricted to the online world. Shame fell on a family when a woman's non-confirming actions were publicly disclosed. Neval was only slightly concerned about the morality of her online interactions. She mostly cared about carefully concealing them from public view. Interactions and communications with strangers remained strictly limited to the online world, with no consequences for her life in the realm of the physical world.

The experiences of Gülüm, Güliz, Sedal, and Neval shed light on local forms of women's agency as "immobile mobility". I use this concept to describe two intertwined forms of mobility. "Immobile mobility" refers to the ways in which homebound women get away into digital places to engage in alternative forms of sociality while sitting at home. This is the mobility from the offline physical place of the home to the online digital place of social media. Second, it refers to the transgression of social norms that comes with their simultaneous reinforcement. The maintenance of forbidden relationships takes place together with the reproduction of the public normative understandings of social relations that is embedded in kin structures. This is the mobility away from the constraints imposed by social norms ruling offline relationships. However, this mobility ends up reinforcing those same norms that limit it in the first place. This understanding of "immobile mobility" reflects Cara Wallis's use of the concept (2011, 2013), which was applied to the study of Chinese young migrant women in Beijing. Wallis defined "immobile mobility" as a form of traversing structural and cultural factors without erasing them, "a means of surpassing, but not erasing, limiting material conditions to gain inclusion in expanded and enriched social networks" (2011: 62). Like mobile phones among Chinese migrant women in Beijing, social media in Mardin alleviate and reinforce the limitations experienced by young women from conservative backgrounds. Social media liberate and constrain at the same time. However, Wallis uses this concept to describe the multiple ways in which the surpassing of structural and social barriers is intertwined with practices of mobile phone usage in the context of physical, social, and economic (im)mobility. Conversely, in Mardin, the concept of "immobile mobility" captures the ways in which the simultaneous transgression and reinforcement of constraints takes place through the movement from physical to digital places. In my case study, "immobile mobility" stresses the role that the new forms of online socialities have in the enactment of agency.

This notion leads to an important guiding concept in the social sciences and humanities, and in the anthropology of the Muslim Middle East, which is that of women's agency (among others, see Kandiyoti, 1988, 1996; Mahmood, 2005; Marsden, 2005; Menin, 2015; Mittermaier, 2012; Sehlikoglu, 2018). Notably, Mahmood (2005) described agency as the capacity for actions that relations of subordination produce and enable. She did not view agency as a resistance to external form of domination, but rather as a struggle toward actively inhabiting hegemonic social norms and values. She aimed at "question[ing] the overwhelming tendency within poststructuralist feminist scholarship to conceptualize agency in terms of subversion or re-signification of social norms" (2005: 14). Anthropologists of the Muslim Middle East have extensively elaborated on women's agency, without particularly focusing on the role of digital media in shaping it. Mardin provides a site from which we can start conceptualizing human agency to take into account the role of digital technologies. The stories of Gülüm, Neval, and many other young women show that digital places play an important role in enabling and constraining possibilities for their actions; digital places facilitate the escape from social constraints, while leaving offline social norms unchallenged. Like the religious practices of pious women studied by Saba Mahmood (2005), the actions of Mardinite women are not an opposition to a social order they do not like and disapprove. Their social media practices are rather one of the possible ways to actively live and inhabit the social norms that limit their lives. Hence, social media practices lighten and reinforce the restrictions that constrain them in the first place.

These findings resonate with Waltorp's (2020) study of Muslim women's use of smartphones in Denmark. Waltorp (2020) builds on Bourdieu (1977) to formulate the concept of "composite habitus" that describes the habitus of second generation Muslim women in Copenhagen. "The term composite habitus points to the ways in which these young women, as most people in the modern globalized world, are informed and formed by *multiple fields* and *multiple forces* during their life trajectory" (Waltorp, 2020: 84). Waltorp (2020) uses the concept of "composite habitus" to explain the paradox of second-generation migrant women who follow – and, at the same time, challenge – different sets of moral norms. In her study, it is mostly the condition of geographic mobility and transnational belonging that creates multiple dispositions. Also in Mardin, the social change brought by modernization, urbanization processes, and digital technologies created multiple fields and dispositions. For example, the use of social media for the creation and maintenance of romantic relations was accepted and valued by the group of female peers and youth subculture. On the one hand, the concept of "composite dispositions" accounts for the forms of socialities enabled by social media in Mardin. But on the other, it does not explain the existing hierarchies between various *forces* and *fields*. In Mardin, one habitus is dominant over others: the new online socialities are generated by one enduring "structuring structure" (Bourdieu, 1977), which is the long-lasting patriarchal organization of family and society that limits women's mobility and social relations. Women have tried to secretly circumvent social norms that limited their socialities and mobility well before the arrival of social media. Secret flirting, pre-marital romantic conversations, and gender-mixed friendships have been part of local practices and imaginations in Turkey for a long time, as documented in the research (among others, see Stokes, 2010). These secret interactions aimed at actively living and inhabiting the social norms that limited women's sociality in the first place. Elderly and middle-aged people in Mardin also recognized that pre-marital and extra-marital romances always existed in the town, but were limited by the lack of physical places in which women and men could secretly meet and talk. The new topography of social media has multiplied private and secret places, as well as anonymous public places where women can privately liberate themselves from social constraints (Costa, 2016). Every self is composite and can be understood as a microcosm inhabited by multiple forces (Strathern, 1988; Hall, 1992). Nevertheless, the experiences of the young women I described in this chapter are the outcome of one main *force*: the patriarchal organization of family and society. This same *force* has produced relational desires (Joseph, 1999) that translate into family commitment, and the production at the same time of highly individualized and highly collective selves (Joseph, 1999 in Deeb, 2011). The concept of "immobile mobility" captures women's movement beyond the social, cultural, and physical limitations created by the patriarchal family, without an open challenge to its founding social norms. "Immobile mobility" sheds light on this specific form of agency that supports structures of male power, on the one hand, while undermining them, on the other hand (Abu-Lughod, 1990), and cannot be reduced to the binary model of subordination or resistance.

Conclusion

This chapter delved into the lived experience of young Muslim women who used social media to create and maintain social relationships that depart from offline forms of sociality. I focused specifically on the case of women who – due to the restrictions imposed by conservative parents or husbands – have limited physical mobility and social opportunities, and spend most of their time at home. They use social media as digital places that allow them to start, live, and end pre-marital and extra-marital romances and friendships. Far from being simple media that mediate relationships between people, social media in this social context are better understood as digital places whose spatial elements determine the ways the relationships are lived and maintained. I specifically refer to the configuration of private and public places, and the boundaries between them, which allowed women to be part of a public space with pseudonymous identities and then easily switch to the private Facebook chat. I then introduced the concept of "immobile mobility" that simultaneously accounts for the mobility from the offline place of the house to the online place of social media, and the mobility away from the constrains imposed by social norms that create gendered inequality, but are not challenged or transformed. The notion of "immobile" accounts both for the stillness of the body and women's unchanging subordinate role within family and society. The concept of "mobility" accounts for the movement from offline to online, and the ability to transgress social and cultural boundaries. Like in Wallis' description of working-class migrants in China, immobile mobility "has a dual logic – it can be liberating and constraining, creating new opportunities for empowerment and disempowerment" (Wallis, 2013: 6). The concept of "immobile mobility" captures the agency enacted by homebound women who use social media to escape the social norms they continue to submit to. A "mobile socialities" perspective allowed me to shed light on questions of human agency which have been at the core of social science's concerns. It foregrounds the importance that the longing for mobility and new socialities has in the life of young women living in a digitally mediated world, and it accounts for the contradictory dynamics associated with it.

Notes

1 To maintain anonymity of the research participants, all names in the chapter are pseudonyms.
2 Clara Wallis adopts the concepts of 'immobile mobility', too. I did not draw on her work, but similarities between her conceptualization and mine will be discussed later in the chapter.
3 The *WhyWePost* project was run by Prof. Daniel Miller and was funded by the European Research Council (Grant ERC-2011-AdG-295486 Socnet).
4 For an anthropology of "strangership" as a distinct form of relationship and the role of social media in shaping it, see McDonald (2019).

References

Abu-Lughod, L. (1990) "The romance of resistance: Tracing transformations of power though Bedouin women", *American Ethnologist*, 17(1): 41–45.

Barthes, R. (1979) "Dining car", in *The Eiffel Tower and other Mythologies*. New York: Hill and Wang.

Boellstorff, T. (2015) *Coming of Age in Second Life: An Anthropologist Explores the Virtually Human*. Princeton: Princeton University Press.

Boellstorff, T. (2020) "The ability of place: Digital topography of the virtual human on Ethnographia Landia", *Current Anthropology*, 61(S21).

Bourdieu, P. (1977) *Outline of a Theory of Practice*. Cambridge: Cambridge University Press.

Costa, E. (2016) *Social Media in Southeast Turkey*. London: UCL Press.

Costa, E. (2018) "Affordances-in-practices: An ethnographic critique of social media logic and context collapse", *New Media and Society*, 20(10): 3641–3656.

Costa, E. (2019) "Location as conspicuous consumption: The making of modern women and consumer culture in southeast Turkey", in R. Wilken, G. Goggin and H. Horst (eds.) *Location Technologies in International Context*. London: Routledge.

Deeb, L. (2011) *An Enchanted Modern: Gender and Public Piety in Shi'i Lebanon*. Princeton: Princeton University Press.

Ellison, N. B. and boyd, d. (2013) "Sociality through social network sites", in W. H. Dutton (ed.) *The Oxford Handbook of Internet Studies*. Oxford: Oxford University Press, 151–172.

Graham, M. and Khosravi, S. (2002) "Reordering public and private in Iranian cyberspace: Identity, politics, and mobilization", *Identities: Global Studies in Culture and Power*, 9(2): 219–246.

Hall, S. (1992) "The question of cultural identity", in Stuart Hall, David Held and Anthony McGrew (eds.) *Modernity and its Futures*. Cambridge: Polity Press.

Hine, C. (2015) *Ethnography for the Internet: Embedded, Embodied and Everyday*. London: Bloomsbury Publishing.

Hine, C. (2017) "Ethnography to the embedded, embodied, everyday internet", *The Routledge Companion to Digital Ethnography*, 21.

Holmes, D. (ed.) (1997) *Virtual Politics: Identity and Community in Cyberspace*. London: Sage.

Jones, S. G. (1997) *Virtual Culture: Identity and Communication in Cybersociety*. London: Sage.

Jordan, T. (1999) *Cyberpower: The Culture and Politics of Cyberspace and the Internet*. London: Routledge.

Joseph, S. (1999) "Introduction: Theories and dynamics of gender, self, and identity in Arab families', in S. Joseph (ed.) *Intimate Selving in Arab Families: Gender, Self, and Identity*. Syracuse, NY: Syracuse University Press.

Joseph, S. (2005) "Learning desire: Relational pedagogies and the desiring female subject in Lebanon", *Journal of Middle East Women's Studies*, 1(1): 79–109.

Kandiyoti, D. (1988) "Bargaining with patriarchy", *Gender and Society*, 2(3): 274–290.

Kandiyoti, D. (1996) "Contemporary feminist scholarship and Middle East studies", in D. Kandiyoti (ed.) *Gendering the Middle East: Emerging Perspectives*. London: I.B. Tauris, 1–27.

Madianou, M. (2016) "Ambient co-presence: Transnational family practices in polymedia environments", *Global Networks*, 16(2): 183–201.

Mahmood, S. (2005) *Politics of Piety: The Islamic Revival and the Feminist Subject*. Princeton: Princeton University Press.

Marsden, M. (2005) *Living Islam: Muslim Religious Experience in Pakistan's North-West Frontier*. Cambridge: Cambridge University Press.

McDonald, T. (2019) "Strangership and social media: Moral imaginaries of gendered strangers in rural China", *American Anthropologist*, 121(1): 76–88.

Menin, L. (2015) "The impasse of modernity: Personal agency, divine destiny, and the unpredictability of intimate relationships in Morocco", *Journal of the Royal Anthropological Institute*, 21(4): 892–910.

Miller, D. and Slater, D. (2000) *The Internet: An Ethnographic Approach*. Oxford and New York: Berg.

Miller, D. et al. (2016) *How the World Changed Social Media*. London: UCL Press.

Mittermaier, A. (2012) "Dreams from elsewhere: Muslim subjectivities beyond the trope of self-cultivation", *Journal of the Royal Anthropological Institute*, 18(2): 247–265.

Schielke, S. (2009) "Being good in Ramadan: Ambivalence, fragmentation, and the moral self in the lives of young Egyptians", *Journal of the Royal Anthropological Institute*. https://doi.org/10.1111/j.1467-9655.2009.01540.x.

Sehlikoglu, S. (2018) "Revisited: Muslim women's agency and feminist anthropology of the Middle East", *Contemporary Islam*, 12(1): 73–92.

Silverstone, R., Hirsch, E. and Morley, D. (1992) "Information and communication technologies and the moral economy of the household", in R. Silverstone and E. Hirsch (eds.) *Consuming Technologies: Media and Information in Domestic Spaces*. London: Routledge, 115–131.

Stokes, M. (2010) *The Republic of Love: Cultural Intimacy in Turkish Popular Music*. Chicago, IL: University of Chicago Press.

Strathern, M. (1988) *The Gender of the Gift: Problems with Women and Problems with Society in Melanesia*. Berkeley, CA: University of California Press.

Ureta, S. (2004) "The immobile mobility: Spatial mobility and mobile phone use among low-income families in Santiago, Chile". Paper presented at the 5th Wireless World Conference, University of Surrey, London, UK, July 15–16.

Vitak, J. (2014) "Connecting in the Facebook age: Development and validation of a new measure of relationship maintenance". Paper presented at the International Communication Association 64th Annual Conference, Seattle, WA, May. Available at: http://goo.gl/rBjPwS.

Wallis, C. (2011) "(Im)mobile mobilities: Marginal youth and mobile phones in Bejing", in Rich Ling and Scott Campbell (eds.) *Mobile Communication: Bringing Us Together and Tearing Us Apart*. New Brunswick, NJ: Transaction, 61–81.

Wallis, C. (2013) *Technomobility in China: Young Migrants Women and Mobile Phones*. New York: New York University Press.

Waltorp, K. (2020) *Why Muslim Women and Smartphone: Mirror Images*. London: Bloomsbury Publishing.

9

MOBILE SOCIALITIES IN BEIJING

Young adult Chinese WeChat users' management of social relations between tradition and modernity

Shuhan Chen and Peter Lunt

Introduction

Young Chinese university students leaving home to study in Beijing present a fascinating and complex case of mobile sociality. They are mobile in three senses: geographically, they have moved from rural and provincial areas of China to the capital city; they are part of the upward social mobility of the emerging aspirational middle class of China (Goodman, 2014); and they are the generation that is enthusiastically adopting mobile and digital technologies, driving the development of this sector of the Chinese economy. These three forms of mobility have significant implications for sociality. Geographical mobility creates distance from family, friends and community of origin, detaching young people from traditional forms of social relations based on hierarchy, patrimony and collective experience immersing themselves in the dynamic modernity of Beijing. Social mobility involves developing instrumental, post-traditional forms of social relations in both friendship and social contacts that potentially support both social and career aspirations. Digital and social media provide potential new forms of self-expression and access to a wide range of social relations based on interest and mutuality. The combination of these three forms of mobility positions our participants as experiencing key social and cultural transformations in sociality related to the meaning of family and friendship, the relation between the collective and the individual, and opportunities for personal expression and new forms of friendship and association.

In this chapter, we explore the mobile sociality of a sample of young adults by exploring their social media strategies and personal reflections on the mediation of their social relations in the context of dynamic social and cultural change, beginning with an outline of the social and cultural transformation of China to provide a background to the study. We then review the potential impact of WeChat on social relations in contemporary China. This is followed by an account of our research method and a presentation of our research findings in four themes: creating virtual co-presence with parents, friends and contacts linked to the aspirations of our participants, and their online rites of sociality.

Economic reform and modernisation

Since economic reform began in the late 1970s, China has been on a rapid path of condensed modernisation that combines transformations that took place over 200 years in Europe: industrialisation, including rapid development of digital technologies; global trade; rapid urbanisation; political transformation (Beck and Grande, 2010; Yan, 2010). Urbanisation has resulted in a sizeable floating population separating people from their families and communities of origin to seek work or higher education in the expanding cities. The floating population initially was mainly comprised of those who moved to work in the factories as cheap labour, driving economic development in China. However, after several decades of economic growth, China has more recently developed an emerging middle class whose members, in the context of highly competitive higher education and job markets, aspire to social mobility through the opportunities available in Chinese cities (Goodman, 2014). They are the children of the one-child policy who have been brought up in ways that are in tension with both Confucian values related to parental – particularly the father's – authority and reciprocal obligations between parents and children. They are often also both first-generation university students and the first from their families to leave their community of origin. We are interested in how they navigate the tensions between traditional social obligations, between collectivism and individualism, and between the demands of modernisation and their own desires for autonomy.

The reform period has mainly been driven by economic change including increasing opportunities for private ownership of businesses and property, and the relaxation of restrictions on mobility. In the Maoist period before 1978, mobility was highly restricted by a strict household registration system that differentiated between rural and urban districts, and tied people to collectives. This period also witnessed a clampdown on social mobility through a state-controlled programme aimed at equalising social class positions or declassification. This transition is not in the distant historical past, but part of the shift in parent/child relations shaping understandings of the nature of family and the associated social obligations (Yan, 2010; Dikotter, 2016).

The children of reform

The current generation of young adults in China (post-1990) is the first generation to grow up in the era both of economic and social reform, increasing opportunities for education and work opportunities, relative freedom of movement and as natives of the rapidly developing digital media landscape. Often the only child and torchbearer of their family, they experienced a very different childhood from previous generations. As individuals, they have access to social and educational resources that were unavailable to their previous generations, but they also face an increasingly competitive education system and dynamic labour market. The government and families see this generation as critical to the future of both the state and the family. A lot is at stake, and much effort goes into creating the conditions that will ensure this generation will develop in a way that serves the interests of the nation, their families and themselves.

Although they have new opportunities and choices, there remain centralised controls that shape their chances. For example, *Gaokao*, the Chinese university entrance examination, plays a pivotal role in navigating the personal development path of Chinese children. In a reflection of the importance of education for social mobility, Chinese media describe *Gaokao* as the most important and fairest opportunity for people to change their lives (Sun and He, 2019). In contrast, other voices suggest that the *hukou* system and the unequal distribution of educational resources between urban and rural districts limits the fairness of the system (Burkhoff, 2014). Whereas those born in the late 1960s and 1970s have witnessed the benefits brought by *Gaokao*, including the potential to migrate to urban centres, obtain an 'iron bowl' job that guarantees job security, and achieve upward social mobility (Sun and He, 2019). The Chinese government supports *Gaokao* as exemplifying the ideology of meritocracy during China's transition to a market economy (Liu, 2013). The idea of educational meritocracy is firmly embedded in Chinese youths' lives, from being merit students, to studying at elite universities. Education is, therefore, an important source of mobility in the attempt to overcome the unequal distribution of educational opportunities in different localities (Burkhoff, 2014).

WeChat and online connection

The floating population contributes to the growth of China, both economically and technically, including boosting demand for mobile phones and online communication. WeChat, called *WeXin* in Chinese, is a mobile-based social media platform developed by Tencent Holdings in January 2011. Through its suite of social media functions, by 2019, WeChat had 1.15 billion monthly users and more than 1.2 million applets on its platform providing services covering almost every aspect of people's lives (Tencent Tech, 2020). For example, WeChat Pay allows people to link their WeChat account with their bankcard and enjoy services like online shopping, taxi services, hospital appointments and tax payments through the WeChat platform. Also, mini programmes allow users to access third-party invented apps to play online games, create online name cards, open online stores, etc. Recently, WeChat launched a channel that encourages users to create and share one-minute short videos on its platform, which is considered a gesture to sharing the short video market of TikTok. Chen et al. (2018) call WeChat a super sticky infrastructure of platforms and applications which constitutes a social space of massive dimensions capturing much of the online activities of internet users across China.

The default architecture of social media reflects networked individualism (Rainie and Wellman, 2012), in which individuals realise their social identity through the construction of personal networks based on aggregations of interconnected pairs of social contacts. In contrast, the default setting of WeChat architecture reflects traditional Chinese conceptions of 'circle culture' in which the foundation of social life is the inner circle of family, beyond which are a set of social relations determined by principles of reciprocal favouritism, and beyond that, the nation or collective. For example, in WeChat Moments, users can only see postings between mutual contacts, creating interconnections between mutual circles of sociality. As social spaces, Western media represent potentially infinite connections based on chains of association, whereas WeChat represents a social space constructed of groups defined by mutual connections. In this chapter, we examine how our participants use the affordances of WeChat to develop and sustain different relationships with family, friends and acquaintances, and how they balance their traditional social obligations with their desire for autonomy and engagement in the new modernity of China.

Relations with family at a distance

Relations with parents in traditional cultures of authority and control focus on *xiaoshun*, which emphasises the blood connection between parents and children, raising children to take care of older parents as part of reciprocal obligations over the lifecycle (Liu, 2015; Tu, 2018). With the one-child policy, family relations have shifted from obligations for mutual support in the context of parental authority to more egalitarian relations with a focus on caring and intimacy (Liu, 2015). In this context, commitment to family obligations depends upon the quality of the relationship between parents and children, and potentially conflicts with children's desire for independence and autonomy. The shift from an authority to an intimate and caring relationship also leads to potential anxiety in parents that distance and independence might reduce children's commitment to family.

In addition to changing family relations, the nature of friendship and acquaintanceship is transformed by the mobility of people in China and the affordances of mobile digital and social media. In his summary of different forms of sociality in traditional Chinese culture, Fei (1948) used the metaphor of the circle of ripples when a stone is thrown into water. The ripples are strongest in the centre, and as they expand, they reduce in energy and size until they disappear. So, it is the same for different forms of sociality with the family at the centre, characterised by proximity and close emotional ties surrounded by more distant relatives and friends, and then looser associations which include both reciprocal ties (*guanxi*) and instrumental social relations. Inspired by Fei's work, Hwang (1987) suggested that different social ties sustain the relations within each social zone of proximity. Expressive ties include family and close friends, and mixed ties among friends are maintained by both personal emotions and reciprocal favours, while instrumental ties are distant social relations based on short term mutual benefits.

A potential consequence of mobility is a shift from these traditional normative assumptions concerning social relations that are tied to family and community of origin. The increasing opportunities for life in the city and access to contacts in social media potentially impact on both the meaning and the form of social relations (Yan, 2010; Chen and Lunt, 2021). There are parallels with late modernity in the West in which increasing individualisation can be read either as an extension of liberalism in neoliberalism or as the emergence of forms of individuality that are embedded in extended relationships of equality and reflexivity (Giddens, 1991; Bauman, 2000). In the Chinese context, we are interested both in how traditional Chinese cultural concepts of sociality are in transition in the context of mobile sociality greater flexibility for geographical and social mobility, and the way this changes the conditions of possibility for social relations.

In the context of urbanisation and modernisation that disrupts and potentially destabilises traditional Chinese social relations based on patriarchy, clan and community, social media attracts users both as a way of sustaining established social relations at a distance and expanding social networks with acquaintances (McDonald, 2016). In Western studies of online sociality, cyber optimists praised social media for allowing social connections regardless of gender, class, age ethnicity and geographical distance (Doyle and Smith, 2002). Mediated public life was understood as a collapsed context that raised concerns over privacy and the effectiveness of social media in managing social relations. Here we suggest that Chinese social media reflects the Chinese cultural context in which potential contradictions relate to the relation between tradition and modernity, between reciprocal social relations and instrumental ones, between hierarchical and equal social relations, and between individualism as independence and as autonomous commitment to social relations.

Method

This research includes 42 university students from six universities in Beijing. Thirty-seven participants are from one-child homes, and 39 participants migrated to Beijing primarily for education. All participants are aged between 18 and 24 their parents are between their late thirties and their early fifties, with most in their mid-forties; and both the parents and children embrace smart phones and social media. Twenty participants are undergraduate students, and 22 are postgraduates. Based on the Chinese university category system, half study at a first-tier university and the others at a second-tier university.

After receiving ethics approval, we conducted semi-structured face-to-face interviews among all participants in Beijing. Inspired by the photo-elicitation research method (Lunnay et al., 2014), following each interview, we conducted a post elicitation method by asking participants to show and reflect on examples of their online posts and explain the stories behind them.

With the help of the data analysis software package NVivo, we conducted a grounded thematic analysis of the qualitative data. Starting with open coding, discarding unimportant codes and grouping similar codes, we developed four themes related to mobile sociality: relations with family, best friends and acquaintances, and new mobile sociality rituals developed through the wide use of WeChat. These four themes shed light on the changing forms of sociality resulting from geographical and social mobility, and the widespread use of mobile digital and social media platforms.

Virtual co-presence with parents

One of the things that people miss when they have to live apart from loved ones is the day-to-day practical, routine things they do together, along with sharing ideas and intimacies. Madianou and Miller (2012) showed that mothers living apart from their families use a variety of digital and social media to create moments of mediated co-presence that mirror some of these routine practices of family life. Our participants discussed several forms of mediated co-presence that they adopt to sustain their relationship with their distant parents. Virtual co-presence helps participants to meet some of their and their parents' emotional needs for intimacy while also fulfilling aspects of their obligations towards their left-behind parents.

When one of our participants (female, aged 22) gained a place at a university in Beijing, her mother obtained two smartphones, one for the daughter, one for herself. She then asked her to install both QQ and WeChat on both phones and asked our participant to teach her how to use these social media applications:

> I installed WeChat and QQ for my mum, created an account for her on both sites, and taught her how to use them. She was amazed by the video chat and that is what we frequently use for daily talks. She can see what I am doing, and I can know how she is doing.

It is interesting that the mother, although with limited digital literacy, understood the potential value of a mobile phone running a social media app as a way of keeping in touch with her daughter while she is away at university. Our participant also accept the importance of keeping in touch with her parents. There were many examples among our participants of regular, sometimes daily, social media contacts with parents to update them on what is happening and spend time hanging out together.

Participants also recognise that social media can become a vehicle for parental smothering, expressions of anxiety and attempts to control. Indeed, some of our participants relished the opportunity to leave the family home as a way of escaping 'helicopter parenting' and traditional forms of parenting based on patriarchal authority:

> I came to Beijing because I wanted to run away from my family. My family knows me as a shy little girl, and they try to arrange everything for me. However, I do not dare to tell them what I think. So, when there was a chance for me to leave, I took it and ran away.
>
> (Participant 11, female, aged 19)

Having to conform to the stereotype of the quiet, dutiful daughter motivated Participant 11 to leave home to develop an individual identity away from both the gender role assumptions and controlling nature of her parents. This account also opens up the distance between the potential for personal growth and development as an autonomous individual that is partly about escaping family constraints but also about the opportunities of sociality and personal development in modern China. Yet, while embracing change and the new opportunities of autonomy and self-development, our participants persist in maintaining close relationships with their parents. For instance, our participants manage parents' online surveillance and retain their need for privacy and autonomy by using the online blocking and self-monitoring features on WeChat.

Blocking: either yours or mine

WeChat block settings allow users to decide who their online posts are available to and which contacts' online posts they will see. For instance, selecting 'Hide his/her Moments' blocks a contact's posts from appearing in a user's WeChat Moments, and 'Block' prohibits contacts from interacting with the user on WeChat either via instant messaging or by replying to posts. Participants use the 'Hide' feature so as not to see their parents' online posts which they regard as either fake news spamming or over-solicitous interventions. For instance, participants are upset when parents share health-related pseudo-information online. To prevent what they regard as over solicitous interventions, they use the 'Block' feature to prevent parents from monitoring their WeChat usage.

> I blocked my mum's WeChat Moments because she always shares useless links. For instance, articles with titles like 'which kinds of food become poisonous when eaten together', 'this food can cure cancer', and parents are very interested in links that contain health information . . . they do not justify whether the information is accurate or not, they just share it on WeChat.
>
> (Participant 30, male, aged 20)

> I make my WeChat moments inaccessible to my parents, because they do not understand me and why I create certain posts online. I told them I have not opened my Friend Circle on WeChat as I do not want to expose my private life to others. They understand me and are quite happy. In fact, I post on WeChat quite often, which they do not know . . . [smile].
>
> (Participant 27, male, aged 22)

Both strategies are a compromise between keeping in touch with parents while retaining a sense of autonomy and privacy. Instead of excluding parents completely outside their online social circles, our participants selectively decide what to make available and what to leave unavailable.

Self-monitoring: limited availability

WeChat users can selectively decide who to share posts with when creating or sharing posts. For instance, they can decide to make a post available to all his/her contacts. However, if they want to keep a post private, they can select who they want to share it with from their WeChat contact list. To ease this process, WeChat users can sort their connections into different groups, then decide which group they want to share a post with. These affordances of WeChat make it possible for participants to decide which posts are available to their parents while keeping open communication with other contacts. Significantly, these strategies overcome the disadvantages of collapsed context experienced in Western social media, as discussed by Marwick and boyd (2010). Most of these ways of presenting are available on Facebook, as well.

A key motivation for using social media is the freedom of self-expression that is available to users. However, participants regard their parents 'following' their online posts as a constraint that affects their online self-presentation. Consequently, they take care not to create or share posts that they think their parents might find 'problematic'. They also note that presenting an image of positive motivation and determination helps to ease their parents' minds more effectively than deleting negative posts. For example, Participant 7 (male, aged 20) presented one of his online posts, which include an image of his handwritten essay papers, accompanied with text 'After tomorrow, I am still a real man', and provided two reasons for creating this post:

> It was the end of the semester; we were all busy writing essays and preparing for the last examination of the semester. I wanted to present to my friends who are still working for the final to make them jealous . . . on the other hand, I know my parents check my WeChat friend circle every day so when they saw my examination paper, they would think 'oh, my kid is so hard-working' . . .[laugh].

As WeChat is often the main mode of online communication with parents, some participants choose to conduct other social relations on alternative social media platforms so that their parents cannot access these posts and they are more able to manage the online impression they give to their parents.

My 'besties', my siblings

In contrast to the complexity of sustaining parental social relations online, participants note that relations with best friends are more open, intimate and based on a wide range of topics and shared concerns. As most participants are from one-child families, they express their desire to have siblings and treat their best friends as siblings. Ling (2008) notes the 'flattening' of social relationships among Chinese experiencing modernisation in which close friends replace kinship as members of their inner circle of associates.

During the interview with Participant 10 (female, aged 21), she spent most of the time talking about two of her best friends who were classmates who sat next to her in high school (*tong zhuo*):

> We are like sisters, even closer than our siblings. We lived in the same accommodation when we were in high school and applied to universities in the same city [Beijing]. . . . We try to meet every week when we are free and text each other every day on WeChat. Whenever one of us feels upset the other two will comfort her by text, voice message, or video chat. If something serious happens, we will see each other immediately.

Friends who remained in their home community help by meeting children obligations on behalf of the person who left for university. Participant 19 (male, aged 24) quited his job as a governmental officer to work toward a master's degree in Beijing. Close friends in his hometown support him by carrying out his obligations toward his parents:

> I have several good *ge men* (brothers) at home. We agreed that we should support each other in the future to take care of our parents, because all of us are the only child at

> home. It would be too much pressure for each [of] us if our parents or grandparents get sick in the future. Therefore, when I left for studying in Beijing, my close friends told me not to worry about my parents as they would be there for them. I talk to my close friends almost every day on WeChat, they also help to celebrate my parents' birthday when I am not at home.

It is a Chinese custom to refer to close friends as brothers or sisters, as it encourages a sense of commitment including responding to each other's emotional and instrumental needs. To sustain these relationships, friends develop online rituals for special friends such as replying immediately to posts. This form of mobile sociality is interesting because it recreates traditional circle culture in an online context by treating friends as siblings. It is as though the traditional notion of a close emotional bond with strong obligations is extended to close friendships as a form of kinship.

Social mobility and aspiration

A final form of mobile sociality that we observed among our participants is their broader social relations in the city and online. McDonald (2016) observes that Chinese teenagers often try out 'messy adding' out of curiosity to meet online strangers. Our participants also discuss having gone through a period of exploring random connections online; however, they admit that they often terminate such online connections as they consider them as not worth sustaining. Participants note they prefer connecting with individuals who they consider 'belong to the same circle', 'are helpful' or 'know things better':

> I added some of my previous primary school classmates as my WeChat friends on a trip back home. Then I deleted or blocked them because they create posts about their kids, share fake news, and posts like 'you will get bad luck if you do not share this'. I just feel we do not belong to the same circle; our interests, field or vision are different. I do not think we have things in common to talk to each other [about], nor do I think we [will] have further positive interaction.
>
> (Participant 3, female, aged 23)

The capacity to connect to large numbers of potential 'friends' is one feature of social media that appears to liberate friendship from the constraints of the offline world. Although our participants start by making a large number of contacts, they eventually move away from that to focus on the quality of their connections. Participant 3's account reflects the idea of online connection as maximising the benefits of receiving useful information and effective social relations rather than friendship. Compared with the traditional friendship that emphasises either shared memories or experience, and inner circle relations based on obligations or reciprocity, many online connections are more pragmatic and instrumentally oriented.

Nevertheless, our participants had interesting ideas about these ephemeral and permissive forms of sociality referring to the cultural tradition of *guanxi* or reciprocal social relations, which are sustained by a sense of shared feeling (*renqing*). Hwang (1987) suggested that *renqing* is like granting a favour which people expect to get back in the future, like the payment of a debt.

Guanxi principles are applied in the context of social aspiration as participants review and let go of their old classmates and friends from their community of origin. They are inhabitants of a dynamic, competitive world, and are increasingly aware of emerging social class differentials in contemporary China (Goodman, 2014). Mobility creates both geographical and social distance between themselves and their old classmates, who represent the past and the location of their childhoods. Therefore, migrant youths may feel nostalgia for old classmates, while their aspiration for life and career trajectory motivates them to be pragmatic. This creates a context in which their social media use has an urgency and a form that differs from the way they sustain their relations with family and close friends. As Pahl and Spencer (2006) note, accompanying the loosening of the bonds of family as a social institution is the rise of flexible and supportive personal relations in which fragile and fractured social bonds are frequently found.

Online interaction ritual

The way our participants talk about social relations both offline and online reflect Hwang's (1987) concept of mixed ties that meet both emotional and instrumental needs, and which are sustained through reciprocal favours. These forms of sociality reflect neither tradition nor collectivism, but the increasing interdependence of people on social relations that are valued in their own right rather than being structured by family, tradition, community or government-sponsored forms of collectivism. By constructing relationships and developing rites within them, people develop mutual knowledge and recognition of each other through social exchanges, such as gifts and words – which, in the online environment, can start from an online 'like'.

From a pat on the shoulder to 'let's have a chat'

Our participants' online rituals reflect Goffman's (1967) notes that routine everyday social relations start from small gestures such as smiles or glances. For instance, an online 'like' is a greeting that is similar to a pat on the shoulder given to a friend (Lim and Basnyat, 2016). Based on the closeness between people, online 'likes' are simple gestures that can have a diversity of meanings and are easy ways to strengthen the familiarity between people. Participants use online 'likes' as a ritual expression of personal empathy and that they expect to be reciprocated:

> I try to like everyone's online posts. Because when I post something, I would expect others' feedback . . . [so] I know they are interested in what I post online. I think others have the same feeling so when they post something, I also give them a 'like'. The implicit rule on WeChat is you like my posts, I like yours, so both of us feel happy.
>
> (Participant 16, female, aged 21)

Participants call this kind of relationship *dian zan zhi jiao* (online likes friendship), which they contrast with 'likes' given to people they do not want to develop a relationship with as a form of ritual politeness without empathy. Participant 4 (female, aged 23) explains:

> Sometimes I post selfies online just to receive online 'likes' and feel good about it. If people comment 'what a beautiful girl' or 'the beauty goddess', I will feel embarrassed and will not know how to reply because I know that I am not that beautiful, and they are just complimenting me.

Commenting is inappropriate in ritual polite forms of sociality, but welcome in the context of ritual empathy. In the case of people who participants would like to interact with, comments provide an interactional opportunity, but for people who they are not interested in developing an online conversation with, online comments are experienced as social pressure or as a breakdown in civil inattention (Goffman, 1966). For instance, Participant 15 (female, aged 23) explains how she develops social relations with her master supervisor through liking and commenting on his WeChat posts:

> When I was preparing for my postgraduate entrance examination, I tried to connect with students at my university. Luckily, my friend's brother had a friend who was a postgraduate here. I got that person's WeChat account and she helped me find a professor's WeChat account, so I connected with the professor on WeChat and told him I wanted to be his student. I clicked like and commented on almost every post that professor created or shared on his WeChat moments. When I took the interview, the professor recognised me and said, 'You are the student who often clicks 'like' on my WeChat, right?' I said yes and he was quite happy. My interview went very well. Even now, I think it is my online 'likes' and 'comments' that secured me a place.

Similarly, Participant 24 (female, aged 21) describes using WeChat to enhance her relationship with her employers:

> I have been an intern in a company for almost a year now. My postgraduate supervisor introduced me to work there and people in the company were nice to me. We all connected on WeChat to organise activities and I paid attention to employers and senior colleagues' online posts as I need to show my respect as well as to *la guan xi* [strengthen social ties] between us.

Social mobility encourages participants to develop instrumental social connections to help them gain a foothold in the dynamic modernity of Beijing. In this context, WeChat provides the opportunity to connect online, while with whom this relationship should be developed further is largely depend on the social resources that one hold and their willingness to offer. To strengthen these social ties, individuals develop interactional rites to distinguish close friends from online acquaintances and to develop and sustain effective reciprocal relations.

Immediate replies

To compensate for the effects of social distance, participants try to make contacts feel special by providing timely responses to online requests, text messages and online posts:

> Some of my good friends are not in Beijing . . . when a friend wants to share some information with me, she will send me an instant message, then we will video or voice chat on WeChat . . . we only pay attention to close friends' online posts, so when she posts and I see it, I will click 'like' or comment on her post to signify that I have seen it. If I realise that she is expressing negative feelings, then I might message her or call her as soon as possible, so we can talk, which makes her feel better.
>
> (Participant 15, female, aged 23)

These examples reflect an alternative form of mobile sociality that reflects Goffman's (1966, 1967) work on *Behaviour in Public Places* and *Interaction Ritual*, in which people expect to both give and receive ritual compliments for recognition of their humanity and individuality. Such public social recognition follows several interaction rituals regulating proximity, ritual forms of acknowledgement and recognition in the form of gestures.

Personalised online gestures

Another mode of mobile sociality that is reflected in our participants' social media practices is the use of mutual knowledge shared with close friends and romantic partners. By embedding mutual knowledge in posts, people in intimate and close relationships both signal their relationship and maintain its secrecy. Participant 17 (male, aged 24) explains that he rarely creates posts on WeChat, except when he wants to 'express thanks' or has something 'special' to share with friends:

> Sometimes, I post something for some special friends, either to say thanks or to share something interesting with them. Sometimes I mention their name in the posts to express thanks. My friends will understand what I want to say in my posts, and even though posts are available to all my WeChat connections, it is not for them, and I do not expect their feedback.

Goffman (1971) claims that such ritual social interaction involves shared momentary moods or feelings and mutual engagement and reinforces participants' sense of worth. Similarly, Ling (2008) notes the importance of mundane ritual in interpersonal interactions that help to develop an individual's sense of pride and honour. Virtual co-presence is becoming so common that people are constructing special online interaction rituals to express and sustain such friendship and intimacy.

Conclusion

Focusing on the social media practices and reflections of a group of young adults who moved to Beijing for education, we explored three types of mediated mobile sociality. Mobile sociality includes creating keeping up with and virtual co-presence with left-behind parents to carry out social obligations, keeping in touch with best friends for social and emotional needs, and the establishment of reciprocal social relations with acquaintances to establish a path to social mobility in the city.

These social media practices provide solutions to tensions in the lives of young people who have left their families and communities of origin to join in the condensed modernisation of contemporary China. Their move to the city for education creates opportunities for upward social mobility and serves as an opportunity for them to develop autonomy through independence, personal expression and the freedom to choose their social contacts. Social media provides ways of sustaining their relationships with parents at home and fulfilling their obligations to them. Virtual co-presence with parents helps them to sustain intimacy and keep in touch, and they develop strategies to limit parents' interventions in their life in the modern city. Social media also provides opportunities for forms of friendship that sustain them in their studies and through periods of difficulty. What seems to be new for them are the opportunities for more pragmatic social relations that also provide a means of connecting with people to achieve their social and professional aspirations. This idea shares similarities with the work of Tönnies and Weber on community and society, in which they mention the different social rules that exist in traditional community and modern society (Bond, 2012). However, what we notice in our research is the complex combination of community that exists in modern China. Once our participants form a pragmatic network, they treat it as a community. Mobile sociality, along with education, provides a route to social mobility.

A particularly interesting feature of the online sociality we have analysed in this chapter is that participants are experimenting with modes of mobile sociality by connecting to a world defined by mutual reciprocity and the exchange of favours (*guanxi* and *renqing*). What emerges is a picture of both the enthusiasm and nuance of mobile sociality as this generation faces the daunting challenges of managing flexible social relations in the context of rapidly expanding opportunities and the challenges of contemporary China. WeChat is used to connect to individuals who have the potential to provide cultural and social resources, to establish networks to meet their instrumental needs through reciprocal social connections, in contrast to their connections with family and friends.

Two kinds of mobile sociality we have discussed are sustained through *guanxi*. The expressive ties with close friends, and the mixed ties with acquaintances with potential for reciprocity (Hwang, 1987). With best friends, who are frequently their old schoolmates, shared collectivist socialisation and school education is in tension with an increasing sense of themselves as individuals seeking to develop their independence while needing the emotional support from peers. However, as Ling (2008) suggests, mobile sociality allows young adults to develop their social network, and the aspiration for a better future drives them to develop a functional social circle, which is sustained by *guanxi* and *renqing*.

We found parallels with the role of facework (Goffman, 1967) in sustaining social interactions in the online environment with online rituals of mobile sociality. We identified three online rites that young adults develop while creating effective social connections. Offering to connect on the WeChat platform is a way of offering friendship and swift responses are politeness gestures to underline the attention given to the other creating a feeling of being special among their contacts is another way to create shared memories or experiences in the online environment.

Migrating to large urban cities and developing mobile sociality through social media platforms like WeChat is a widespread phenomenon among Chinese youth. In this process, although we notice the increasing need for developing pragmatic social connections for their city lives this does not replace family-oriented collectivism. Embedded norms of filial piety continue to define responsibilities between parents and children. These traditions are reinforced by law, in which each migrant youth is expected to visit their parents a few times a year to provide emotional and financial support to their ageing parents. This traditional model of obligation is reflected at the level of the state and the nation in China as the term 'family' is used to describe Chinese citizens' relationship with the country through notions like motherland, daughters and sons of China suggesting that individuals have social responsibilities to contribute to the nation modelled on their obligations to their family. Complementing this, our participants carefully select what to make available online to their parents and friends practising self-censorship as a response to the government's call for creating a clean and clear online environment.

References

Bauman, Z. (2000) *Liquid Modernity*. Cambridge: Polity Press.

Beck, U. and Grande, E. (2010) "Varieties of second modernity: The cosmopolitan turn in social and political theory and research", *British Journal of Sociology*, 61(3): 409–443.

Bond, N. (2012) "Ferdinand Tönnies and Max Weber", *Max Weber Studies*, 12(1): 25–57.

Burkhoff, A. (2015) "'One exam determines one's life': The 2014 reforms to the Chinese national college entrance exam", *Fordham International Law Journal*. Available at: https://ir.lawnet.fordham.edu/cgi/viewcontent.cgi?article=2411&context=ilj (accessed June 5, 2020).

Chen, S. H. and Lunt, P. (2021) *Chinese Social Media Face, Sociality and Civility*. Bingley: Emerald.

Chen, Y. J., Mao, Z., and Qiu, J. (2018) *Super-sticky WeChat and Chinese Society*. Bingley: Emerald.

Dikotter, F. (2016) *The Cultural Revolution: A People's History 1962–1976*. London: Bloomsbury Publishing.

Doyle, M. E. and Smith, M. K. (2002) "Friendship: Theory and experience. The encyclopaedia of informal education". Available at: https://infed.org/mobi/friendship-some-philosophical-and-sociological-themes/ (accessed February 2020).

Fei, X. T. (1948) *Xiang Tu Zhong Guo* (in Chinese). Beijing: San Lian Shu Dian.

Giddens, A. (1991) *Modernity and Self Identity*. Cambridge: Polity Press.

Goffman, E. (1966) *Behavior in Public Places: Notes on the Social Organization of Gatherings*. New York: The Free Press.

Goffman, E. (1967) *Interaction Ritual*. New York: Pantheon Books.

Goffman, E. (1971) *Relations in Public: Microstudies of the Public Order*. London and New York: Routledge.

Goodman, D. S. G. (2014) "Middle class China: Dreams and aspirations", *Journal of Chinese Political Science*, 19: 49–67.

Hwang, K-K. (1987) "Face and favor: The Chinese power game", *American Journal of Sociology*, 92(4): 944–974.

Lim, S. S. and Basnyat, I. (2016) "Face and online social networking", in Sun Sun Lim and C. R. R. Soriano (eds.) *Asian Perspectives on Digital Culture: Emerging Phenomena, Enduring Concepts*. London: Routledge.

Ling, R. (2008) *New Tech, New Ties. How Mobile Communication is Reshaping Social Cohesion*. London: MIT Press.

Liu, S. (2015) *Identity, Hybridity and Cultural Home: Chinese Migrants and Diaspora in Multicultural Societies*. London and New York: Rowman & Littlefield.

Liu, Y. (2013) "Meritocracy and the Gaokao: A survey study of higher education selection and socio-economic participation in East China", *British Journal of Sociology of Education*, 34: 5–6, 868–887.

Lunnay, B., Borlagdan, J., McNaughton, D. and Ward, P. (2014) "Ethical use of social media to facilitate qualitative research", *Qualitative Health Research*. Available at: https://journals.sagepub.com/doi/abs/10.1177/1049732314549031 (accessed May 2015).

Madianou, M. and Miller, D. (2012) "Polymedia: Towards a new theory of digital media in interpersonal communication", *International Journal of Cultural Studies*, 16(20): 169–187.

Marwick, A. E. and boyd, d. (2010) "I tweet honestly, I tweet passionately: Twitter users, context collapse, and the imagined audience", *New Media & Society*, 13(1): 114–133.

McDonald, T. (2016) *Social Media in Rural China*. London: UCL Press.

Pahl, R. and Spencer, L. (2006) "Personal communities: Not simply families of 'fate' or 'choice'", *Current Sociology*, 52(2): 199–221.

Rainie, L. and Wellman, B. (2012) *Networked: The New Social Operating System*. Cambridge, MA: MIT Press.

Sun, J. and He, M. J. (2019) "How Chinese Gaokao changed many lives in China (in Chinese)". Available at: http://politics.people.com.cn/n1/2019/1007/c429373-31385971.html (accessed June 5, 2020).

Tencent. Tech. (2020) "WeChat released data report: By the end of the third quarter of 2019, WeChat monthly live account were 1.15 billion". Available at: https://tech.qq.com/a/20200109/051470.htm (accessed February 2020).

Tu, M. W. (2018) *Education, Migration and Family Relations between China and the UK: The Transnational One-child Generation*. Bingley: Emerald.

Yan, Y. (2010) "The Chinese path to individualization", *The British Journal of Sociology*, 61(3): 489–512.

PART III

Working with mobile socialities

INTRODUCTION PART III

Erika Polson

A man hired as a day labourer to move equipment at a small regional airport is invited to climb inside a bright red airplane and has his photo taken in the cockpit – an artefact he cherishes as a memory to show others on his smartphone. A new mother uses a web platform to find foreign visitors to lodge in her home in exchange for helping with childcare and providing the kind of cultural interaction she values and knows will be in short supply in her temporarily less-mobile life. A remote tourism destination enables digital visitors to command the movement of networked, Go Pro-outfitted tour guides as a way of staying connected to the world in real time during a pandemic that halts travel. A food courier leans his bike against a bench and sits down to call his wife while waiting for the next delivery order to pop up on screen. A young woman fluffs up her hair, raises her arms to the sky and says "Now!" to the stranger taking her photo, which will be cropped, filtered, posted and then hopefully liked thousands of times on her Instagram page. And an exhausted Queensland mine worker defies the no-phones rule on her break to text a quick picture of a kangaroo to her daughter back in Brisbane.

As the time and space of labour and leisure become increasingly blurred, examples such as these illustrate from varying perspectives how socialities enabled by digital media are interwoven into the work of everyday life. The chapters in Part III of the book address socialities at the centre of multiple forms of mobile work, highlighting day labourers, "fly-in-fly-out" workers, food courier 'gigs', housing-for-labour cultural exchange, building the empowered self through travel, and selling live, digital "moments" across geographic space.

The ability to use networked devices to both maintain and create social connectivity while on the move has corresponded to the increasing collapse of boundaries between work and leisure – whether for the worker who can connect to friends, family, chat rooms or games during breaks, or the traveller who can extract added value from a vacation by maintaining productivity and/or accruing symbolic capital. As is explored in Part III, mobile media play an articulating role in the work of everyday life, tuned by users as they seek to maintain just the right balance around various dialectical socialities: family and individual, sociality and solitude, self and community, task and relationship, absence and presence, now and later.

As the editors described in the Introduction to this volume, the concept of 'mobile socialities' is multivalent and contradictory, seeking to understand the layered ways in which mobile people might experience a sort of groundedness and belonging through relating to others. Knitting these chapters together are threads of space, place and time in the fostering of mobile socialities, particularly considering the changing spatial nature of the 'workplace' in mobile work, and the fluctuating tempo of mobile labour practices. In Magnus Andersson's Chapter 12, bicycle couriers tap into the familiar comforts of digital media to become ensconced in place during unscheduled moments of downtime, experienced as breaks in random corners of the city. The social use of time when 'waiting' is a salient aspect of courier work, and also for the day laborers in Carlos Jimenez's Chapter 13. Jimenez demonstrates how waiting structures the day labour experience in many ways, from waiting along the street for a job offer, waiting to get to the job site and sometimes waiting to be paid; finding themselves in entirely new places (sometimes quite far-flung) each day, the consistency provided through their mobile communication devices provides spaces of comfort. In my Chapter 10 on solo travellers, people can feel more comfortable embarking on an epic self-building journey to unknown places because their mobile phones enable them not only to keep in touch with loved ones, but also to upload photos on-the-go, hashtagged in a way that connects their place-based moments to those of others on similar journeys. Deniz Neriman Duru in Chapter 15 describes how transnationally mobile people can engage in placemaking in their destination country through the "convivial media experiences" created in the more successful scenarios of the home-share platform, Workaway. Key to the sociality accessed in Duru's chapter is the slowing down that can occur when people agree to dwell together over time in place. Roger Norum in Chapter 11 confronts how when the world seemed to be slowed down by a global pandemic, some places continued to assert themselves through creating digital experiences built to make visitors feel as-if they were there, *now*. Although the regional commuters in Jolynna Sinanan's Chapter 14 work away from home for a week or more, sacrificing time and regular communication with their families (due to time differences, shift schedules, and spotty wi-fi), Sinanan also finds that the new placed-based relationships that emerge in the sites of this mobile work "influence how people may imagine their future". Indeed, for all the spatial and temporal implications of the chapters in this section, the role of the future is the more latent, yet perhaps most underlies the nature of working with mobile socialities. Aspirations and imaginative futures are negotiated through socialities that are profoundly linked through mobile labours.

Although there are many reasons to commend how social connectivities have enabled and empowered mobile work, in closing, I cannot help but wonder about how the personal usefulness of these socialities might be considered critically in relation to the extractive value they provide. Seeing the many commonalities across these chapters, it is notable that socialities are often used as moments of coping, distracting or dreaming. Thus, it is certainly worth asking to what extent the ability to bring sociality into experiences that are, in fact, a form of labour, is one of the latest versions in ongoing tensions between demands for expanding market productivity and meeting human needs.

When read as a whole, Part III asks us to reflect on the *work* that socialities can do, and points to future lines of critical inquiry into emerging forms of mobile labour. As the authors in this section mine how sociality intersects with both mobile labour and the labours of mobility, we see that socialities themselves are mobilised and mobilising – accessed as a form of transport, out of work and into another place and time. As a migrant worker in Carlos Jimenez's chapter takes a photo of beautiful snow-capped mountains from a job site he has been trucked to for a day of labour, and as this photo takes on the role of a vacation picture in his photo album, from which it will be texted to friends and uploaded to Facebook, we can both celebrate his ability to find this joy at work and wish he had a job that would provide that vacation. This is the contradictory nature of working with mobile socialities.

10
THE SOCIALITY OF #SOLOTRAVEL

Erika Polson

A woman wrapped in flowing bronze silk stops along the stone walkway in front of an ornate Thai temple to turn and smile at the camera behind her. With windswept amber hair tucked back behind one ear, her sandal-clad feet in mid-stride, she beckons over her shoulder to the viewer. "Have you ever travelled solo?" she asks.

The Instagram post continues:

> In my life I've never experienced the feeling of freedom more than when I'm travelling alone. Nobody to wait around for you when you want a jam packed and proactive day, nobody to make you feel guilty for sleeping in and lazing by the pool drowning in martini's [sic], nobody to tell you what to do!
>
> (Instagram, 2019)

Through a self-branded website containing links to a YouTube channel, magazine features, blog posts, and a collection of carefully edited digital travel stories, the self-described "travel content creator" who created this post is just one of the many entrepreneurs endeavouring to grow one of the tourism industry's hottest new trends: solo travel. The post deftly wields the individually focused language of self-care that has become prevalent in advertising discourse, crafted with a friendly and intimate tone. Readers are assured they need not feel guilty about wanting some alone time; and considering they are among the author's 13,500 followers on social media, plus whoever else follows the hashtags she has attached to the post, they can assume to be in good company.

The international tourism industry, which contributed 8 trillion euros annually to global gross domestic product (Chakrabarti and Pololikashvili, 2019) and exhibited steady growth before the COVID-19 pandemic put much global travel temporarily on hold in 2020, has become increasingly segmented by niche to account for growing consumer desires to satisfy their "bucket list" items in the quest to accumulate unique memories (rather than things) as part of the burgeoning "experience economy" (Pine and Gilmore, 1998). Within travel, the search for singular, tailor-made experiences is increasingly located along an independent itinerary that entails doing exactly what one wants, on one's own schedule. Indeed, multiple surveys conducted by industry groups in recent years (e.g. AirBnb, 2020; Fah, 2020; Brown, 2019) claim that travelling alone has lost its stigma to become a highly sought-after experience, and according to Google Trends, use of the search term "solo travel" doubled between 2014 and 2019 (Sovich, 2019). Considering that international travel is among the status-bearing practices of global middle-class consumers (Polson, 2018; Jansson, 2019), and embedded in a cultural context where *self-actualization* is increasingly "packaged and distributed according to market criteria" (Giddens, 1991: 198), it is not surprising that the growing popularity of solo travel cuts across age groups, ethnicity, gender, marriage status, and nationality (e.g. Mackay, 2018; "Solo travel," 2018; "Don't just visit," 2016; Pillai, 2013). The always-innovating tourism boards are suggesting that this market could further flourish post COVID-19 as people seek a safe, socially distanced break from being shut in with others at home (e.g. "The Rise," 2020).

Mobile media and communication affordances are integral to contemporary travel trends, with social media increasingly used for choosing destinations and planning trips (Xiang and Gretzel, 2010), and data-connected smartphones considered vital for keeping in touch, posting pictures, real-time mapping, and accessing local information, resources, and connections while travelling (Parra-López et al., 2016). As an indicator of the connective value of these affordances, an Expedia poll from July 2019 found that 33 per cent of respondents said they would rather go on vacation with their smartphone than with a companion ("Could 'mobilemoons,'" 2019). And while a desire to leave friends and family behind to enjoy a vacation alone (with wi-fi) may initially sound like the sort of self-focused behaviour that upholds early concerns about individualism and alienation in relation to mobile communication devices (c.f. Willson, 2006), the analysis conducted for this chapter provides a case that supports and adds to scholarship recognizing how digital media may expand social relationships (Miller, 2011), creating intimacies beyond the domestic sphere (Chambers, 2013).

In this chapter, I examine the Instagram hashtag #solotravel as a means of critically investigating emerging interconnections between individual mobility and sociality. The hashtag itself is fundamental, not merely in its substance but also its form. Although situated within a discourse of escape from social obligation, #solotravel as an index creates a supportive and collective space where travellers can experience a form of what Maffesoli calls "undirected being-together" (1996: 79) through a mobile sociality generated via the linking function of the hashtag – what I refer to here as "hashtag sociality." To frame my analysis of these posts, I review scholarship on social connectivity through mobile devices, and in particular the implications these affordances have had for travellers. I then consider the online context of discussion forums devoted to solo travel to situate how these provide a social space to support and encourage solo mobilities, including before and after travel. Finally, with an examination of #solotravel posts, I argue that by adopting similar stylistic representations of their solo travels, as well as liking, sharing, and posting supportive messages to one other, solo travel *is* social travel. I conclude by interrogating how the empowerment message of these mobilities is significant for creating solo travel as a marketable experience.

Sociality in mobility

In the face of the growing popularity of mobile communication, Wellman (2004: 22) wrote that the individual nature of access could lead to alienation from grounded groups as people began to "connect in individualized and flexible social networks"; at the same time, Castells and colleagues seemed to celebrate that such traditional ideas of sociability could become transformed, as people joined "elective social networks" (Castells et al., 2004: 237). Wittel's (2001) idea of "network sociality" separated flexible social connection from more stable notions of community and belonging, arguing that in a mobile, networked society, sociality is "fleeting and transient" and consisting of "ephemeral but intense encounters" (Wittel, 2001: 51). Yet, rather than recognizing the support that such encounters might provide, Wittel saw them as beset by a "commodification of social relationships" (Wittel, 2001: 53). Postmodern era social theory addressed the growing flexibility of sociality as relating to how individuals increasingly see themselves as the authors of their own lives: the self of today is an entrepreneur (Beck and Beck-Gernsheim, 2002) who must create their life as a work of art (Bauman, 2008). However, considering the possible societal effects of this individualization, van Dijk argues that "technology to make us more independent from others [has] not made us less social human beings" (2006: 1), and has in fact developed to serve societies whose social relations are increasingly stretched across large distances. Similarly, Sheller and Urry (2006) place mobile technologies as central to the emergence of increasingly independent/individual mobile subjects, claiming that the 21st century will be "organized around new 'machines' enabling 'people' to be more individually mobile through space, forming small world connections 'on the go'" (Sheller and Urry, 2006: 221). They suggest mobile subjects are able to tap into their network capital to produce community on an as-needed basis which, as Urry (2007: 196) notes, points to "the real and potential social relations that mobilities afford." He claims these "social consequences of mobilities" (Urry, 2007: 196) allow people to create and sustain social relations with others (people and places) who are mostly not physically proximate. Interestingly, along these lines, Urry (2003) also observes that travel has *grown* alongside the expansion of mobile communications devices, with desires for face-to-face interaction actually increasing with (rather than being replaced by) the ability to engage in long-distance electronic communication.

Some have been concerned about how mobile media might dislocate users from the public spaces they dwelt in or passed through, with "public space" considered as part of the social (e.g. Meyrowitz, 1985; Turkle, 2011). Yet as Halegoua (2019: 2–3) points out, although that earlier research saw public uses of media as "enacting processes of alienation, abstraction, and distraction," more recent work demonstrates that many actors are working *with* digital technologies, using them to "re-embed themselves within urban space and to create a sense of place." Key to these perspectives is the recognition of the extent to which "digital and material realities dialectically co-construct each other" (Jurgensen, 2011: para 3), which points to the relevance of social media as a site for understanding the contemporary experiences and practices of travellers.

Researching the impacts of networked mobile media on connectivity during travel, Molz (2006: 377) finds that "mobile phones, the Internet, wireless devices, email, laptop computers, and websites have become increasingly significant not just as technological objects, but as social objects and as sites of sociality." Because they can send email, post photos, and search for necessary logistical information while on the go, she refers to these practices as "interactive travel" – a term intending to capture the idea that people are "travelling *with* the Internet as well as *on* it" (Molz, 2006: 378). Interactive travel involves not only frequent communication with faraway friends and family; it also speaks to how digital media help travellers to find, communicate, and even meet up face-to-face with others nearby. For example, Mascheroni (2007: 532) explores how digital media practices help create "mobile spaces of sociality, where mediated interactions at a distance overlap and intersect with face-to-face interaction with other travellers and, sometimes, local people." Many social networking sites or groups are built around fostering these traveller networks, including dedicated organizations such as Internations and Couchsurfing (which offers online chat forums and offline events, in addition to free lodging), as well as topical discussions occurring across Reddit, Facebook groups, and other social media. As I argued in Privileged Mobilities (Polson 2016: 163), by using such networks, mobile actors can develop and maintain affiliations and attachments, configuring "their networks as communities tailored to their needs."

Molz (2006) also points out that interactive media open up situations akin to surveillance, where travellers know they are being watched by their audience and perhaps even alter their behaviours accordingly. She cites Urry's (1990) notion of the "tourist gaze" to point out that the gaze "becomes more complicated when the gazing tourist is also the object of the gaze" (Molz, 2006: 378). By sharing personal experiences online, Molz claims, interactive travellers "insert themselves into surveillance relationships with friends and family members back home, with other travellers they meet on the road, and with the host of strangers who come across their websites and follow their trips" (Molz, 2006: 383). Molz's study on travel blogs and other web-based personal travel writing only just predated the ubiquitous uptake of social media use, yet it is easy to see how the ability to connect with personal contacts as well as other travellers would only increase with the availability of a common platform such as Twitter or Instagram, as well as the use of hashtags to help #solotravellers associate with each other.

Many people find personal and social value in signalling their network associations through social media. For example, in laying out the contours of the "networked self," Papacharissi (2011: 304, 307) points out that social network sites provide "a stage for self-presentation and social connection" by both enabling public displays of one's connectivity as well as promoting the "exchanges and ties" that help foster and maintain relationships. Social media allow for placing oneself as a traveller in relation to a network of friends and family, larger social networks, and other travellers who, because of the way hashtags work to create publics (e.g. Papacharissi, 2015) need not already be contacts of one another. Indeed, as Azariah (2012) documents, even for travellers who mainly use travel blogs, social media have become useful places to demonstrate the connections they have made with other individuals as well as to post references to their travel writing. Although posting while on the move will undoubtedly exacerbate the previously mentioned audience surveillance aspects of interactive travel, Kim and Tussyadiah (2013) also claim the tourism experience is improved by the social support received from social media use during travel. Magasic (2018) finds that although most support comes from connecting to loved ones, it can also be impersonal, derived through online feedback from strangers. As sociality becomes networked and mobile – and thus less dependent on geographic location – travellers are able to keep in touch with close ones and even connect with fellow travellers and locals using a variety of social and location-based media. The social support provided through use of these communication technologies means that solo travellers need not feel completely alone. This, as is discussed in the section that follows, is one of the factors that enables or even motivates solo travel.

Solo, but not alone

The mystique of setting out alone to explore foreign territories has roots in the travel writings of the 18th and 19th centuries, as white upper-class Europeans (usually men) set out on grand tours around Europe, or more distant travels around East Asia and Africa, producing "knowledge" about other parts of the world suffused with nationalism, colonialism, and fantasy (Bishop, 1989; Buzard, 1993; Festino, 2010; Norum, 2006). Although steeped in traditions of reproducing the European upper classes through a first-hand cultural education and social networking (Conway, 2011), such travels eventually became a storied means of building one's character through facing obstacles and solitude in foreign lands (e.g. Iyer, 2000), and as such became a practice taken up by upwardly mobile middle classes (Munt, 1994). Today, as solo travel grows in popularity, researchers find this empowerment narrative to be a dominant factor in motivations for undertaking journeys alone (Abbasian, 2019). Interestingly, however, Bianchi (2016: 205) notes that although solo travellers in her study reported they gained satisfaction through feeling "free, independent, brave and relaxed, and [that they] were doing something new and different," she also found that "the most salient single source of satisfaction for solo travellers was meeting new and interesting people." Indeed, as much as the term "travelling alone" sounds like a solitary endeavour, research does not bear this out; just as the historical trope of the solo traveller as lonely sojourner in historical travel writings was mostly a myth, so too is any notion that today's growth in travelling alone is about leaving the world (and its socialities) behind.

A great deal of current research on solo travellers comes from applied scholarship in the tourism or vacation marketing sector, where researchers aiming to learn more about this emerging consumer group claim its growth stems from rising rates of consumerism and a plethora of travel options in conjunction with increasing individualism (Laesser et al., 2009). According to Laesser and colleagues, the substantial increase in solo travel today can be attributed to "an active, aging population, the phenomena of childless couples and later marriage, and especially a rising population of singles, [which have all] combined to produce a substantial change in travel and leisure demands" (Laesser et al., 2009: 218). Although tourism scholars have begun to study the trend, much of this work has been segmented by gender, with a great deal of research looking at motivations and barriers faced by women who travel alone; for example, female travellers' safety and risk (Jordan and Aitchison, 2008; Yang et al., 2018), their motivations (Wilson and Little, 2005; Chiang and Jongaratnam, 2006), and their use of solo travel as a search for meaning (Wilson and Harris, 2006). The specific experiences and practices of male solo travellers have been explored. as well, particularly in relation to sex tourism (Hall, 1996) and the vacationing practices of single men (Heimtun, 2015). There is also abundant research on specific genres of travel, such as backpacking, where Riley (1988: 313) finds that "the average traveller prefers to travel alone, is educated, European, middle-class, single, obsessively concerned with budgeting his/her money, and at a juncture in life." However, as much as studies are often delineated by gender, age group, nationality, or genre, the wide-angle view indicates solo travel is popular among a broad and growing swath of people. Indeed, Bianchi (2016) has lamented that earlier research on solo travellers has focused too much on gender and tourism type, while general experiences with and motivations for solo travel remain markedly understudied.

One of the benefits of studying an issue through its appearance in conjunction with a hashtag is that the researcher does not need to decide in advance how to hone in on a gender, nationality, age, location, or other categorizing factor. And although analysis of a hashtag will obviously not return information on all solo travellers, such a study does produce a sense of the wide variety of people who choose to attach this descriptor to their social media posts. The very act of posting with this hashtag, of course, separates the #solotravellers from those who travel solo either without branding themselves as #solos or without posting to social media at all. In fact, this performative use of the hashtag can be seen as another indicator of how travel is a status-bearing practice within the middle classes.

In the following section, I examine how participants on a Reddit discussion forum interact around topics relating to "solo travel." Next, I analyse a selection of Instagram posts that contain the hashtag #solotravel, paying attention to visual framing, text, comments, and additional hashtags used to understand how the solo travel identifier works to create a sense of connectedness among travellers (if it does so at all). Taken together, these various sites of interactive communication give a sense of how solo travel is constructed as a method of self-empowerment through mobility – a lifestyle practice that is much more about social belonging than about shutting oneself off from the world.

Producing the solo travel lifestyle

In a study of *Cosmopolitan* magazine from five countries, including India and China, Machin and van Leeuwen (2005: 583) argue that new global lifestyle identities have emerged around "shared consumer behaviours (shared taste), shared patterns of leisure time activities (e.g. an interest in similar sports or tourist destinations), and shared attitudes to key social issues." Cohen (2011) suggests these lifestyle practices can at the same time provide a unique sense of personal identity at the individual level and a recognizable collective identity at the group level. Important to the expression of adherence to a lifestyle is the notion of "style," as people can announce their affiliations through their appearance (Machin and van Leeuwen, 2005). Yet, Machin and van Leeuwen (2005: 584) also call attention to how these outward signifiers of participation are increasingly produced as consumer goods, "globally disseminated through the experts and role models featured in the global media." Their analysis predates but also points the way to the rise of "influencers" through social media.

The values of solo travel are extolled across social media, advertisements, travel magazines, and online discussion boards, with inspirational language, stunning photographs, and a series of hashtags used to broadcast and unite enthusiasts. The Instagram post described at the beginning of this chapter is one of nearly six million photographs on that platform with the hashtag #solotravel attached (as of June 2020). When considered alongside related tags such as #solotravels, #solotraveler, #travelsolo, #solobackpacker, #solofemaletraveler, and #solotravelgirl, that number reaches around ten million posts. Over a quarter of a million online videos on YouTube and other video platforms share stories about solo travel or tips for solo travellers, some professionally produced by travel or media companies, others created by influential vloggers or posted online diary-style by users sharing updates from their trips. A variety of other social media sites have discussion or support groups for solo travellers, as well. As examples: the Reddit group r/solotravel has over one million "wanderers" (members), there are a number of high-membership Facebook groups for solo travellers, and a variety of websites (such as solotravelerworld.com) are devoted to information-sharing, discussion groups, and travel writing. Taken together, these can be seen as a site where solo travel as a lifestyle is constructed, circulated, and reproduced.

Online, where the solo meets the social

A collection of online forums on platforms such as Reddit, Facebook Groups, and a variety of travel blogs facilitate interactive and in-depth communication among travellers and aspiring travellers. Discussions on these sites are characterized by an overall theme that solo travel is a means of personal empowerment and fulfilment, and much of that fulfilment is seen as relating to the social connections that can happen – whether in the serendipitous moments of travel or from being part of a network of other solo travellers online. The prevailing message across these forums is that solo travel is an experience where the solo meets the social. For example, the Reddit group, r/solotravel, has as its headline, "Where travelling solo is travelling social!" (Reddit.com, 2020). Although that group says it is a place to "share experiences and stories," it is much more active as a place to give suggestions and lend support to those either planning or currently in the midst of a solo journey, with advice and reflection coming from those who have made such journeys previously. Posts can be long and quite personal, and indeed these seem to generate the most engagement. For example, one woman writes that she is planning her first solo trip – to her dream location, Japan – after having divorced a year and a half earlier. She explains that this is a trip she has always wanted to take, had been hoping to do it with her partner, but is excited anyway and posting to ask for advice on solo travel to Asia. Her request receives 148 responses, a number of which include a back-and-forth discussion with specific ideas about where she should go and what she should do, as well as reassurance about the rewards that await in her new solo life.

Over one hundred current discussions on the solo travel Reddit focus on how to meet people during a solo trip – some are specific to places, activities, or cultural biases (e.g. "how to meet people in Thessaloniki"; "how to meet people without drinking"; "most friendly countries for Black travellers"), but most are generally focused on sharing tips and encouragement. One user received almost 150 responses to the question, "If you're looking to meet new people in your travels, what are some activities that don't involve clubbing?" After a first round of responses suggesting that staying in hostels is the best way to meet fellow travellers, someone points out that a lot of solo travellers are older than the average hostel-goer. This leads to a very interactive discussion about alternatives, such as meetups and walking tours. When one user suggests asking simply, "Food, anyone?" to the group at the end of a walking tour, many others start weighing in with their qualms about doing something so "bold," with others providing reassurance. For example, someone says they would love to do this but fear they would simply hear "crickets" in response, and another responds, "Great! You can eat the crickets!" Other users jump in to share their appreciation for this advice and their excitement at trying it out in the future; it quickly becomes one of many empowering life lessons that can come from solo travel, as seen in this exchange:

Post A: This is the best advice! And seriously, what's the worst case scenario when you ask something like this? People say no, or maybe they say nothing and it's a bit awkward – so what? It's alright. Just ask people for food or drinks or whatever and see what happens. :)

Post B: This is brilliant!

Post C: In so many situations in life, most people are waiting for somebody, ANYBODY to take the lead. Learning to smoothly do things like this in such situations is a tremendously useful life skill to learn no matter what age you are.

The idea that solo travel is empowering, and teaches important life lessons and skills, comes out as part of general advising in discussions like this exchange, but is also addressed specifically in some posts centred on those very skills. For example, one discussion asks: "What is the greatest thing . . . you have gotten out of solo travel?" Answers balance self-growth and social rewards. Participants report learning how to take care of oneself, stay safe, maintain a budget, trust their instincts, and embrace the liberation and opportunity to know oneself better, nicely summarized by a post that states simply, "Solo travel offered me growth." Mixed among these responses are comments on the joys of "staying up all night in a café with people I met only hours earlier" and the importance of recognizing what it means to have been brought "into someone's social circle just for the night" in order to eventually pass on such a favour to solo travellers one might meet back home.

The Reddit group has a high level of interaction, which stems in part from its format as a discussion board rather than a social media feed; the platform also benefits from group rules that specifically forbid posting of self-promotional content and warn that "low-effort posts will be removed without warning." In this way, it is distinct from the communicative culture on Instagram, where interactions among strangers rely much more on visual representations, nods of approval through likes, and brief (although supportive) messages, and where organic user-posts mix with those created by advertising-sponsored influencers.

#Solotravel: hashtag sociality

Photos using the #solotravel hashtag on Instagram are notable for the fact that despite the geographic, ethnic, and linguistic diversity of participants (there is also some gender and age diversity, although more women, with the dominant age group being around mid-twenties to early forties), there is a striking similarity in the visual framing and style of the solo travel photos posted, which speaks to the geographic spread of this lifestyle practice. Users posting in English, French, Russian, Portuguese, Spanish, Arabic, Bahasa Indonesian, Malay, Hindi, and many more languages add #solotravel to the comments sections below their photos. Most have included #solotravel in a list of other tags related either to travel in general (e.g. #wanderlust) or to personal interests and affiliations (e.g. #travelphotographer, #melanintravel, #muslimtraveller) – connective strategies that serve to spread one's message across as many hashtag searches as possible. As Papacharissi (2015: 318) points out, when read together, posts using the same hashtag can produce "a collective chorus" that reinforces a theme and creates an overall narrative.

On the photo-centric Instagram platform, a few thematic devices become key to the production of a narrative that solo travel is an empowering journey into the self. First, a similar aesthetic runs through the images, with a few overall strategies dominating how the joys of travelling alone are portrayed. The photos are set in locations around the world, highlighting the stunning *cenotes* of Mexico, palace courtyards in Morocco, Indonesian hostel lounges, Parisian balconies, long-tail boats in Thailand, fjords of Norway, and street markets in India, to name a few. The shots are often highly stylized, combining content and design techniques with lens filters and other editing tools to create the effect that Manovich (2016) calls "Instagrammism"; many seem as if they could be advertisements for the locations and activities presented.

Although occasionally a picture contains only a landscape or cultural vignette, an overwhelming majority features just one person (photographed either as a selfie or presumably by a passer-by): the traveller, alone. If other people are in the photo, they are often blurred and treated as background. The solo traveller might be pictured staring at the camera while immersed in an aesthetic scene indicative of the location's uniqueness (sometimes posing, model-style; other times smiling broadly, tourist-style). One of the most common tropes of solo travel photography, though, is for the traveller to have their back to the camera, so that viewers can picture the traveller in motion – walking or moving into the scene – or sitting/standing quietly while staring out at an infinite expanse (See Figure 10.1). The captions added to these images often point to the metaphors of the photo's staging, containing messages about finding oneself through the contemplative nature of travel or losing oneself through immersion into local scenes. These techniques can be seen in two examples shown here: First, under a photo of himself standing still and staring out at the reclaimed former site of an open-pit mine in Malaysia, a man writes, "Travel far enough to meet yourself" (See Figure 10.2); next, a woman walks into an empty plaza, explaining that, "Exploring new cities on my own, I imagine a new life for myself in each one" (See Figure 10.3). Both of these techniques echo the discourse of mindfulness and self-discovery that has been a growing mantra of educated middle classes (Wilson, 2014) – pursuits that are increasingly fed by the experience economy and its focus on travel-related "quests" (Mansfeldt, 2015).

The visual metaphors of solo travel also include single props related to either eating, drinking, working, or being in transit alone. Common among these representative images are items such as a motorcycle or bicycle, a single pair of shoes, a wine or champagne glass, a coffee mug, or an elegant table set for one (always framed by an alluring background). Despite the visual of framing of their solitude through pictures, the #solotravellers are clearly engaging in a collective practice. This occurs not only through the captions they add to their "Comments" boxes, which often personally address viewers (e.g. "Turkish tea is famous . . . *you'll* be able to get it in all Turkish restaurants"; "when *we* can pick up *our* bags . . ."), but also through the act of recreating similar tropes through their posted photos, such as those mentioned previously. MacCannell (1999) argues that the sharing of similar images can play a role in producing collective experiences and identities around and through travel. More recently, I have written that

Figure 10.1 Solo travellers contemplate the view.

Source: Collage of Instagram screengrabs by author

as travellers share photographs tagged with similar hashtags on social networking sites, they can "connect their experiences to those of others through similar visual representations and emotional responses to travel" (Polson, 2018: 166). This speaks to how travellers incorporate awareness of the surveillance aspects of interactive travel, discussed earlier, into practice: signalling to and communicating with both a known and imagined audience.

Beyond using shared photographic aesthetics, solo travellers on Instagram also use a similar style of language to describe their journeys. Many photos are accompanied by words equating solo travel with empowerment, describing difficulties and fears that lead to inner-peace, strength and self-realization. Some examples include:

> Sometimes a path calls you to walk alone. And still, it is beautiful.

> I'm just going with the flow and it's a little bit scarier than I'd like it to be, but I'm staying in this place and making a choice to be here.

Figure 10.2 Contemplation: "Travel far enough to meet yourself."

Source: Instagram screengrab

Figure 10.3 Immersion: "Exploring new cities on my own, I imagine a new life for myself in each one."

Source: Instagram screengrab

> Pushing myself outside of my comfort zone is something I try to do everyday and this experience has allowed me to feel more comfortable: 1) Being myself, 2) Enjoying the photography process, 3) Embracing imperfection.
>
> If you don't go after what you want, you'll never have it.
>
> I am a firm believer that every person who loves to travel should take at least one solo trip in a lifetime.
>
> And in the end, all I learned was how to be strong Alone.

Such comments reinforce the narratives of self-improvement and discovery through hardship that are central to discourses of upward mobility for the middle classes. The notion that each person must travel alone at least once is reminiscent of how middle classes use certifications to classify and rank social status, and points to how the "bucket list" has become a new version of previous self-improvement mechanisms (such as educational achievement) as life skills and consumption merge in the experience economy. Of course, the value in such attainment comes through awareness of it by others (another way in which these are not solitary journeys), and posts about hardship serve both to broadcast self-growth and invite support and acknowledgement. Indeed, such posts from travellers are met with supportive feedback from viewers, who include friends as well as observers who follow the hashtag. Replies tend to be short, supportive, and enthusiastic, such as: "You've got this, friend!" or "You're amazing!"

Finally, a number of remarks shared in the Comments sections of #solotravellers include a networking angle, where a photographer or travel "influencer" might compliment the photo and then specially request that the traveller follow them on Instagram. For those who are using their accounts to build an audience for extended solo travels, the Comments section is also a place to improve their post's placement on Instagram (which is boosted by demonstrating high "engagement," such as a lot of comments) and/or to send viewers to other platforms where they have a presence, such as YouTube or a longer-form travel blog. In fact, many #solotravellers have attempted to monetize their travel (+empowerment) experiences by getting product placement deals or sponsorship (on their social media, blogs, or YouTube channels), or by becoming personal travel consultants. The consultancies in particular tend to be produced by and aimed at helping people from groups who have traditionally been more at risk when travelling alone, such as female travellers and Black travellers.

Overall, the #solotravel hashtag on Instagram serves both to connect solo travellers in a stylized network (which happens through a common photographic aesthetic and through text that constructs the solo traveller as a brave, contemplative individual who becomes empowered through the difficulties of being alone on the road) and as potential consumers of content created by the producers among them. A "hashtag sociality" comes both from the sense of having an audience, which is borne out through shared aesthetics, "likes," and short, supportive comments (much more than through two-way conversation), and through returning that participation by making comments to other #solotravellers.

Conclusions: a market for mobile sociality

This chapter has argued that the growing trend in solo travel is not an indication that people are becoming more anti-social or are increasingly alienated from one another. Instead, considering that today's travellers use a variety of mobile devices to stay in touch with geographically dispersed friends and family members (Molz, 2012), and that digital media expands the access that foreigners have to the places they visit (Jansson, 2007), an increase in solo travel seems to stem from the ability to be socially connected rather than an indication of a decreasing interest in such connectivity. Indeed, simply by framing their travel as "solo," people are able to access and generate new forms of sociality before, during, and after travel.

As seen through social media posts and online communities for travel discussion and support, while solo travellers extol the virtues of controlling their own schedules, making plans on the go, and having time for "number one," they also point out that it is easier to meet locals or other travellers when on their own. Many also say they could not travel solo if not for the ability to keep in touch with friends and family on their phones. And it is noteworthy that one of the fast-growing sectors for solo travellers is *group travel* – whether at the scale of organized day trips or even entire group tours where travellers prefer the combination of social connection without obligation, plus a bedroom to themselves. This tension – between the outstanding growth in the popularity of solo travel, on the one hand, and the crucial roles that socially connective media play in enabling such mobilities, on the other – provides a useful case for considering how self-social dynamics are evolving with and through mobilities, and indeed for understanding solo travel as a mobile sociality. Consistent across the platforms studied is that solo travel is a form of self-empowerment, one whose value is predicated on: a) others knowing about and recognizing it; and b) the fact that solo-ness can be shared with others, by forming a community of people out there on their own, connected through media.

Duru and Trenz (2016: 94) have suggested that digital and social media function as "difference media," which encourage individual self-representation and new ways of defining and expressing the self, and point out that by using networks formed through social media, mobile people constitute themselves as "new active publics." Although the discourse around solo travellers is one that portrays an active (and empowered) public, it is worth considering the extent to which this public is tied to the development of a new niche market.

As friction between "self-seeking through solo experience" and the need for "social connection and community" plays out in this growing mobility practice, entrepreneurs are ready with a vast assortment of handbooks, seminars, Instagram feeds, group tours for singles, sponsored travel blogs (and YouTube channels), and an array of socializing opportunities to access. In fact, a number of websites and blogs specializing in solo travel combine resources and information along with long-form travel writing, shorter blog posts, "tips" lists and sponsored advertising which all continue the main message of solo travel as providing both the opportunity for self-growth and for expanding one's social horizons. Content and communication on these sites are part of a larger solo travel economy; for example, the founder of solotravelerworld.com (which includes articles from a numerous travel bloggers and boasts 150,000 monthly readers) also publishes a *Solo Traveller Handbook* (now in second edition), sells "how to travel solo" online courses, and manages a Facebook group for solo travellers. Much of that site is devoted to helping solo travellers figure out how to be safe, how to budget, and how not to be lonely while travelling (much like in the Reddit discussion forum, suggestions include talking to strangers, taking a group tour or day tour, and sitting at communal tables), and although presented as helpful advice, the site is monetized from many directions.

This particular form of mobile sociality is connected to the rise of consumption-based social practices that connect people through lifestyle identities – the signifiers of which, as Machin and van Leeuwen (2005) explain, are in part produced as consumer goods and disseminated through social and other digital media by role models and experts. In 1997, Cova posited that in a post-modern culture, people would not only look for "products and services which enable them to be freer, but also products and services . . . which can link them to others, to a community, to a tribe" (Cova 1997: 311). In this way, the growth of niche mobilities such as solo travel relies both on the discursive construction of the identity as an empowering one – to be attained through practice – and on the social support that comes from the symbolic and connective value of the digital moniker, the #solotraveler.

References

Abbasian, S. (2019) "Solo travellers to city destinations: An exploratory study in Sweden", *International Journal of Tourism Cities*, 5(1): 35–50.

AirBnb (2020) "What's next in travel: Nature treks, trips to the past and conscious dining", Press Release, February 11. Available at: https://news.airbnb.com/2020-travel-trends/.

Azariah, D. R. (2012) "Beyond the blog: The networked self of travel bloggers on Twitter", *Platform: Journal of Media and Communication*, 4(1): 63–78.

Bauman, Z. (2008) *The Art of Life*. Cambridge: Polity Press.

Beck, U. and Beck-Gernsheim, E. (2002) *Individualization: Institutionalized Individualism and Its Social and Political Consequences*. London: Sage.

Bianchi, C. (2016) "Solo holiday travellers: Motivators and drivers of satisfaction and dissatisfaction", *International Journal of Tourism Research*, 18(2): 197–208.

Bishop, P. (1989) *The Myth of Shangri-La: Tibet, Travel Writing and the Western Creation of Sacred Landscape*. London: Athlone.

Brown, R. (2019) "What billions of GDS data signals tell us about traveler booking behavior", *Travelport*, October 7. Available at: www.phocuswire.com/Travel-booking-behavior-data-signals.

Buzard, J. (1993) *The Beaten Track: European Tourism, Literature, and the Ways to 'Culture'*. Oxford: Clarendon Press.

Castells, M., Tubella, I., Sancho, T., de Isla, M. I. D. and Wellman, B. (2004) "Social structure, cultural identity, and personal autonomy in the practice of the internet: The network society in Catalonia", *The Network Society: A Cross-Cultural Perspective*, 233–248.

Chakrabarti, S. and Pololikashvili, Z. (2019) "Encouraging sustainable and inclusive tourism2", *UNWTO*, October 8. Available at: http://www2.unwto.org/content/encouraging-sustainable-and-inclusive-tourism.

Chambers, D. (2013) *Social Media and Personal Relationships: Online Intimacies and Networked Friendship*. Basingstoke: Palgrave Macmillan.

Chiang, C. Y. and Jongaratnam, G. (2006) "Why do women travel solo for purposes of leisure?" *Journal of Vacation Marketing*, 12(1): 59–70.

Cohen, S. A. (2011) "Lifestyle travellers: Backpacking as a way of life", *Annals of Tourism Research*, 38(4): 1535–1555.

Conway, S. (2011) "The grand tour", in *Britain, Ireland, and Continental Europe in the Eighteenth Century: Similarities, Connections, Identities*. Oxford: Oxford University Press, 189–213.

"Could 'mobilemoons' be the newest trend in travel?" (2019) *PRNewswire*, July 31. Available at: www.prnewswire.com/news-releases/could-mobilemoons-be-the-newest-trend-in-travel-one-third-of-americans-would-choose-a-solo-trip-with-only-their-phone-instead-of-taking-a-travel-companion-300894044.html.

Cova, B. (1997) "Community and consumption: Towards a definition of linking value of products and services", *European Journal of Marketing*, 31(3): 297–316.

"Don't just visit, live it: Travel solo on your next trip: 80 percent of Homestay.com's bookings focus on booming solo travel sector" (2016) *PR Newswire*, May 31. Available at: www.prnewswire.com/news-releases/dont-just-visit-live-it – travel-solo-on-your-next-trip-300277086.html.

Duru, D. N. and Trenz, H. J. (2016) "Diversity in the virtual sphere: Social media as a platform for transnational encounters", in H. G. Sicakkan (ed.) *Integration, Diversity and the Making of a European Public Sphere*. Cheltenham: Edward Elgar Publishing, 94–115.

Fah, E. G. (2020) "Solo travel and its promising business opportunity", *Klook*, January 6. Available at: www.phocuswire.com/Solo-travel-deep-dive-Klook.

Festino, C. G. (2010) "Hybridity and travel writing in English: The grand tour and the imperial frontier", *Uberlandia-MG*, 26(2): 325–343.

Giddens, A. (1991) *Modernity and Self-Identity: Self and Society in the Late Modern Age*. Cambridge: Polity Press.

Halegoua, G. (2019) *The Digital City: Media and the Social Production of Place*. New York: New York University Press.

Hall, C. M. (1996) "Gender and economic interests in tourism prostitution: The nature, development and implications of sex tourism in South-East Asia", in Y. Apostolopoulos, S. Leivadi and A. Yiannakis (eds.) *The Sociology of Tourism: Theoretical and Empirical Investigations*. London: Routledge, 265–180.

Heimtun, B. (2015) "Midlife single men and the solo holiday", in G. R. Ricci (ed.) *Travel, Tourism, and Identity*. London: Routledge, 191–210.

Instagram (2019) "Username kept anonymous", October 4.

Iyer, P. (2000) "Why we travel", *Salon*, March 18. Available at: www.salon.com/2000/03/18/why/.

Jansson, A. (2007) "A sense of tourism: New media and the dialectic of encapsulation/decapsulation", *Tourist Studies*, 7(1): 5–24.

Jansson, A. (2019) "The mutual shaping of geomedia and gentrification: The case of alternative tourism apps", *Communication and the Public*, 4(2): 166–181.

Jordan, F. and Aitchison, C. (2008) "Tourism and the sexualisation of the gaze: Solo female tourists' experiences of gendered power, surveillance and embodiment", *Leisure Studies*, 27(3): 329–349.

Jurgensen, N. (2011) "Digital dualism versus augmented reality", *Cyborgology*. Available at: https://thesocietypages.org/cyborgology/2011/02/24/digital-dualism-versus-augmented-reality/.

Kim, J. and Tussyadiah, I. P. (2013) "Social networking and social support in tourism experience: The moderating role of online self-presentation strategies", *Journal of Travel & Tourism Marketing*, 30: 87.

Laesser, C., Beritelli, P. and Bierger, T. (2009) "Solo travel: Explorative insights from a mature market (Switzerland)", *Journal of Vacation Marketing*, 15(3): 217–227.

MacCannell, D. (1999) *The Tourist: A New Theory of the Leisure Class*. Berkeley, CA: University of California Press.

Machin, D. and van Leeuwen, T. (2005) "Language style and lifestyle: The case of a global magazine", *Media, Culture & Society*, 27(4): 577–600.

Mackay, Q. (2018) "Solo travel the 'in' thing", *The Daily Mercury*, 5: 2, July.

Maffesoli, M. (1996) *The Time of the Tribes: The Decline of Individualism in Mass Society*. London: Sage.

Magasic, M. S. (2018) "Connected tourists: From sightseeing to sitesharing", PhD Dissertation, Curtin University. Available at: http://hdl.handle.net/20.500.11937/75446.

Manovich, L. (2016) "Notes on Instagrammism and mechanisms of contemporary cultural identity", *Public Access*. Available at: http://manovich. net/index.php/projects/instagram-and-contemporary-image.

Mansfeldt, O. K. (2015) "The airbnb experience and the experience economy: The spatial, relational and experiential in-betweenness of airbnb", in A. Lorentzen, L. K. Tops and L. Schroder (eds.) *Spatial Dynamics in the Experience Economy*. London: Routledge, 103–116.

Mascheroni, G. (2007) "'Global nomads' network and mobile sociality: Exploring new media uses on the move", *Information, Communication & Society*, 10(4): 527–546.

Meyrowitz, J. (1985) *No Sense of Place: The Impact of Electronic Media on Social Behavior*. Oxford: Oxford University Press.

Miller, D. (2011) *Tales from Facebook*. Cambridge: Polity.

Molz, J. G. (2006) "'Watch us wander': Mobile surveillance and the surveillance of mobility", *Environment and Planning A*, 38: 377–393.

Molz, J. G. (2012) *Travel Connections: Tourism, Technology and Togetherness in a Mobile World*. London: Routledge.

Munt, I. (1994) "The 'other' postmodern tourism: Culture, travel and the new middle classes", *Theory, Culture & Society*, 11: 101–123.

Norum, R. (2006) "Popular pleasures and discursive moralities: 'Sex tourism', the Middle East and the (re)orienting of desire", unpublished M.Phil Thesis, Institute of Social and Cultural Anthropology, University of Oxford, Oxford.

Papacharissi, Z. (2011) "Conclusion: A networked self", in Z. Papacharissi (ed.) *A Networked Self: Identity, Community and Culture on Social Network Sites*. New York: Routledge, 304–318.

Papacharissi, Z. (2015) "Affective publics and structures of storytelling: Sentiment, events and mediality", *Information, Communication & Society*, 19(3): 307–324.

Parra-López, E., Gutiérrez-Taño, D., Díaz-Armas, R. J., and Bulchand-Gidumal, J. (2016) "Travellers 2.0: Motivation, opportunity, and ability to use social media", in M. Sigala, E. Christou and U. Gretzel (eds.) *Social Media in Travel, Tourism and Hospitality: Theory, Practice and Cases*. London: Routledge, 171–188.

Pillai, R. C. (2013) "Bitten by Wanderlust, Kerala women travel solo", *The Times of India*, October 17. Available at: https://timesofindia.indiatimes.com/Bitten-by-wanderlust-Kerala-women-travel-solo/articleshow/24242176.cms.

Pine, B. J. and Gilmore, J. H. (1998) "Welcome to the experience economy", *Harvard Business Review*, July–August.
Polson, E. (2016) *Privileged Mobilities: Professional Migration, Geo-Social Media, and a New Global Middle Class*. New York: Peter Lang.
Polson, E. (2018) " 'Doing' local: Place-based travel apps and the globally networked self", in Z. Papacharissi (ed.) *A Networked Self: Platforms, Stories, Connections*. London: Routledge, 159–174.
Reddit.com (2020) "r/Solotravel". Available at: www.reddit.com/r/solotravel/.
Riley, P. (1988) "Road culture of international long-term budget travellers", *Annals of Tourism Research*, 15: 313–328.
Sheller, M. and Urry, J. (2006) "The new mobilities paradigm", *Environment and Planning A*, 38: 221.
"Solo travel awards winners announced – best practices for market revealed: Multiple reports show solo travel continues to rise" (2018) *PR Newswire*, November 5. Available at: www.prnewswire.com/news-releases/solo-travel-awards-winners-announced-best-practices-for-market-revealed-300743696.html.
Sovich, N. (2019) "For a truly indulgent vacation, fly solo", *Wall Street Journal*, September 19. Available at: www.wsj.com/articles/for-a-truly-indulgent-vacation-fly-solo-11568925999.
"The rise of the solo traveler" (2020) *Sawubona*, June 18. Available at: https://saasawubona.com/the-rise-of-the-solo-traveller/.
Turkle, S. (2011) *Alone Together*. New York: Basic Books.
Urry, J. (1990) *The Tourist Gaze: Leisure and Travel in Contemporary Societies*. London: Sage.
Urry, J. (2003) "Social networks, travel and talk", *British Journal of Sociology*, 54: 155–175.
Urry, J. (2007) *Mobilities*. Cambridge: Polity.
van Dijk, J. (2006) *The Network Society: Social Aspects of New Media*. 2nd edition. London: Sage.
Wellman, B. (2004) "Connecting communities: On and offline", *Contexts*, 3(4): 22–28.
Willson, M. A. (2006) *Technically Together: Rethinking Community Within Techno-Society*. New York: Peter Lang.
Wilson, E. and Harris, C. (2006) "Meaningful travel: Women, independent travel and the search for self and meaning", *Tourism*, 54(2): 161–172.
Wilson, E. and Little, D. (2005) "A 'relative escape'? The impact of constraints on women who travel solo", *Tourism Review International*, 9(4): 155–176.
Wilson, J. (2014) *Mindful America: The Mutual Transformation of Buddhist Meditation and American Culture*. Oxford: Oxford University Press.
Wittel, A. (2001) "Toward a network sociality", *Theory, Culture & Society*, 18(6): 51–76.
Xiang, Z. and Gretzel, U. (2010) "Role of social media in online travel information search", *Tourism Management*, 31(2): 179–188.
Yang, E. C. L., Khoo-Lattimore, C. and Arcodia, C. (2018) "Constructing space and self through risk taking: A case of Asian Solo female travelers", *Journal of Travel Research*, 57(2), 260–272.

11

TIME FOR REPRESENTATION

Mediating the moment in a mobile space

Roger Norum

Introduction

Slow food. Fast Company. Slow travel. QuikStop. Stop action. Live-stream. Slow boat. Speed Queen. Slofie. Real time. The forms, tempos and rhythms of time are ever-present in contemporary socio-cultural life, as the manifold sensibilities characterised by these common phrases show. Associations with the temporalities of contemporary experience, which often make their appearance heavily mediated by socio-technical intervention, are ubiquitous. Time is an entity that, via a range of human and digital practices, can now be readily contested, enhanced, stretched, shrunk, played with and reconfigured. Time is a fulcrum through which we both experience *and* represent such experiences – or through which those experiences are represented to us.

A key point of departure of this chapter is that the temporalities of mobile tourism experiences are today fundamental to shaping these experiences. Within mobile mediated tourism, location-based social media platforms connect people as much to particular *places* as to specific *times*. While not excluding spatial considerations in mobile practice, I seek here to theoretically explore how inflections of time give form to momentary human experience. Framed by two online mobile travel experiences which came about during the COVID-19 pandemic, the chapter investigates how time is (re)presented in momentary and mediated tourism experiences. The core question I seek to answer in this chapter is this: how does the digital mediation of individuals' temporal experiences alter the socialities of being mobile or immobile? In showing the social and temporal possibilities of both mobility and immobility, I argue that digital practices do not merely alter how we experience time and space, but that these practices *make* human experience *in se*.

By shifting how present, in situ moments are lived and represented, digital practices reconfigure how humans experience the expansions and limitations of time and space. However, in order to be ushered into being, this present must be *presented*, which is to say that in contemporary social life, lived experiences are turned into present moments worthy of distribution and dissemination through their re-presentation. In this sense, real-time mobile media practices that have emerged during recent decades are shown to be panaceas or workarounds to immobilisation (whether created by global pandemics or other challenges to mobility), encouraging people to not simply feel, but to be, mobile – to be and to feel (in) the present.

The chapter is structured as follows. First, I will present a brief background of the ways in which travel or tourist practice has come to dominate much of contemporary socio-cultural life in the Euro-American context. This is followed by an overview of how these shifts have proceeded alongside key changes in the travel media content industry. Then, I draw attention to two live-streamed mobile media broadcasts, each of which offers an example of how temporality is deployed through mobile media experiences. The Faroe Islands' "Remote Tourism" initiative and Saint Lucia's "#7MinutesinSaintLucia" campaign, two distinct commercial initiatives conceived of and launched within weeks of the global lockdown, became ways for would-be (but immobilised) travellers to experience place through online mobile experience. Having set the scene with these two cases as a backdrop, I then discuss four specific instantiations of the signification "representation" through which experience is deployed via a temporally mediated mobility. Specifically, I explain the ways in which: a) experience is represented, depicted or signified; b) experience is *re*-presented, or presented again, for consumption; c) experience wields the present moment in new ways, re-*presenting* it; and d) experience is subsequently gifted – given as a *present* – through reciprocal social action. Through this archaeology of understandings of "representation," I demonstrate how time becomes a pervasive, defining entity in the context of mediated mobility, showing how online sociality is mediated by the particular temporalities afforded by various platforms. By way of conclusion, I argue that digital mobile practices today do not merely alter how the temporal and the spatial are encountered and perceived; they can in fact *make* human experience itself, giving a valorisation to human experience that equates it with the experience of time passing.

Mobilities and socialities: some context

For all the importance that is (rightfully) drawn to place in mobile practice, time is of tantamount consequence to the experience of both being mobile and being in a given location (cf. Schwartz and Halegoua, 2015). This chapter builds on previous work on global expatriates which has considered the integral links between mobility, temporality and what I have termed the "hypersociality" that persists among particular forms of ephemeral community (Norum, 2015, 2016). Hypersociality is just one example of the way in which new forms of mobile practice can at once amplify and throttle modes of human social engagement. The framework of mobile socialities encourages thinking about the temporalities presented by the mediated, mobile distribution and dissemination of such sociality – and the real-time, discursive engagement around such texts that connects disparate individuals separated by distance but brought together by the time/space-bending affordances of digital technologies.

Mobilities have long been understood by scholars across disciplines to play key roles in social relationships formed among individuals and communities (Hannam et al., 2006). Invoked alongside mobility, sociality becomes a conduit for remaking individual, individualised mobile practices as intertextual practice. I understand sociality as the propensity to associate and engage with others and form social groups, a "distinctly human dynamic relational matrix within which human subjects are constantly interacting in ways that are coproductive, continually plastic and malleable" (Long and Moore, 2012: 41). This chapter brings together understandings of sociality and mobility as a way of enabling us to better understand the importance of the structures, processes and practices of mediation (Keightley and Reading, 2014). The imagined communities that technological innovations such as the printing press (Anderson, 1983) and the internet (Miller and Slater, 2000) have generated are now fundamental to the ways in which we are, and are able to become, social. Through processes of mobilisation and digitisation, both individual and community locate new roles and find new relationships forged between them.

New socio-technological interventions facilitate and encourage humans to expand our actual and imaginative capacities, enabling us to creatively envision new modes of sociality that may not yet be present in our lives or worlds.

As spelled out in the Introduction to this volume, the concept of mobile socialities speaks to the obfuscation and complexification of particular social practices in the context of mobility and mediated mobile communications. A specifically mobile sociality serves us well here as a frame for considering the temporalities of reproduction, distribution and dissemination of media texts, and the real-time, discursive engagement around such texts that connect disparate individuals who are separated by distance, but brought together by the temporally altering connectivities of digital technologies. In this chapter, I inflect temporality through the mobile socialities frame as a means of bridging the temporal structuring processes and everyday lives of transient individuals and communities with shifts in the relevance and importance of place and place-making concomitant with mobile practice. I aim to draw attention to these particular social practices and the temporal structuring processes implicit in them.

Mobilities becomes tourism: travel background

In December 2012, the United Nations World Tourism Organization announced that the one billionth tourist that year had arrived at her destination (UNWTO, 2012). Seven years later, the number of international arrivals had increased by 50 per cent to 1.5 billion per year (UNWTO, 2020a). That same year, the tourism and travel industry made up some 10 per cent of global GDP (contributing approximately $9.26 trillion to the global economy), and was the world's third largest export sector (after fuels and chemicals) (Statista, 2020). The COVID19 pandemic changed things dramatically. During the pandemic, borders were closed, hotels were shut and air travel dropped dramatically, with international tourist arrivals decreasing by 56 per cent in the first five months of 2020, resulting in $320 billion in exports lost (UN, 2020). By the end of the year, international tourist arrivals will likely have plunged by up to 80 per cent and visitor spending is estimated to have dropped to nearly $300 billion (from $1.5 trillion in 2019) (UNWTO, 2020c). A proper recovery of the industry is not foreseen until 2024, putting as many as 120 million jobs at risk (UNWTO, 2020b).

And yet, what mobility, travel and tourism have come to mean in both the human imagination and in human practice, has not altogether (yet) been changed by the pandemic. Such shifts had been long in the process of becoming. The twenty-first century demarcated a watershed for mobile technology, introducing a sea change in the ways in which people think, move and communicate across space and time (see e.g. Goggin, 2006; Wilken and Goggin, 2013). The new opportunities for and forms of travel and tourism, and the new affordances of mobile media technology, have together led to dramatic shifts in the practices of mobile individuals, expanding and speeding up connections between people travelling and at home. As mobile practices change, they lead to new ways in which movement is mediated, concomitantly enabling shifts in what it means to have, communicate and remember experience both on and off the road.

James Clifford (1986: 15) has noted that travel is not merely about going places; it is a conceptual and geographic displacement intended to catalyse consciously enriching experiences. While some people may travel in order to have a memorable bucket list adventure in a given part of the world, others might do so in order to experience an experience in a place and then immediately thereafter (or indeed, simultaneously) share it, making use of now-ubiquitous practices of documenting, narrativising and broadcasting momentary experiences for consumption by others. Such shifts in the production and consumption of events, and the subsequent capturing and fixating of "moments" of experience, can also change how experiences, and the subjective experience of the present, are *experienced* (Simanowski, 2018). Along with the recent growth in the forms of experiential (Skift, 2014) and transformative (Skift, 2018) travel practices, changes in the production, consumption and mediation of mobile experience highlight the importance of temporal considerations for how place is experienced.

Today, the forms of travel – and of mobile practice more broadly – obscure the seemingly stable coordinates which have long demarcated the sedentary life and the mobile life. People, things, data and ideas are on the move now more than ever, and the forms of such movement are continually re-situated in theoretically interesting ways. The shift to the experience economy more generally has spawned novel "tourism practices that resist orthodox or 'sedentary' definitions [and] explanations" (Gale, 2009: 130). Moments that once might have been considered specifically touristic experiences, framed by a search for the disappearing authentic or the exotic other, are now collapsed into our quotidian lives, with new lines between the holiday and the everyday being simultaneously drawn and blurred (Mostafanezhad and Norum, 2018). For example, contemporary mobile and "touristic" experiences often extend well beyond consumption of place during temporary sojourns, as intensified transnational connectivity and regimes of flexible, de-situated employment also contribute to a withering of distinctions between categories that may have once seemed to be rather fixed, such as home and away, quotidian and extraordinary, and labour and leisure (cf. Urry, 2002). While the effects of the COVID19 pandemic have put much global travel on hold, the immobilisation of many once-hypermobile individuals has not made such categories any less fixed, binary or clearly defined.

New (travel) media production

In order to more fully appreciate these changes in how tourism is done, it is critical to understand the broader commercial and industrial context in which it is sold and represented. Such socio-cultural shifts are facilitated by the changing media industries: online and content production practices and new forms of circulation of media texts have radically transformed what it means to be a "content producer." With the advent of mobile media and the decline in traditional modes of publishing, recommendations for (and ideas about) specific travel destinations are found less frequently on the printed page of a dog-eared travel guidebook or travelogue than on the glowing screen of a mobile device. The traditional published narratives of travel – evocative, long-form texts that once communicated the value and meanings of tropes such as foreign, exotic and other – are no longer produced exclusively by professional, rarified or full-time writers. Nor are they written as evocatively or lengthily. Today, content arrives round the clock to mobile devices from the keystrokes of other travellers. Blog posts, tweets, top-ten "listicles," advertorials, Instagram images, Periscope feeds, Tripadvisor ratings and Facebook statuses today comprise the now-ubiquitous short-form journalism of the travel industry, bite-sized for short attention spans and on-the-go consumption.

In previous work, I have considered the roles of professional travel journalists and other industry actors in the creation and dissemination of travel narratives (see Abram and Norum, 2016). But new relationships in new media production merit looking at the role played by travellers themselves in generating "content." If travel writing helped produce the rest of the world for a Western audience (Pratt, 1992; Said, 1979), the forms of contemporary travel media production can expose the powerfully colonising force of Western romantic imaginaries of place (Holland and Huggan, 2000). Investigating these new mobile subjectivities reveals how the desire for travel is being created and mediated (Franklin, 2008), and how such travel is consumed, experienced and then re-narrativised through a range of mobile media practices.

Contemporary mobile practice is heavily mediated by the geolocative natures and multiple layers of interaction of mobile devices. Hundreds of travel-related apps are marketed at a location-aware generation to "make every stage of your trip easier and more enjoyable" (Buxton, 2016) and to "keep you connected and informed" (Coldwell, 2016). These apps range from currency converters, boarding pass generators and flight trackers to travel-focused dating and mapping systems that track one's every movement, enabling comparisons with the trajectories of other travellers. If, as Zygmunt Bauman (2005: 67) put it, mobile individuals today are "people of many places but of no one place in particular," then mobile media facilitates exploring the self across these many places. Indeed, Kannisto (2014) has observed that the new subjectivities of mobile practice are often typified by location independence.

These narratives often make experience appear to be accessible to everybody in much the same way, unhindered by the vaunted gates of the grand publishing houses. Travellers now no longer just consume travel experiences: they co-produce, co-design and co-exhibit (and co-consume) them, often simultaneously. Implicit in these processes is a wider mesh of reciprocal transactions and relationships of exchange between actors that include tourism boards, destination management organisations, public relations firms, publishers and editors, all of whom enter into negotiation with each other (and with potential travellers) in the mediation of travel content production to contribute to what Simone Abram and I have referred to as the "negotiated creation of a destination" (Abram and Norum, 2016). The advent of new media and mediated experience has brought unexpected changes to the relationships between producers and consumers (and distributors and regulators). Once clear distinctions between amateur and professional content producers (and public and private domains) are shifting, blurring and in many cases dissolving, since people who at one time made up an "audience" are now performers and viewers, producers and consumers, at the same time. These mentioned new media practices transform not just the production and consumption practices that define the political economy of the media system; they also transform the pleasure and meaning of these practices and associated socially significant, bodily and emotionally lived practices (Ardèvol et al., 2010).

The mediated mobility forms which I discuss in this chapter speak to recent work by Axel Bruns regarding the involvement of end-users in processes of production in the context of digital media (Bruns, 2008, 2016). Building on concepts first put forth by Toffler (1980), and later elaborated by others (see e.g. Ritzer et al., 2012; Charitsis, 2016) the concept of *produsage* – a somewhat clunky portmanteau of "production" and "usage" – is helpful for describing the new commercial constellations that have evolved following the development of platforms and technologies which enable the creation of user-generated content. A political economy of produsage might be explained thus: as the separation between consumer consumption and production becomes clouded, the traveller evolves into a "produser" (Bruns, 2016) comprised of a "produsing" (or "prosuming") self who creates value broadcasted experience, and a "prodused" (or "prosumed") self, an active and entrepreneurial subject governed to produce data and content that bring value for large media corporations through engagement with various platforms. This "communicative capitalism" (Dean, 2010) relies on the appropriation and exploitation of the active participation of users who create content, interact with each other, develop and cultivate relationships, and provide suggestions, recommendations and solutions. Now, new modes of produsage are emerging as "smart" tracking technologies become embedded in the everyday mobile practices of travellers, offering numerous possibilities to users to track, quantify, arrange and disseminate their activities, contributing to what has been termed the "quantified self" (Wolf, 2011). *Produsive* labour, as well as being an opportunity for capital accumulation, also governs people's lives and subjectivities. Time is, of course, central to what it means to do *produsage*. In a clear temporal cycle, the production of content for an experience is pre-meditated;

the experience is had, the content is produced; the content is then shared, then consumed; then content based on this consumption is reciprocally produced, shared, consumed; and so on. The agentive and participatory production, consumption and evaluation of content in relationships of *produsage* is particularly visible in the travel media industry, as I will argue in this chapter. In the section which follows, I briefly describe two novel travel campaigns launched during COVID19 which used live-streaming video feeds to connect otherwise immobilised end-users to new places in real time.

Present imperfect: from emplaced immobility to displaced mobility

The COVID19 virus outbreak had impact on a wide range of aspects of human mobility through the multiple lockdowns, border closures, restriction of services and extended periods of immobilisation under one roof that were imposed on people by state agencies. Already by early March 2020, the virus had swiftly obliterated the demand for travel all over the planet. With the world's piazzas denuded, biennales postponed and flights grounded, the doomsday economic predictions were quick to come: losses of €11 billion between March and May in Italy alone; a drop in global tourism arrivals of double-digit percentages. The bankruptcy in March 2020 of the UK-based discount airline FlyBe was just one poignant reminder that much of our seemingly robust hypermobile world operates on a knife's edge by the receipts generated from business meetings, technology conferences, music festivals, city breaks and honeymoons. Within a few weeks, the tourism industry responded with a litany of hopeful hashtags (e.g. #throwbackthursday, #bucketlisttravel, #traveltomorrow) to encourage people to believe that, with patience, their persistent wanderlust would one day be rewarded. Jocular reminders in the form of "coronavirus travel memes" quickly flooded Instagram and Twitter feeds, aiming to reassure people that we would all at some point travel again. Companies such as Airbnb turned their locally hosted offline tours into virtual "online experiences" held over Zoom, to which would-be travellers could purchase access. A number of cultural organisations, destination management organisations and tourist boards themselves capitalised on the large number of people in lockdown at home, hoping both to entertain people who had time on their hands and win loyal future visitors for once travel again became possible. The Smithsonian National Museum of Natural History established virtual tours. The Monterey Bay Aquarium and San Diego Zoo set up live-streaming webcams and pointed them at their sea otters, jellyfish, baboons, penguins and owls. Clubs in Berlin and Ibiza set up live-streaming DJ sets (Mitchell, 2020). Perhaps somewhat ironically, the worldwide travel ban inspired a wide range of new travel experiences.

Launched on 15 April, the Faroe Islands' "Remote Tourism" campaign allowed virtual travellers to log on and take part in live, interactive online tours of the islands with local guides. In mid-March, the Faroe Islands tourist board commissioned a local tech firm to swiftly build a platform that enabled end-users to "control" the movements of a local tourist board employee. The local wore a helmet mounted with a GoPro camera, which was connected to the web by mobile phone. The video feed streamed the feed from the camera overplayed with an on-screen gaming controller (see Figure 11.1), which offered the opportunity to suggest movement in any of four directions, as well as "run" and "jump" options. Once logged into the platform, users could watch the streamed video feed and sign up to join a queue that would enable them for several minutes to be in charge of the controller, which directed the local with voice commands played through their mobile phone's earphones, to perform a certain type of action. Tourist board employees held several tours weekly, during which they walked across various parts of the island, rode horses and flew in helicopters, throughout the tours responding to the directional commands of various users. The platform streamed on a newly established URL, www.

Figure 11.1 Screenshot of archived Faroe Islands Remote Tourism live-stream video tour #7.
Source: www.facebook.com/VisitFaroeIslands/videos/255546842247527/

remote-tourism.com, archiving each tour on the tourist board's Facebook page. Over two months, until 17 June when the campaign ended (and the country could open again to visitors from several countries), 22 separate live-streamed tours were carried out across the islands' waterfalls, fjords and mountain paths, receiving close to 10,000 viewers (Globetrender, 2020).

On 26 March, Saint Lucia Tourism Authority launched its virtual live moments campaign under the hashtag "#7MinutesinSaintLucia." For three months, every Tuesday and Thursday at 6pm EST, the tourist board broadcast seven minutes of live-streamed video showcasing picturesque scenes, interviews and tours that highlighted local memorable experiences. Shot in portrait mode with a mobile phone, the tourist board filmed yoga sessions (see Figure 11.2), DJ sets, cooking classes, sunset scenes (see Figure 11.3) and lessons in SocaFit, a renowned cardio dance workout to the region's Soca music, which originated in Trinidad and Tobago during the early 1970s. Hosted across the tourist board's Instagram, Facebook and Twitter accounts, the video broadcasts were streamed over Instagram Live and later archived on IGTV. Comments and graphical multicoloured hearts posted by users scrolled up across the videos during broadcasts, through which users voiced their opinions on, for example, the stream, the presenter or the landscape. The welcome text on each video read: "Welcome to the beautiful Island of Saint Lucia 🌴 While travel may not be possible at this time we want to immerse you in our culture, music, dance, food and more, from your home." Tourism minister Dominic Fedee explained in an interview that the social media series was intended to "virtually share Saint Lucian culture and natural beauty in an authentic way while people stay safe at home, to keep in touch with future visitors and our valued travel agents – hopefully providing a bright spot in their days" (Plates, 2020).

These two brief examples allude to the ways in which flexible understandings of time – understandings which new media technologies in past decades have afforded us – have altered how we now imagine and experience place. Scholars of the human have long seen the experience of time to be a core determinant of how humans make sense of the world (see e.g. Evans-Pritchard, 1969;

Figures 11.2 and 11.3 Screenshots of @travelsaintlucia Instagram Live streams of yoga and a sunset scene.

Sources: www.instagram.com/stories/highlights/18143550433003197/?hl=en; www.instagram.com/stories/highlights/17856090475972043/?hl=en

Mauss, 2004 [1905]). While anthropologists have acknowledged that the passage of time and the modes and forms in which it is experienced may be organised distinctly across different cultural contexts (Herzfeld, 2009; Frederiksen, 2008), we have also been accused of perpetuating allochronicity – fixing our subjects in the historical past via the practice of writing in the ethnographic present (Fabian, 1983). Philosophers, for their part, have found multiple ways of explaining how

Figure 11.2 and 11.3 (Continued)

humans subjectively experience time. They have often focused on perceivable phenomena such as motion, succession and persistence, incurring debates as to whether how we experience eventful time is necessarily devoid of movement or change (seeing time akin to individual frames in a movie, in which movement from one moment to the next is only based on our memories of previous positions) or whether our ability to experience events is in fact temporally extendable, consisting of much more than isolated, durationless moments (Artsila, 2016).

The recent attention of media and cultural scholars to time (see e.g. Barad, 2010; Keightley, 2012; Sharma, 2014; Kaun et al., 2016; Bastian, 2017; Bødker, 2017; Pentzold, 2018;

Hartmann et al., 2019) has been a breath of fresh air for understandings of temporality in contemporary socio-cultural life. By focusing on lived, vernacular temporal subjectivities, such work has thankfully sidestepped anthropologists' navel-gazing self-consciousness about ethnographic presentism and avoided philosophers' preoccupation with rhetorical categorisations. Such work cuts to the core of how media and mediation shape the human experiences of time, and how these experiences in turn shape our subjective perception of the world. While Larissa Hjorth has argued that much early research on mobile media practice reasserted the importance of place as a "socio-cultural precept" (Hjorth, 2007), Hartmann et al. (2019: 3) point out that time has in fact been fundamental to early discussions of electronic media, in which temporal and spatial phenomena in the context of digital mediation were frequently understood to comprise malleable categories (see e.g. Harvey, 1989; Robertson, 1992; Relph, 2008 [1976]). Yet, as Adam has noted, temporal notions are themselves hardly straightforward considerations, in part due to the "incommensurable ideas about the source of our experience and concept of time" (Adam, 1990: 15). Indeed, the times we experience may be multiple, distinct and even conflicting – biological or chronological, linear or cyclical, industrial or technological, momentary or routine, communal or national, shared or imagined. Importantly, too, times are at once individual *and* social. And the ways in which we perform and enact time are never independent of some form of mediation (Hartmann et al., 2019: 4). Following Sharma (2014), then, my focus on individual, quotidian experiences of lived time is a way of counterweighting the emphasis on space prevalent in many studies of e.g. globalisation, migration and locative media.

In this vein, Annette Hill proffers the metaphor "roaming audiences" as a way of accounting for the "dynamic practices of people as they experience storytelling that takes place across dispersed sites of production, distribution and reception" (Hill, 2019: 1). Through a consideration of the spatio-temporal practices of audiences, roaming speaks to audiences' access to media across multiple devices and platforms, and from different places and at different times. This topographic focus is decidedly spatial, as the viewing, consuming audiences Hill describes are very much "roaming around storytelling," engaging with media content not just by "following pathways but acting as pathmakers in their media experiences" (Hill, 2019: 2). But, as Hill alludes to, these metaphorical pathmaking practices are temporal, as well, connecting consumers and producers *with* each other and *to* place, across multiple mediated times. Tourists themselves *roam*. This is in fact what historical and contemporary tourism has often been – particularly if we follow the New Oxford American Dictionary's definition of the verb: "to move about or travel aimlessly or unsystematically, especially over a wide area" (McKean, 2005). During the global lockdown, people's inability to travel in any physical sense led them to satiate their wanderlust through seeking out the possibility of roaming mediated place – in the Faroese case through the act of controlling the movements of a mobilised, datafied guide; in the Saint Lucia example through interacting with live-streamed broadcasts of various touristic activities. In so doing, these virtual travellers, each individually experiencing place on their own in front of their computers or mobile devices, also became – together – an audience.

The "representation" with which I have entitled this chapter speaks to this this temporal mediation in four distinct ways. Before each following explanation, I have enumerated the general meaning and italicised the stressed syllables in an attempt to convey the various emphases and subtleties of meaning:

1 repre*sent* (the primary definition, e.g. to depict or signify)
First, and perhaps most straightforward, a subjective entity (in our case, a travel experience) is *represented* through some form of mediation. This entity to be communicated might be a particular experiential moment/momentary experience (e.g. my time bungee jumping off

a bridge in West Bengal) or a particular relationship to or understanding of a place (what I like about a flower market).

2 *re*present (presenting again)
Second, the entity is *re*presented, or *presented once again* to be perceived and consumed by another actor. This typically occurs within a generally quite short temporal span and speaks to the continual mediation of many (e.g. social) media which function across real-time broadcast or transmissive technologies. Such mediation takes place on (at least) two levels. In one, the entity is represented to the self: our body presents an experience to our senses both for us to enjoy its momentary sensory-ness (e.g. pleasure, pain) *and* for it to be written into our own memory banks. On a second level, the moment is once again presented back to both ourselves (e.g. in order to jog our memories, our own personal media content storage devices, to recall the physical experience that has already happened to us) and to others (e.g. to share with friends and followers what the experience was like for us).

3 re*pre*sent (producing a new present, e.g. "to present" as a verb)
Quite distinct from the second definition is my third understanding of "representation," which is perhaps the most challenging to clarify. Here, the *temporal entity known as the present* is itself redeployed as a mobile, mediat-able actor. In this, I re-envision the term "present" with a non-standard verbal sense (e.g. imagine a new infinitive, "to *pre*sent") that facilitates speaking about forms of engendering the present moment – we might say that it herein formulates practices of *presenting*. I take here "*pre*sent" to speak to both a social-individual sensibility that is attuned to a momentary instantiation of an experience (e.g. "living in the present") as well as a generalised sense of bringing the immediacy of momentary lived experience always, already into the present through practices of documenting, saving and archiving. As such, the present is itself refashioned as a new temporal instantiation, given that it can be kneaded and moulded through its mediations across various digital forms.

4 re*pre*sent (giving "the gift" of reciprocity, again)
Finally, fundamental to understandings of sociality given its inherently reciprocal nature, a fourth understanding of "representation" embodies the sense of "the gift" (Mauss, 2004 [1905]), or in more familiar English, a present. In the context of *produsage* embedded within the circulation and recirculation of media content, the gift of re*present*ation is really fully reciprocal: it is at once the gift to others, which becomes something that is then always already gifting, or being gifted, back to ourselves (for more on which, see Coates, Chapter 7 of this volume). It is at once a present to the self and a present to the other.

I will now discuss the ways in which each of these representations are taking place. Note that for purposes of clarity and to pay homage to a key concept from early tourism anthropology (see Smith, 1977), I use the term "host" here to refer to the person who would in the traditional tourist encounter be understood to be the producer (or creator) of content, and the term "guests" to refer to the people who remain on the other, consuming (or viewing) end of the communication. However, as I have stated earlier in this chapter, these categories are now – and indeed, always have been – far from fixed, and now can readily blend into one another over the course of a single streamed broadcast or other intertextual communicative act.

First, all of the examples presented above are instantiations of (1) a representation, "the action of standing for, or in the place of, a person, group, or thing, and related senses" (OED Online, 2020). They are textual or visual renditions that aim to, affectively or substantively, communicate some a given experience. Second, the example of the live-streamed, interactive programme, travel content is also (2) *re*presented – presented once again – since each communicative act on the part of the host and the guest is a recursive presentation of the experience being had by each of them. It is important to stress that "content" is being produced and consumed on both sides of the connection, as well: the guests are taking in the video feed, while the host is receiving the hearts and comments. The act of consumption, because it is taking place in real time, is changing each subsequent performative, (re)productive act. The content is further presented later on, too, since the stream will be video archived and can be viewed at a later point by guests – although at that point, the feed is no longer live, and is somewhat less interactive. Third, the experience as a singular entity is (3) represented, or brought back into the present moment for both host and guest, with each passing frame of mutual communication on either side of the connection. Finally, end-users are repeatedly gifted or (4) re*pre*sented, not merely with "the gift of travel," but with the reciprocal interaction that comes from real-time circulation and recirculation of data and content.

Such multimodal phenomena exemplify Barthes' paradoxical (spatial) experience of "transported immobility," in which a mobile device and an immobile subject are brought together in a paradox of experiential reality (1979: 143; see also Hill, Chapter 18 of this volume). In Barthes' example, the mobile device is a moving railroad dining car; the immobile subject is Barthes himself, sat before an array of fine silverware on a white tablecloth about to tuck into a fine dining experience in a fixed restaurant space, which he perceives as illusory when juxtaposed with the fast-moving train. For Barthes, "each constraint seems to produce its contrary freedom, each gesture is a denial of its original limits" (Barthes, 1979: 141). Hill has suggested that this conflict generates a mirage that "conceals contingency, displacement, and movement" (Hill, 2019: 38). Indeed, this paradigmatic example itself dines out on the dialectic between mobility and immobility, and, penned in the last 1970s, it portends a multitude of similar experiential paradoxical confrontations and juxtapositions that were to grow exponentially in number as technology progressed and globalisation expanded post-Barthes. Yet this dialectic of movement and locatedness is, in fact, a defining problematic of the travel experience.

When tourist experiences were first sold to the masses in the mid-nineteenth century, their aim was to package a form of transported immobility within the narrative of a "staged authenticity" (MacCannell, 1999). Friedberg (1994) points out that this was part of an evolution in the gaze of public viewership from panorama to diorama to cinema, through which, as the gaze itself became more "virtually" mobile, the traveller/spectator was empowered to "travel" while remaining more physically immobile. As she writes, tourism typically

> produces an escape from boundaries, it legitimates the transgression of one's static, stable, or fixed location. The tourist simultaneously embodies both a position of presence and absence, of here and elsewhere, of avowing one's curiosity and disavowing one's daily life.
> (Friedberg, 1994: 59)

However, I would suggest that while both selves and imaginations in the presented cases are indeed transported, as Barthes' phrasing suggests, their (re)presentations fix to enchant the consumer. I thus suggest that these are perhaps less fully encapsulated by the phrase "transported immobility" and are more indicative of what Chandler (2009) has called a "trans*fixed* immobility" [emphasis mine]. Here, inherent to the perceived immobility is a sense of amazement, of awe, of enrapture, speaking directly to the effect of mechanised travel on a mobilised body. If we consider the temporal aspects of this "fixation" (Vium, 2017), it becomes clear that mutations of mobility previously described do not just work on the plane of the spatial; they are also all inherently operating on temporal level as well.

Conclusion: mediated sociality and social media

In his critique of Facebook's "global community" and of Western society's media practices more broadly, Roberto Simanowski (2018) argues that many social media platforms encourage an uncritical narrativisation of the self that privileges mechanical presentation (e.g. through automated sharing) over mindful representation (e.g. through textual statements). He suggests that the "automatic autobiography" culture of overshared self-expression is characterised by "spontaneous, episodic" production, rather than "deliberate, coherent" narrative, and by storytellers that are not humans but rather networks and their algorithms (Simanowski, 2018: 185). Simanowski is less than sanguine on live-streaming practices which enable responsive, real-time broadcasting of pastel-coloured hearts, thumbs-up likes and comments alongside audio-video transmission – such as the examples on which I draw in this chapter – arguing that "spontaneity and immediacy function here as synonyms for authenticity" (Simanowski, 2018: 189). Through such practices, Simanowski posits, new media platforms and their affordances destroy the experience of the present by their very endeavour to permanently capture it, replacing "real experience" with reflexive, unreflecting documentation of the lived moment. He even marshals to his defence Agamben's observation that true contemporaneousness is impossible in a mode of absolute immediacy (Agamben, 2009); Agamben himself built on the work of Benjamin in his 1933 essay "Experience and Poverty" (1999) earlier notion of the replacement of longer-term experience (Erfahrung) with incidental lived experiences (Erlebnisse).

In this chapter, I have considered the role of time in particular mobile media practices, in which location-based social media platforms connect people not to particular *places* but to specific *times*. I discussed the ways in which digital technology co-opts time and the ways in which time concomitantly co-opts human sociality, pointing to the emergence of two online mobile travel experiences during COVID19 to show how online sociality is directly mediated by the particular temporalities afforded by these techno-social interventions. I have argued, *pace* Simanowski (2018), that the *immediatised* temporality of such mobile experiences are not merely "synonyms for authenticity." Rather than simply changing the perception of time and the relationship to physical spaces (or the digital representation of such spaces) encountered, digital or otherwise, the digital mobile practices of travellers "in" these experiences *make* experience itself. In this way, time is neither abstract idea, nor fleeting presence, nor even physical barrier. Time is not just experiential; it *is* experience, defining how we see, feel and be in the world, and how we make others do the same.

In the tourist landscape, the souvenir and the camera have long been pointed to as possible proxies for experience (see e.g. Silberman and Bensen, 1987; Peters, 2011; Stylianou-Lambert, 2012). So, the argument goes, a culture of compulsion that privileges documentation and only the surfeit of communication through the presentation of what one experiences to the world is exactly what hampers many people from experiencing the present in the first place. In contrast, as I have attempted to argue in this chapter, real-time mobile media practices can in fact be just what enable individuals, immobilised by global pandemics or other challenges to mobility, to not merely *feel* mobile but to *be* mobile. In effect, hosts and guests, together, become time travellers, wielding the spatio-temporal affordances of devices and platforms as tools for playing with and contesting what we typically feel to be the physical limitations of time and space. I argue that these mediated experiences have a positive, nurturing effect on people's everyday lives, unlike Simanowski, who would argue that such mediations distract people from actually experiencing life. Indeed, such practices connect hosts and guests not merely to particular places and times – imagined and real, virtual and lived, past, present and future – but to each other.

In tourism's early years, the days of Thomas Cook's Nile journeys in the mid-nineteenth century, travel was predicated on a physically mobile subject, one who would have pre-planned experiences. Indeed, the dimension of time in (offline) travel writing is essential for the way in which readers refigure, through a travelogue's discursive configuration, a mental image of an actual journey (Korte, 2008). After all, travelogues typically recount actual, physical travel – which is subject to time's arrow. But in the virtual travel practices discussed in this chapter, space and time themselves may be traversed and refashioned; a refiguring takes place through performed, digital texts. While the techno-socially mediated real-time-ness of such broadcasted acts impart upon end-users a chronotope that "bears a strong resemblance to what people know actual time spaces for travel to be" (Korte, 2008: 48), they also contest their own understandings of the temporal and spatial phenomena – both their opportunities and challenges – that characterise their lives. Following the example of Barthes, in whose example "travel itself imparts a psychic status" (Friedberg, 1994: 59), the immobile travellers who connect to the Faroese and Saint Lucian virtual experiences are both able to experience known and familiar times and places, while also being able to navigate and negotiate new understandings of, and experiences in, time and space. As Hill has put it, while new technologies and concomitantly novel modes of production and consumption may well "complicate, frustrate and outpace traditional media," new media innovations in the ways in which place and time are communicated remind us that "people feel they have a right to experience content in their own way" (Hill, 2019: 7). Indeed, as actors who determine which media, content and platforms succeed and fail in the global travel marketplace, such travellers will invariably express – through their mobile imaginations and practices – such prerogative.

Coda

The complexities, connectivities and – at times – surprising juxtapositions that bind together mobility, sociality and time are of no secret to the internationally networked scholar. When I settled on the Northern Fennoscandian Peninsula a while back – in part because it seemed to be the only place on the continent where I could afford even a small plot of land, in part because much of my fieldwork kept me returning to the North anyway – the story I would tell myself was that I was doing so largely to "get away from it all." As I quickly discovered, if you are busy being away from it all, but still allow yourself to be tied into buzzing global networks, you will feel compelled to *go* to and *be* in places in order to engage in some form of presentist sociality, whether or not you suffer from chronic FOMO, the fear of missing out (I leave the obvious issues posed by COVID19-era "presentism" for another time). So within just a few weeks of my move into an old red gingerbread-style homestead, I was back out on the road again and into my old, well-seasoned habits: scuttling through airport lounges, waiting for delayed trains while doom scrolling mindlessly through news updates, scarfing down croissants on my way to can't-miss academic workshops. Even when in my new rural home – quiet, forested and, for all intents and purposes, devoid of transport connections – I could not help but be drawn back into timing practices. For example, I would catch myself not merely regularly checking the local bus schedule on my phone, but relying on the municipal transport agency's GPS live-tracking app so as to time my on-foot arrival to the bus stop with the arrival of the bus itself – aiming to maximise use of the time I had in the middle of nowhere while also using my time to track the most efficient use of it. Even then, I still often ran for and frequently missed the once-an-hour bus connections anyway. Time was something I had convinced myself I could somehow create more of just by virtue of living off the map and off the grid. It soon became clear to me that I was relying just as much on the tools of temporal mediation as when I lived in global hubs of speedup and hypersociality such as New York and London. It is true that without peripheries, such as the one in which I now spend much of my time, the "centerness" of centres may itself be meaningless. But the ways in which technology mediates both our movements and our imaginations of mobility keep us always already bound to some centre, wherever we find ourselves.

Acknowledgements

I remain extremely grateful to Annette Hill and Jolynna Sinanan for their insightful comments on previous versions of this chapter, to Katie Ritson and Alessandro Rippa for last-minute editorial acuity, to Jamie Coates for his affectionate nudges to remain critical at every turn, to the three editors of this collection for their patience, generosity and encouragement, and to the entire family of Mobile Socialities authors for being such a inspiring and welcoming group of scholars.

References

Abram, S. and Norum, R. (2016) "An archaeology of Arctic travel journalism", *Studies in Travel Writing*, 20(3): 272–284.

Adam, B. (1990) *Time and Social Theory*. Cambridge: Polity.

Agamben, G. 2009 "What is the contemporary?", in D. Kishik and S. Petadella (eds.) *What Is an Apparatus? And Other Essays*. Stanford: Stanford University Press, 39–54.

Anderson, B. (1983) *Imagined Communities*. London: Verso.

Ardèvol, E., Roig, A., San Cornelio, G., Pagès, R. and Alsina, P. (2010) "Playful practices: Theorising 'new media' cultural production", in B. Bräuchler and J. Postill (eds.) *Theorising Media and Practice*. Oxford: Berghahn.

Artsila, V. (2016) "The time of experience and the experience of time", in B. Mölder, V. Arstila and P. Øhrstrøm (eds.) *Philosophy and Psychology of Time*. Berlin: Springer, 163–186.

Barad, K. (2010) "Quantum entanglements and hauntological relations of inheritance: Dis/continuities, spacetime enfoldings and justice-to-come", *Derrida Today*, 3(2): 240–268.

Barthes, R. (1979) *The Eiffel Tower and Other Mythologies*. New York: Hill and Wang.

Bastian, M. (2017) "Liberating clocks: Developing a critical horology to rethink the potential of clock time", *New Formations*, 1(92): 41–55.

Bauman, Z. (2005) *Liquid Life*. Cambridge: Polity Press.

Benjamin, W. (1999) "Experience and poverty", in M. W. Jennings, H. Eiland and G. Smith (eds.) *Selected Writings, Vol. 2, Part 2: 1931–1934*. Cambridge, MA: Harvard University Press, 731–736.

Bruns, A. (2008) *Blogs, Wikipedia, Second Life, and Beyond: From Production to Produsage*. New York: Peter Lang.

Bruns, A. (2016) "Prosumption, produsage", in K. B. Jensen and R. T. Craig (eds.) *The International Encyclopedia of Communication Theory and Philosophy*. New York: John Wiley & Sons.

Buxton, M. (2016) "How to travel like a pro", *Refinery29*, December 22. Available at: www.refinery29.com/travel-apps#slide (accessed December 27, 2016).

Bødker, H. (2017) "The time(s) of news websites", in B. Franklin and S. Eldridge (eds.) *The Routledge Companion to Digital Journalism Studies*. London: Routledge, 55–63.

Chandler, K. (2009) "The Eiffel Tower and Ferris wheel: Transfixed immobility and the performance of the modern in the late-nineteenth century", unpublished Master's Thesis, Royal College of Art, London.

Charitsis, V. (2016) "Prosuming (the) self", *Ephemera: Theory & Politics in Organization*, 16(3): 37–59.

Clifford, J. (1986) "Introduction: Partial truths", in J. Clifford and G. E. Marcus (eds.) *Writing Culture: The Poetics and Politics of Ethnography*. Berkeley, CA: University of California Press, 1–26.

Coldwell, W. (2016) "10 of the best travel apps . . . that you'll actually use", *The Guardian*, January 2. Available at: www.theguardian.com/travel/2016/jan/02/10-best-travel-apps-free-paid-for (accessed December 27, 2016).

Dean, J. (2010) *Blog Theory: Feedback and Capture in the Circuits of Drive*. Cambridge, MA: Polity.

Evans-Pritchard, E. E. (1969) *The Nuer: A Description of the Modes of Livelihood and Political Institutions of a Nilotic People*. Oxford: Oxford University Press.

Fabian, J. (1983) *Time and the Other: How Anthropology Makes Its Object*. New York: Columbia University Press.

Franklin, A. (2008) "The tourism ordering: Taking tourism more seriously as a globalising ordering", *Civilisations*, 52: 2–39.

Frederiksen, M. D. (2008) "Temporality in participation and observation: Perspectives from Albania and Georgia", *Anthropology Matters*, 10(2).

Friedberg, A. (1994) *Window Shopping: Cinema and the Postmodern*. Berkeley, CA: University of California Press.

Gale, T. (2009) "Urban beaches, virtual worlds and 'the end of touris'", *Mobilities*, 4: 119–138.

Globetrender (2020) "Faroe Islands welcomes thousands of virtual tourists via live video stream", *Globetrender*, June 7. Available at: https://globetrender.com/2020/06/07/faroe-islands-virtual-travel-remote-tourism.

Goggin, G. (2006) *Cell Phone Culture: Mobile Technology in Everyday Life*. London: Routledge.

Hannam, K., Sheller, M. and Urry, J. (2006) "Editorial: Mobilities, immobilities and moorings", *Mobilities*, 1(1): 1–22.

Hartmann, M., Prommer, E., Deckner, K. and Görland, S. O. (2019) "Introduction", in M. Hartmann, E. Prommer, K. Deckner and S. O. Görland (eds.) *Mediated Time. Perspectives on Time in a Digital Age*. London: Palgrave Macmillan, 1–21.

Harvey, D. (1989) *The Condition of Postmodernity: An Enquiry into the Origins of Cultural Change*. Oxford: Blackwell.

Herzfeld, M. (2009) "Rhythm, tempo and historical time: Experiencing temporality in the neoliberal age", *Public Archaeology: Archaeological Etnographies*, 8(2–3): 108–123.

Hill, A. (2019) *Media Experiences: Engaging with Drama and Reality Television*. London: Routledge.

Hjorth, L. (2007) "The game of being mobile: One media history of gaming and mobile technologies in Asia-Pacific", *Convergences*, 13(4): 369–381.

Holland, P. and Huggan, G. (2000) *Tourists with Typewriters: Critical Reflections on Contemporary Travel Writing*. Ann Arbor: University of Michigan Press.

Kannisto, P. E. (2014) *Global Nomads: Challenges of Mobility in the Sedentary World*. Ridderkerk: Ridderprint.

Kaun, A., Fornäs, J. and Ericson, S. (2016) "Media times | mediating time – temporalizing media, Introduction to the special issue", *International Journal of Communication*, 10: 5206–5212.

Keightley, E. (ed.) (2012) *Time, Media and Modernity*. London: Palgrave Macmillan.

Keightley, E. and Reading, A. (2014) "Mediated mobilities", *Media, Culture & Society*, 36(3): 285–301.

Korte, B. (2008) "Chrono-types: Notes on forms of time in the travelogue", in J. Zilcosky (ed.) *The Poetics and Politics of the Modern Journey*. Toronto: University of Toronto Press, 25–53.

Long, N. J. and Moore, H. L. (2012) "Sociality revisited: Setting a new agenda", *Cambridge Anthropology*, 30(1): 40–47.

MacCannell, D. (1999) *The Tourist: A New Theory of the Leisure Class*. Berkeley, CA: University of California Press.

Mauss, M. (2004 [1905]) *Seasonal Variations of the Eskimo: A Study in Social Morphology*. London: Routledge.

McKean, E. (2005) *The New Oxford American Dictionary*. New York: Oxford University Press.

Miller, D. and Slater, D. (2000) *The Internet: An Ethnographic Approach*. Oxford: Berg.

Mitchell, S. (2020) "The best live streams from around the world, to cure your cabin fever", *Escape*, March 27. Available at: www.escape.com.au/travel-advice/the-best-live-streams-from-around-the-world-to-cure-your-cabin-fever/news-story/33ea0c9505226290b14bcc86f94263ed.

Mostafanezhad, M. and Norum, R. (2018) "Tourism in the post-selfie era", *Annals of Tourism Research*, 70: 131–132.

Norum, R. (2015) "The hypersocial: Transience, privilege and the neo-colonial imaginary in Expatria, Kathmandu", unpublished D.Phil Thesis, University of Oxford, Institute of Social and Cultural Anthropology, Oxford.

Norum, R. (2016) "The 'nextpat': Towards an understanding of contemporary expatriate subjectivities", in J. Rickly, M. Mostafanezhad and K. Hannam (eds.) *Tourism and Leisure Mobilities: Politics, Work and Play*. London: Routledge, 91–105.

OED Online (2020) *Representation*. Oxford: Oxford University Press. Available at: www.oed.com/view/Entry/162997?rskey=yCdUBR&result=1&isAdvanced=false#eid (accessed June 28, 2020).

Pentzold, C. (2018) "Between moments and millennia: Temporalising mediatisation", *Media, Culture and Society*, 40(6): 927–937.

Peters, K. (2011) "Negotiating the 'place' and 'placement' of banal tourist souvenirs in the home", *Tourism Geographies*, 13(2): 234–256.

Plates, A. (2020) "7 minutes in Saint Lucia", *Global Traveler*, March 30. Available at: www.globaltravelerusa.com/7-minutes-in-saint-lucia/.

Pratt, M. L. (1992) *Imperial Eyes: Travel Writing and Transculturation*. New York: Routledge.

Relph, E. (2008 [1976]) *Place and Placelessness*. London: Pion.

Ritzer, G., Dean, P. and Jurgenson, N. (eds.) (2012) "The coming of age of prosumption and the prosumer", *American Behavioral Scientist*, 56(4): 379–398.

Robertson, R. (1992) *Globalization: Social Theory and Global Culture*. London: Sage.

Said, E. W. (1979) *Orientalism*. New York: Vintage.

Schwartz, R. and Halegoua, G. (2015) "The spatial self: Location-based identity performance on social media", *New Media & Society*, 17(10): 1643–1660.

Sharma, S. (2014) *In the Meantime: Temporality and Cultural Politics*. Durham, NC: Duke University Press.

Silberman, R. B. and Bensen, J. (1987) "Tourist photographs as souvenirs", *Prospects*, 11: 261–272.

Simanowski, R. (2018) *Facebook Society: Losing Ourselves in Sharing Ourselves*. New York: Columbia University Press.

Skift (2014) "The rise of experiential travel". Available at: https://skift.com/wp-content/uploads/2014/06/skift-peak-experiential-traveler-report1.pdf.

Skift (2018) "The rise of transformative travel". Available at: https://skiftx.com/wp-content/uploads/2018/04/The-Rise-of-Transformative-Travel.pdf.

Smith, V. (1977) *Hosts and Guests: The Anthropology of Tourism*. Philadelphia: University of Pennsylvania Press.

Statista (2020) "Tourism worldwide". Available at: www.statista.com/study/9996/tourism-worldwide-statista-dossier.

Stylianou-Lambert, T. (2012) "Tourists with cameras: Reproducing or producing?" *Annals of Tourism Research*, 39(4): 1817–1838.

Toffler, A. (1980) *The Third Wave*. New York: Bantam.

UN (2020) "Policy brief: COVID-19 and transforming tourism", August. Available at: www.un.org/sites/un2.un.org/files/sg_policy_brief_covid-19_tourism_august_2020.pdf.

UNWTO (2012) "International tourism hits one billion" [Press Release], 12 December. Available at: https://www.unwto.org/archive/global/press-release/2012-12-12/international-tourism-hits-one-billion (accessed January 12, 2021).

UNWTO (2020a) "World tourism barometer", *World Tourism Organization*, 18(1). Available at: https://www.e-unwto.org/toc/wtobarometereng/18/1.

UNWTO (2020b) "International tourist numbers could fall 60–80% in 2020, UNWTO reports", *World Tourism Organization*, May 7. Available at: www.unwto.org/news/covid-19-international-tourist-numbers-could-fall-60-80-in-2020.

UNWTO (2020c) "World tourism barometer", *World Tourism Organization*, 18(3).

Urry, J. (2002) *The Tourist Gaze: Leisure and Travel in Contemporary Societies*. London: Sage.

Vium, C. (2017) "Fixating a fluid field: Photography as anthropology in migration research", in A. Elliot, R. Norum and N. B. Salazar (eds.) *Ethnography and Experiment: Methodologies of Mobility*. Oxford: Berghahn, 172–194.

Wilken, R. and Goggin, G. (2013) *Mobile Technology and Place*. London: Routledge.

Wolf, G. (2011) "What is the quantified self?", *Quantified Self*, March 11. Available at: http://quantifiedself.com/2011/03/what-is-the-quantified-self (accessed December 27, 2016).

12

THE FOOD COURIER AND HIS/HER MOBILE PHONE

Magnus Andersson

The interlinked relationship between communication and transport has a long history. It is today further strengthened by the platform-based gig economy whereby transport services of different kinds (for example food delivery and taxi) can be ordered via applications on our mobile phones. In this chapter, I will delve into communication technologies as an essential part of such transport work and what meaning it has for individual bicycling food delivery workers. I will do this in order to illuminate not only mobile socialities as a concept, but also the (distinct) research areas the concept stems from: mobilities and mobile communication, using sociality, social interaction and social relations as a lens. An elaboration of this kind shed light on foundational questions within mobilities and mobile communication where, for example, the dialectic relationship between mobility and immobility (Adey, 2006) stands out as a paramount one, and where (related) questions regarding waiting and time, work and dwelling are others.

The aim of this chapter is to pursue a theoretical enquiry and thereby elaborate the multifaceted and ambiguous role of the mobile phone for bicycling food couriers, and how this device is *both* a prerequisite for the mobile work *and* crucial for the comfort and social interaction between the rides. The socio-material conditions in this new type of work environment, I argue, constitute a kind of mobile sociality, where the two 'entities', mobility and sociality, are articulated in new configurations. These conditions make the delivery work different from traditional transport work, which will be discussed and theorized in this chapter.

The focus on bicycle couriers as a transport work is justified by at least three aspects. First, in the literature of mobilities, there are many studies of transnational professionals criss-crossing the globe while conducting their work on laptops and mobile phones in cafés and business lounges (Nowicka, 2006; Favell, 2008; Elliott and Urry, 2010; Andersson, 2013; Jansson, 2018). The mobility of blue-collar workers – mobility from below – has not received the same attention (although see the chapters by Jimenez [Chapter 13], Hartmann [Chapter 17] and Sinanan [Chapter 14] in this handbook).[1] In line with the politics of mobility, we need to broaden the scope of the study of mobility and work. There is, as Doreen Massey (1991) has argued, a "power geometry of mobility", which in this context means that the mobilities of professionals are *supported* by and *dependent* on workers providing mobility – for example, taxi drivers, food couriers, messengers, network technicians. Hence, one type of mobility – or someone's mobility – is always interlinked with other types of mobility – or someone else's (im)mobility.

Second, there are theoretical motivations to focus on food couriers. A closer look at their work reveals that mobility is not all there is; immobility, i.e. waiting (to be mobile) is an essential part of transport work in general and food delivery work in particular. This dialectic relationship, how mobilities and immobilities are interdependent, are discussed by several scholars within the studies of mobilities, and studying the working conditions for food couriers may contribute to this larger discussion on the dialectics of mobility.

A third reason to scrutinize the work of bicycling food couriers is the very particular organizational and material conditions of their work: there is no present boss who gives orders, no colleagues, no physical workplace to retreat to between jobs – and thus neither restroom nor room for rest. There is nothing but the mobile phone (and a bicycle). It stands for the boss who gives order and the sociality while waiting. As a food courier in an Australian study puts it: "you're really just following what the app tells you, so you don't have to think much. You just do what you're told by the app" (interviewed bicycle courier in Goods et al., 2019: 513).

The research behind the chapter started with some initial observations and informal talks with some bicycle couriers. Based on that, I have been studying theory with a particular focus on mobility, dwelling and the mobile phone as central phenomena and also analyzed previous research from this specific angle. The character of the chapter is hence an empirically induced theoretical enquiry in order to shed light on mobile socialities related to transport work, especially the temporal and spatial conditions of work and in-between times.

The chapter is structured as a funnel: first a descriptive part, contextualizing transport work in general and describing the context work of food delivery in particular. This is followed by a section on waiting; the time between jobs with an ambivalent relation to the jobs per se. In order to fully grasp this ambivalence, waiting is reflected and elaborated theoretically. The next step in the theoretical enquiry of in-between time is to relate waiting to dwelling; the appropriation and personal shaping of waiting into something one can feel at home in. This is then further explored through the study of the mobile phone and the difference it makes. The final stage is to open up the context again, focusing the context and how work as a food courier is about the dialectic of mobility and dwelling, and the constant switching between them.

Transport work and communication technologies

Transport work, i.e. bringing material things and/or people and from point A to point B, comprises everything from container shipping between Seoul and Sevilla to home delivery of pizza. Transport – which includes infrastructures, means and workforces – is a vital aspect of contemporary societies. And even if we all are dependent on transports, we seldom give them a thought. Today, transports and logistics are digitalized to a great extent. Nevertheless, as, Morley reminds us, even in a digital age, global transport is dominated by shipping and containers (Morley, 2009, 2017). The entanglement of physical and digital dimensions within transport is illustrated by e-commerce, i.e. 'electronic commerce'. For the customer, the whole process appears as 'electronic': the purchase, the economic transfer and the monitoring – via an app – of the commodity's geographic journey. However, not all of the stages between producer and customer are 'electronic'. Many of the things shopped online are manufactured in China and shipped from there in a container to harbours in the West; trucks (or trains) transport them from harbours to logistic nodes/warehouses in the outskirt of cities, from which couriers bring them to our homes.

The geographer Nicky Gregson has conducted a very interesting ethnographical study of truck driving in UK that gives insights into the hidden backstage of our material everyday life in which the supply of goods is taken for granted. This is some of the routinized and arduous aspects of the movement of stuff from point A to point B:

> Behind the cab, on a trailer, is a container empty; on the job sheet clipped onto the dash board is our destination for loading the container – a self-storage warehouse in Slough – and the port to which it is to be taken for export (London, Tilbury). In some ten minutes, we pass or are passed by, hundreds of other LGV [large goods vehicle] cab and trailer units, doing exactly the same job of goods transportation. . . . This cab unit, then, as with every other LGV on the road, is an economic space, dedicated to the work of driving, and driving, the act, when coupled up to a loaded trailer, is a means to realise money, through the movement of goods.
>
> (Gregson, 2018: 295)

The movement of the truck is choreographed by traffic planners in a national office via digital and non-digital media technologies like Bluetooth, a hands-free mobile technology, and printed job sheets. This complex spatial and temporal coordination of a fleet of LGVs between warehouses, ports, traffic jams and slots of loading and unloading makes the space of the cab a "part of a distributed logistical network that is in continual motion. Cab and trailer units on the roads are the visible endpoints of that network" (Gregson, 2018: 295). It is not difficult to imagine the equivalent mediated administrative coordinating work both regarding container shipping between Seoul and Sevilla and the home delivery of pizzas.

If truck drivers represent traditional transport work, bicycle couriers are executing transport work under new conditions. Their work, delivering food from restaurants to workplaces and private homes regulated by an app, is something else than transporting goods between logistic hubs following a job sheet. Their working conditions are also very different. It may look like this:

Two men in the same blue jackets with logotypes are sitting on a bench outside a shopping mall in the centre of a Swedish city. They have their blue square bags, specially designed for transport of warm food, beside them. They small talk, commenting passing people, the weather, the condition of their bikes, and tell each other memorable delivery jobs. Meanwhile, they keep an eye on their mobile phones, checking Facebook as well as the app where new deliveries are announced and up for grabs. A buzz signals a potential job; a delivery from a hamburger chain restaurant in the adjacent mall to a coworking space in a now gentrified, former industrial, area of the city. One of them immediately pushes the 'accept' button on the app, but not quick enough; another courier, in the same blue jacket, grabbed it before him. Ten minutes later, he is luckier; three portions of sushi to a home address just five blocks away.[2]

An individual courier tells me about the condition of his work: first of all, he has to 'check in' in a management app on the mobile phone, confirming that he is present in a particular zone of the city, ready to accept a job. Hence, he cannot check in and accept a job while still at home, which he thinks is a shame, since he – just like many of his colleagues – does not live nearby, but in a suburb. There are three shifts to choose from: daytime (over lunch), afternoon/evening and evening. Double shifts are also an opportunity. Besides being present in the zone, the courier has to be attentive and quick to get an order. The delivery jobs on offer are shown on an app (another one than mentioned previously) where they are up for grabs; the first courier who pushes 'accept' gets the job. This implies that the couriers have to monitor the app all the time. It also means that the other couriers are not your colleagues, but competitors. The experienced couriers gather several jobs before they go, something less experienced couriers do not dare to do since it takes longer time to deliver, which might lead to unsatisfied customers and poor ratings, which again have consequences for your work opportunities (in what follows). The number of jobs may vary between two and 25 per day, meaning cycling between 20 km and 100 km, always carrying the special square, branded backpack (that you have paid for yourself). To always keep an eye on the app, be prepared to work on demand is stressful and tiring, according to the couriers I have talked to; it is not like a proper break, more like non-work or 'waiting-and-hoping-for-work'. In addition, you are monitored via GPS, gig allocation and performance monitoring (Goods et al., 2019). Experienced couriers may have the 'confidence' to turn off the app and a have a real break.

Just like taxi drivers, the food couriers learn to know the streets of their city, their field of work. When delivering to unfamiliar neighbourhoods they use Google Maps on their phone, attached to the handlebar, to find the quickest way. Most of the couriers stay in the city centre between jobs, not only because of the numerous restaurants there, but also due to the many wi-fi spots, and the opportunity for sociality, chatting with other food delivery workers. Couriers need phones with a lot of data and battery power – a flat battery or broken connection to the apps would be a catastrophe. It makes cafés (charging opportunity) and shopping malls (free wi-fi spots) popular sites. Still, many of them have an extra battery in their pocket.

The contrast between truck drivers and bicycle couriers is not only a matter of what they transport and with what kind of vehicle they have; it is also, as hinted, a question of working and employment conditions. The combo of contemporary capitalism and digitalization has created what has become known as the 'gig economy'. It signifies labour turned into discrete one-off tasks, and consequently one-off payment, without further commitments (Prassl, 2018). The presence of the gig economy is visible in the transport sector through companies like Uber, and through many food delivery platforms. A lot has been written on labour in a gig economy, and this is not the place for delving into that literature.[3] Nevertheless, a few things about the working conditions are also of importance in this context. Companies like Uber or within the food delivery sector are not transport companies with employees; they are tech companies offering a technology to *individual independent contractors* that enable these drivers and couriers to get in contact with customers. The network service is thus used for connecting 'buyers' and 'sellers', i.e. those in need of transport and those who can offer it (Malin and Chandler, 2017; Rosenblat and Stark, 2016). Being tech companies delivering a network service implies that they do not have any responsibility as employers (Prassl, 2018; Webster, 2016). For the individual contractor, it means that that there is no health insurance, no paid vacation or any other form of social security traditionally provided by an employer. And as mentioned previously, as individual contractor, you do not have any colleagues – only competitors. Nevertheless, the competition is not that tough. For example, it does not prevent them from hanging out together when they are between jobs.

Another significant feature of this kind of work is the disciplinary function of algorithms; how it surveils drivers, shapes their behaviour and polices their performance. Customers' rating of the workers is at the heart of these functions (Rosenblat and Stark, 2016; Goods et al., 2019).

To sum up, for transport workers in general, communication technologies (maps, traffic information on the radio, GPS, Google Maps, printed job sheets, barcode readers, the mobile phone, apps, etc.) are crucial for the performance of work. A particular feature of working as a food courier, in relation to platform-based company, is the concentration of all kinds of interaction to the mobile phone. All other media technologies are absent. Governance, management and surveillance of work are executed solely via the mobile phone.

Doing waiting

It goes without saying that movement is crucial in transport work. Every interruption, every moment of stillness, means economic setback for the logistic company – or for the individual courier. Time is the currency, the 'product'. What the courier is offering is: 'if you don't have time to get your food, I'll get it for you and I'll do the waiting'. This is in accordance with the foundational principles of time management, set already in the beginning of the 20th century: "effective time management is made possible by delegating work to others" (Gregg, 2018: 57). In order to fulfil these requirements, unpaid waiting, the absence of a workplace or a place to hang out, and the irregular supply of jobs are parts of the work. These insecure conditions are, however, advertised as 'flexibility' and freedom by the companies. Since the waiting is such a large part of food couriers' work, it is interesting to take a closer theoretical look at this human aspect that everyone is familiar with.

A lot has been written about waiting as a phenomenon (see Schweizer, 2008; Hage, 2009; Farman, 2018; Bisell, 2018: ch. 6; Highmore, 2011: ch. 4), a term which is extremely stretchable: we may wait for the bread to pop up from the toaster as well as wait (and have been waiting for thousands of years) for the Messiah to show up. In waiting, "we encounter a time structure that is imposed on us", as Schütz and Luckman argue, and exemplify:

> First I must take care of this (in other respects unimportant, subordinate) matter, so that I can then turn to a more significant concern (I have to shave, and so I must first wait until the water is hot before I go to the conference whose result may be a turning point in my life).
>
> (Schütz and Luckman, 1973: 48)

Waiting time is thickened by emotions, often by impatience and frustration. As Ben Highmore poetically puts it: "The looping rhythms of time as it is pulled into new shapes by resentment and guilt or debilitating forms of boredom" (Highmore, 2011: 111). Waiting is also political, as Massey reminds us: "Much of life for many people, even in the heart of the First World, still consists of waiting in a bus-shelter with your shopping for a bus that never comes" (Massey, 1994: 163). Waiting can be emotional and political at the same time. That is the reality for many refugees in the world today. Digital technologies play a crucial role for their situation, for bridging and bonding, building hopes for the future, while remembering the past. Waiting and poor digital connection feed into each other, as Alexandra Greene notes in her ethnographic research on refugee women in Greece. It marks the whole life from the trivial to the existential; from the waiting for the signal of "message successfully delivered" and for the stop of audio and image lagging to the waiting for family reunification and a better life (Greene, 2020: 742–43).

In a parallel Western world, waiting for many people is closer to trivial (rather than existential). As such, it is malleable time; it can be used by individuals in creative and unexpected ways: drawing squiggles on a left-behind newspaper on a long train journey, texting in a traffic jam, browsing a glossy magazine, knitting, playing solitaire on the phone or just letting the mind drift away. The Swedish ethnologists Billy Ehn and Orvar Löfgren link this to the split consciousness of waiting; being mentally here and in the future at the same time (Ehn and Löfgren, 2007: 74–75). This, in turn, can be related Ben Highmore's interesting discussion of the ambiguous and contradictory term 'distraction' that he suggests should be understood as "the mobility of attention" (2011: 118).[4] Distraction should thus not only be associated with interruption in a negative sense; it is also, according to Highmore, an opportunity – something one can utilize.

To be distracted is to lose one's focus of attention, to be drawn into another reality, or another dimension of reality, something all commuters are familiar with. Waiting may appear as a very passive practice. However, that passivity is a delusion; the inner life of the one who waits is vivid – his/her attention is on the move in time, as well as space. Yet, symbolic and material input from the outside may often facilitate the drifting. Supported by material culture, waiting can be a personal space that allows me-time.

When the truck in which Gregson is a passenger has to wait at a hub, she notes how a metamorphosis is taking place. The cab stops being an economic and logistical space, as mentioned in a previous section, and becomes a personal space. She notices the hidden domestic utensils and related practices that turn the cab to a kitchen/diner:

> Under the lower bunk is a small fridge. A plastic crate sits beside it. Inside this there is: a camping gas stove with canister attached, a kettle, plates and cutlery, mugs, a 5 l drum of water, instant coffee, tea bags, dried milk and food. . . . I am shown how a tray extends out from the dashboard in front of my seat, and am put in charge of making the breakfast tea, whilst sitting down – the stove is balanced on the tray, the kettle filled from the water drum, a lighter lights the stove. . . . Meanwhile Driver 1 contorts himself to extract the milk from the fridge, and then starts preparing the breakfast pastries on a plate on his lap, whilst sitting at the wheel.
>
> (Gregson 2018: 296)

Gregson notices how the daily routines are synchronized with the other drivers: a nap in the cabin's bed after breakfast and social chats with fellow drivers between 2pm and 3pm via handsfree mobile phones (Gregson 2018: 297). This underscores that the truck is not only an economic space, but also space of dwelling and of sociality. A lot of domestic practices take place in the cab: cooking, watching TV, browsing magazines, cleaning, mobile banking and the crafting of "social relationships by talking on their mobiles" (Gregson 2018: 298). There are hence functional dimensions of the use of the waiting – eating, sleeping, interacting with authorities – but that is not all. There is also something comforting: a personal time-space with social dimensions.

Just like the truck drivers, food couriers are doing a lot of waiting. However, they have to make the most of the time with neither a cabin nor any material gadgets. Can they still create a personal space that allows 'me time'?

Dwelling

A personal time-space is a strategy of making oneself at home, creating at-homeness out of practices and accessible material culture. This is not necessarily related to a home, in the sense of a fixed location; this is rather a temporary colonizing of a slice of time and space to create a sense of home for a moment. Empirically, it can be related to how generic places, like café chains, can provide comfort for mobile professionals, as they "offer a blank canvas for people to lightly personalize and thus achieve a feeling of home" (Ley-Cervantes and Duyendak, 2017: 74). For a courier, a bench outside a shopping mall, shared with other couriers, can be that blank canvas. Theoretically, it can be understood within a phenomenological perspective where dwelling is not a fixed entity or locality, but repetitive, habitual practices; not a specific site but an accomplishment in a practical and experiential sense, emotionally felt (Moores, 2012: 27–31).

This moulding of space and waiting time to something personal, to a temporary cocoon of ontological security (Giddens, 1991), so to speak, is closely linked to dwelling. Dwelling, according to Peter King (2004), means living on earth and thus inhabiting space, in its broadest meaning. In contrast to housing that is related to physical structures, dwelling links to intangibles of inhabiting space and to intense emotional relations (Meier and Frank, 2016: 368). Dwelling is thus about emotions which form a bridge between the *social* exchange of ideas, feelings and experiences and a small world of our own (King, 2004: 22). Dwelling, and the conditions of dwelling, are things we can create: it is to establish and operate conventions, to build structure for oneself, as King, drawing on Heidegger, puts it (King, 2004: 21–22). In more concrete terms, to dwell is "to curb, to discipline, a piece of space by providing it with our personal things and marking our presence" (Andersson, 2006: 182). The key here is that dwelling is based on our practical relationship to the world, not cognitive (Ingold, 2000). Repetitive practices, as well as social relations, provide environments with meaning – and they become familiar to us, and thereby associated with dwelling (see Moores, 2012: 28–29). Dwelling as independent of a spatially fixed home is also brought to the fore by mobile work and long-distance commuting that affect workers' residential arrangements. In this context, Gorman-Murray and Bissell (2018: 233) suggest "multilocal dwelling" as a concept to grasp dwelling-in-mobility and thereby getting away from ideas of home and dwelling as something singular, place-based, stationary and immutable.

There are many overlaps between dwelling and the phenomenological concept of the lifeworld, i.e. the meaningful world of taken-for-granted things, ideas and experiences. As a routinized and taken-for-granted thing, dwelling is difficult to grasp:

> Our everyday experience – which is of dwelling on the earth – hides the significance of dwelling. We are too busy dwelling – concentrating on those matters close to us – to see the full significance of what these very acts enclose.
>
> (King, 2004: 25)

These aspects may seem difficult to grasp in empirical studies (we cannot ask people how they dwell). Yet, the practices, what people do when commuting, when they wait for unloading a trailer, waiting for three portions of sushi to be prepared for delivery, is a way to understand dwelling; a temporary, practical and personal – although often social – time-space in the midst of mobilities. This is, I argue, how we can understand couriers' waiting; they use it to dwell. They socialize with their 'competitors' and thus create space of comfort to temporary inhabit, built on sociality and mental distance to work. They also communicate with close friends and loved ones – practices which also are a way of dwelling; an emotional and social structure for oneself. The latter point highlights the role of the mobile phone.

Mobile phones and the making of dwelling

Instead of a physical workplace, which usually holds spaces for work and for rest, bicycle couriers have mobile phones. It means that they, while waiting, can chat with family and friends, play games, do social media, text message, do shopping, pay bills – and make phone calls. As Hjorth argues, today's mobile phones are "miniature caravans", accommodating all aspects of personal life and thus functions as kind of home (2012: 199). The dwelling based on mobile phone practices, if we follow Moores (2012: 27), is a kind of dwelling accomplished through practices and experiences. The range of afforded practices mentioned previously shows that the mobile phone provides opportunities to create at-homeness beyond the spatially fixed home. A personal enclosed space can be created everywhere if you have a mobile phone.

The mobile phone is an "environment of communicative opportunities" (Madianou, 2014), and in this section I will delve into two related ways that the mobile phone contributes to the construction of a personal structure for dwelling beyond the stationary home. These build upon the mobile phone as: (a) a source of habitual practices and distraction (i.e. mobility of attention); and (b) a tool for communication, social belonging and the meaning of socialities. The individual key words for these are 'practice' and 'social interaction'.

Dwelling through practice

It happens that the restaurant is not ready with the food when the courier gets there, which means a short wait. At these occasions, it is common that the courier picks up the phone for a casual game, a quick glance at Instagram or just for fiddling about with it. Familiar mobile phone practices become a tool to shut out the surroundings, letting the attention drift away to make the time pass. This is a kind of dwelling-in-mobility that couriers are practicing between their jobs, carving out a bit of personal time and space in the middle of a crowdy and hectic place. Hence, they can feel at home not in (relation to) the place, but in the familiar practices.

These mobile phone practices become a way to not only build, but also tailor structures for oneself. In addition, it signals to the surroundings that I am busy; my attention is on the move, to speak with Highmore (2011: 118). This kind of phenomenon has been studied among commuters, where individuals are strangers to each other, involuntarily in each other's vicinity. Watts (2008) discusses this phenomenon as a way to "colonize a space for oneself". In a similar vein, although through the use of mobile phones, Bengtsson describes how mobile phones can be used to create "a personal micro-space" in obtrusive environments of public transport (Bengtsson, 2018: 6–7). In a larger study of commuters in three global cities, Ito and colleagues identify the meaning of mobile media for similar encapsulating phenomena, describing them as "micro-places built through private, individually controlled infrastructures, temporarily appropriating public space for personal use" (Ito et al., 2010: 74). Beside a way to shut out the surroundings, Ito and colleagues argue, cocoons have meaning as " 'filling' or 'killing' in-between time when people are inhabiting or moving through places within where they are not interested in fully engaging" (Ito et al., 2010: 74). The common denominator of these examples is that the practice is more important than the content of the interaction.

Personal micro-places can also be linked to games, like casual mobile games, location-based games and social media games (Hjorth and Richardson, 2014). They provide instant distraction; mobility for the attention and thus a personal cocoon while waiting. 'Ambient' is term that is often used in relation to games and play. Ambient games in this context refer primarily to the blurred boundaries between the game and its physical context, like for example Pokémon Go. However, Hjorth and Richardson extend the notion of ambient games. For them, 'ambient' play highlights the embeddedness of games in everyday life and the blurred boundaries to other practices – "play as a part of a background to life" (Hjorth and Richardson, 2014: 73).[5]

Dwelling through practice can be related practice theory, stressing that the construction of meaning is more related to 'the doing' than to representation or cognition (Ingold, 2000; Shove et al., 2012). An early illustrative example from media studies is Herman Bausinger (1984), who challenged media studies in the 1980s through arguing, almost provocatively, that the importance of TV derives from not the content but watching as a practice (and thereby as a ritual).

Dwelling through social interaction

The mobile phone's affordance of traditional phone calls, i.e. real-time communication, is sometimes forgotten in the current era, dominated as it is by other forms of mediated social interaction. Nevertheless, for some people, it is a very valuable function. It is tempting to say that the voices make it to the most intimate form of mediated communication. However, such an argument neglects the complexity of mediation in relation to authenticity and intimacy – a handwritten letter or an email may very well be received as very intimate and authentic (Milne, 2010). It is important to note that the character of the interaction is shaped by the way the interlocutors use them (Licoppe, 2004). It means that the forms of social interactions afforded by a mobile phone – phone calls, text messages, mails, social media apps, direct messaging, etc. – receive their meanings by the social use, and also in relation to each other (Madianou and Miller, 2012: 137).

A courier tells me that the "phone-function" of the mobile is vital for him; he spent lot of his waiting time between jobs speaking on the phone with his wife and parents. It is a nice way to wait and it makes time go faster, he says. It may be related to how mundane conversations between close interlocutors are framed by and reconfirms a common taken-for-granted world (Berger and Luckmann, 1966: 172). The social interaction dissembles the courier from the immediate physical context, the public bench outside the shopping mall, and moves him somewhere else, to the company of family and friends in a socio-emotional layer of reality. Put differently, he is temporarily dwelling in an intersubjectively maintained reality, feeling at home not due to the place but in the conversations with significant others.

Theoretically, dwelling through social interaction can be linked to symbolic interactionism, ethnomethodology and phenomenology, for example in Erving Goffman's discussion about "communication boundaries" (Goffman, 1963) and Alfred Schütz' argument for multiple realities (Schütz, 1962). The Swedish sociologist Johan Asplund talks about "social responsivity"; how social interaction is spatial in the sense that it erects immaterial boundaries around the interlocutors that expel others in some ways (Asplund, 1987). It can be illustrated by being late to a mingle party, and the feeling of being outside all small groups of ongoing 'social responsivity'. Episodes like this happen all the time; in lunch rooms, cafés and bars, and when someone uses his/her phone in public spaces.

Related to this micro-sociological field is a discussion within media studies about different forms of presence, for example illustrated by Esther Milne's book *Letters, Postcards, Email: Technologies of Presence*. Presence is not (only) about being close in physical space. It is, as Milne points out, "achieved in communication" and "depends on the complex interaction between cultural practices and technological infrastructure" (Milne, 2010: 2). Christian Licoppe (2004) also discusses presence via communication technologies. According to him, the act of calling can be more important for the experienced presence than what is said. Even more interesting in this context is his argument that such conversations are almost geographically detached: "The relationship seems to be reduced to the connection allowing this mediated presence" (Licoppe, 2004: 147). The consequence of this is blurred boundaries between absence and presence, something I will come back to in the final section.

Conclusions and reflections

There is (always) an ongoing discussion on the future of work, which takes many expressions. I have seen jokes on Twitter about a future that holds two kind of working people: digital knowledge workers working from home and those who deliver food and coffee to them in their homes. It is possible to add a twist to that joke: it is not impossible that the home-working knowledge workers are designing apps that regulate the work of the food couriers delivering their food and coffee: hospitality workers as "the flipside of creative labor", as Melissa Gregg (2015) puts it in a similar context. This chapter started off in the awareness of precisely that power geometry where "some are more in charge of [mobility] than others" (Massey, 1991: 26), in order to identify and theorize the fine-tuned circumstances, the role of the communication technology, the practices of mobility and sociality that also are part of the differentiation.

What has been established in regards to bicycle couriers is that immobility and waiting are integrated parts of their working life, and that the mobile phone is crucial for them. The work is regulated and governed via apps. Other apps help them find the quickest way to the delivery address. Furthermore, the mobile phone accommodates other realities in between jobs: worlds of games, music and sociality. The absence of a physical workplace and work organization with superiors giving orders underscore the centrality of the phone, a device which here represents a spectrum of practices as well as states of mind.

This spectrum of states of mind can be related to Alfred Schütz's ideas about multiple realities (Schütz, 1962). According to him, we experience the world as plural, as constituted by "finite provinces of meaning"; for example, "the world of dreams, of imageries, of sacral experiences, of art or of play" (Schütz, 1962: 232).[6] Each of them is characterized by their own 'logics' of interpretation, meaning-making, relationships and communication. Such finite provinces of meaning are not something rare; we leap between them all the time in the course of a day. It means that we constantly adapt our meaning-making capacity to the specific context at hand. The food courier with his/her phone are continuously switching between provinces of meaning: the set of apps that regulates food delivery jobs, biking in the city while being guided by Google Maps, the sociality with other bike couriers (competitors), a world of a casual game, the recurrent communication with significant others, etc.

In the chapter, I also argued that food couriers use waiting time for dwelling. The important point about this seemingly trivial argument is the particularity of their *unpaid* waiting; it is not so much about rest and relaxation as it is about waiting for a chance to get a paid job while being 'forced' to be physically present in a specific zone of the city. It is thus a 'break from work' under very special preconditions, including restricted freedom. These circumstances make the dwelling a challenge. Dwelling should here be understood as mobility of attention, creating a personal capsule out of familiar practices and sociality – and thereby appropriating a slice of public space and shut out the surroundings.

The phenomenon, using breaks and micropauses for letting the attention move, is not something unique to food couriers. Similar examples among truck drivers were discussed in the chapter. Moreover, moving in and out of concentration is also a feature of knowledge work which has been studied in a range of contexts (Wajcman and Rose, 2011; Broadbent, 2016; Pink et al., 2018). Yet, beside the huge difference in control and autonomy in relation to work and dwelling, it is once again important to stress that the food courier just has his/her mobile phone (and bicycle). There is no physical shelter to create a goffmanesque backstage for withdrawing; the dwelling has to take place in the midst of the crowdy public space.

Dwelling is an aspect of a dialectic between mobility and immobility, or between work and dwelling as it is expressed here. If the constant oscillation between work and dwelling is seen through the lens of mobile socialities, one could almost say that it is a dialectic between mobility and sociality; mobile work and sociality as dwelling. A closer look reveals that the work implies corporeal mobility; most of all biking, but also running on stairs of apartment buildings, balancing food delivery boxes – everything executed with a focused attention. In contrast, the dwelling is based on "nimble forms of attention", as Highmore calls it (Highmore 2011: 120), which also can be understood as leaps to other – often social – provinces of meaning. These leaps are combined with corporeal immobility; often, they take place at a public bench.

The interrelationship between mobility and immobility (mobility and dwelling), and the necessity to understand them in relation to each other, is a recurrent theme in social theory in general and mobility studies in particular. In 1992, almost 30 years ago, James Clifford discussed this dialectic relationship and argued for studies of "traveling-in-dwelling, dwelling-in-traveling" (Clifford, 1992: 108). At an early stage of mobility studies, John Urry argued that people "dwell in various mobilities" (Urry, 2000: 157). Since then, other scholars have talked about the complex relationships between mobility and immobility (Adey, 2006), mobility and moorings (Sheller and Urry, 2006), nomadism and sedentarism (Cresswell, 2006), mobility and place (Easthope, 2009), and mobile dwelling (Bissell and Gorman-Murray, 2020).

The present theoretical enquiry may contribute to this ongoing discussion, especially when it comes to the role of the mobile phone. Drawing on the discussion of ambient character of phenomena (Hjorth et al., 2012; Hjorth and Richardson, 2014; Madianou, 2016), we may talk about *ambient practices* and *ambient communication*, i.e. practices and communication that blur boundaries between online and offline, absence and presence, and individuality and sociality.[7] To reiterate: as for other gig workers, food couriers' work is governed through apps, which require certain familiarity with the logics of apps. When biking, the courier needs to keep one eye on Google Maps (with a mobile phone attached to the handlebars) and the other eye on the street, including the road signs. There are ongoing leaps between the online environment on the screen and the physical surroundings. Hence, these working practices are ambient practices; neither the online practices nor the offline practices are complete – their meaning is mutually co-constructed. Similarly, as discussed earlier, dwelling (through practice and through social interaction) also has an ambient character. I am thinking of ambient games and other distracting mobile phone practices and how they become, in the words of Hjorth and Richardson, "part of a background to life" (Hjorth and Richardson, 2014: 73). I am also thinking of the regular conversations with close ones and how they blur the boundary between presence and absence and between different forms of communication (Licoppe, 2004; compare Madianou, 2016; Bourdon, 2020). Mediation of this kind justifies the concept of ambient communication – that is, communication that moves between and conflates the background and foreground of attention. This ambient character, and how it relates to the dialectic between mobility and immobility in general and in work life in particular, is a potential avenue for further studies of mobilities, socialities – and mobile socialities.

Notes

1 A similar tendency can be seen in the study of digitalisation and work, where white-collar workers have received most of the scholarly attention. See Schäfer et al. (2018) for an alternative.

2 This fictive account is based on a conversation I had with two bicycle couriers in August 2019. It took place in a café and lasted for 70 minutes, when the last of them had to accept a job – the first one left for a job after 25 minutes. A prerequisite for having a chat with me was that they did not have to turn off their work app. It is understandable, given their work conditions. The conversation is also a source for many of the basic facts concerning food couriers work described in the chapter. Hence, I describe the system for a specific company; different companies use different systems and are not organized exactly in the same way, for example regarding the competition for rides (see Veen et al., 2020). However, the basic organizational principles are similar.

3 See for example Rosenblat and Stark (2016), Webster (2016), Malin and Chandler (2017), van Doorn (2017), Rosenblat (2018), Veen et al. (2020), Griesbach et al. (2019).

4 The point of departure for Highmore is Siegfried Kracauer and Walter Benjamin's work on "Zerstreuung", which has no direct English translation, but can mean anything from amusement via diversion

to absentmindedness. It is this ambiguity that Kracauer and Benjamin take advantage of. I would argue that the Swedish term "förströelse" has the same ambiguous meaning.

5 An interesting note: game studies has borrowed the term ambient from Brian Eno, who coined the term 'ambient music' in relation to his record *Music for Airports* (1978), with the motivation "I wanted to make a kind of music that would actually reduce your focus on this particular moment in time that you happened to be in and make you settle into time a little bit better" (in Eyles and Eglin, 2008, no pagination). This is actually a very good summary of the couriers' dwelling while waiting.

6 In an essay on utopia, Ghassan Hage (2011) discusses multiple realities as an ontological alternative to the dominant "mono-realism" within social science, although without mentioning Schütz. For Hage, it is an approach that "invites us to think of ourselves as always inhabiting a multiplicity of intersecting spatialities and realities" (Hage, 2011: 8).

7 I would like to thank Annette Hill, who led me in to this way of thinking.

References

Adey, P. (2006) "If mobility is everything then it is nothing: Towards a relational politics of (im)mobilities", *Mobilities*, 1(1): 75–94.

Andersson, M. (2006) "Flexible home", in J. Falkheimer and A. Jansson (eds.) *Geographies of Communication: The Spatial Turn in Media Studies*. Gothenburg: Nordicom.

Andersson, M. (2013) "Cosmopolitan practices in social contexts: Swedish skilled migrants in the Netherlands", *Crossings: Journal of Migration and Culture*, 4(2): 155–173.

Asplund, J. (1987) *Sociala Livets Elementära Former*. Göteborg: Korpen.

Bausinger, H. (1984) "Media, technology and daily life", *Media, Culture & Society*, 6(4): 343–352.

Bisell, D. (2018) *Transit Life: How Commuting Is Transforming Our Cities*. Cambridge, MA: MIT Press.

Bissell, D. and Gorman-Murray, A. (2020) "Mobile dwelling", *Applied Mobilities*, 5(1): 1–5.

Bengtsson, S. (2018) "Sensorial organization as an ethics of space: Digital media in everyday life", *Media and Communication*, 6(2): 39–45.

Berger, P. and Luckmann, T. (1966) *The Social Construction of Reality: A Treatise in the Sociology of Knowledge*. London: Penguin.

Bourdon, J. (2020) "From correspondence to computers: A theory of mediated presence in history", *Communication Theory*, 30: 64–83.

Broadbent, S. (2016) *Intimacy at Work: How Digital Media Bring Private Life to the Workplace*. Walnut Creek, CA: Left Coast Press.

Clifford, J. (1992) "Traveling cultures", in L. Grossberg, V. Nelson and P. Treichler (eds.) *Cultural Studies*. London: Routledge.

Cresswell, T. (2006) *On the Move*. London: Routledge.

Easthope, H. (2009) "Fixed identities in a mobile world? The relationship between mobility, place and identity", *Identities: Global Studies in Culture and Power*, 16(1): 61–82.

Ehn, B. and Löfgren, O. (2007) *När ingenting särskilt händer*. Stockholm, Stehag: Brutus Östlings Bokförlag Symposion.

Elliott, A. and Urry, J. (2010) *Mobile Lives*. London: Routledge.

Eyles, M. and Eglin, R. (2008) "Ambient games, revealing a route to where work is play?", *International Journal of Computer Games Technology*, article ID 176056.

Farman, J. (2018) *Delayed Response: The Art of Waiting from the Ancient to the Instant World*. New Haven: Yale University Press.

Favell, A. (2008) *Eurostars and Eurocities: Free Movement and Mobility in an Integrating Europe*. Oxford: Blackwell.

Giddens, A. (1991) *Modernity and Self-Identity: Self and Society in the Late Modern Age*. Cambridge: Polity Press.

Goffman, E. (1963) *Behavior in Public Places*. New York: The Free Press.

Goods, C., Veen, A. and Barrat, T. (2019) "'Is your gig any good?' Analysing job quality in the Australian platform-based food-delivery sector", *Journal of Industrial Relations*, 61(1): 502–527.

Gorman-Murray, A. and Bissell, D. (2018) "Mobile work, multilocal dwelling and spaces of wellbeing", *Health & Place*, 51: 232–238.

Greene, A. (2020) "Mobiles and 'making do': Exploring the affective, digital practices of refugee women waiting in Greece", *European Journal of Cultural Studies*, 23(5): 731–748.

Gregg, M. (2015) "The doublespeak of the gig economy", *The Atlantic*. Available at: www.theatlantic.com/business/archive/2015/09/gig-economy-doublespeak-new-labor/404779/ (accessed June 15, 2020).

Gregg, M. (2018) *Counterproductive: Time Management in the Knowledge Economy*. Durham, NC: Duke University Press.

Gregson, N. (2018) "Mobilities, mobile work and habitation: Truck drivers and the crisis in occupational auto-mobility in the UK", *Mobilities*, 13(3): 291–307.

Griesbach, K., Reich, A., Elliott-Negri, L. and Milkman, R. (2019) "Algorithmic control in platform food delivery work", *Socius: Sociological Research for a Dynamic World*, 5: 1–15.

Hage, G. (ed.) (2009) *Waiting*. Melbourne: Melbourne University Press.

Hage, G. (2011) "Dwelling in the reality of utopian thought", *Traditional Dwellings and Settlements Review*, 23(1): 7–13.

Highmore, B. (2011) *Ordinary Lives: Studies in Everyday*. London: Routledge.

Hjorth, L. (2012) "iPersonal: A case study of the politics of the personal", in L. Hjort, J. Burgess and I. Richardson (eds.) *Studying Mobile Media: Cultural Technologies, Mobile Communication and the iPhone*. London: Routledge.

Hjorth, L. and Richardson, I. (2014) *Gaming in Social, Locative and Mobile Media*. Basingstoke: Palgrave Macmilan.

Hjorth, L., Wilken, R. and Gu, K. (2012) "Ambient intimacy: A case study of the iPhone, presence and location-based social media in Shanghai, China", in L. Hjort, J. Burgess and I. Richardson (eds.) *Studying Mobile Media: Cultural Technologies, Mobile Communication and the iPhone*. London: Routledge.

Ingold, T. (2000) *The Perception of the Environment: Essays on Livelihood, Dwelling and Skills*. London: Routledge.

Ito, M., Okabe, D. and Anderson, K. (2010) "Portable objects in three global cities: The personalization of urban spaces", in R. Ling and S. Campbell (eds.) *The Reconstruction of Space and Time: Mobile Communication Practices*. New Brunswick, NJ: Transaction.

Jansson, A. (2018) *Mediatization and Mobile Lives: A Critical Approach*. London: Routledge.

King, P. (2004) *Private Dwelling: Contemplating the Use of Housing*. London: Routledge.

Ley-Cervantes, M. and Duyendak, J. W. (2017) "At home in generic places: Personalizing strategies of the mobile rich", *Journal of Housing and the Built Environment*, 32: 63–76.

Licoppe, C. (2004) "'Connected' presence: The emergence of a new repertoire for managing social relationships in a changing communication technoscape", *Envirnonment and Planning D: Society and Space*, 22(1): 135–156.

Madianou, M. (2014) "Smartphones as polymedia", *Journal of Computer-Mediated Communication*, 19(3): 667–680.

Madianou, M. (2016) "Ambient co-presence: Transnational family practices in polymedia environments", *Global Networks*, 16(2): 183–201.

Madianou, M. and Miller, D. (2012) *Migration and New Media: Transnational Families and Polymedia*. London: Routledge.

Malin, B. and Chandler, C. (2017) "Free to work anxiously: Splintering precarity among drivers for Uber and Lyft", *Communication, Culture and Critique*, 10(2): 382–400.

Massey, D. (1991) "A global sense of place", *Marxism Today*, 24–29, June.

Massey, D. (1994) *Space, Place and Gender*. Cambridge: Polity Press.

Meier, L. and Frank, S. (2016) "Dwelling in mobile times: Practices and contestations", *Cultural Studies*, 30(3): 362–375.

Milne, E. (2010) *Letters, Postcards, Email: Technologies of Presence*. New York: Routledge.

Moores, S. (2012) *Media, Place and Mobility*. London: Palgrave Macmillan.

Morley, D. (2017) *Communications and Mobility: The Migrant, the Mobile Phone and the Container Box*. Oxford: Wiley-Blackwell.

Morley, D. (2009) "For a materialist, non-media-centric media studies", *Television & New Media*, 10(1): 114–116.

Nowicka, M. (2006) *Transnational Professionals and Their Cosmopolitan Universes*. Frankfurt: Campus Verlag.

Pink, S., Hjort, L., Horst, H., Nettheim, J. and Bell, G. (2018) "Digital work and play: Mobile technologies and new ways of feeling at home", *European Journal of Cultural Studies*, 21(1): 26–38.

Prassl, J. (2018) *Humans as a Service: The Promise and Perils of Work in the Gig Economy*. Oxford: Oxford University Press.

Rosenblat, A. (2018) *Uberland: How Algorithms Are Rewriting the Rules of Work*. Oakland: University of California Press.

Rosenblat, A. and Stark, L. (2016) "Algorithmic labor and information assymetries: A case study of Uber's drivers", *International Journal of Communication*, 10: 3758–3784.

Schäfer, S., Andersson, M., Bjarnason, E. and Hansson, K. (eds.) (2018) *Working and Organizing in the Digital Age*. Lund: The Pufendorf Institute for Advanced Studies, Lund University. Available at: www.pi.lu.se/sites/pi.lu.se/files/workingandorganizinginthedgitalage.pdf (accessed September 22, 2020).

Schütz, A. (1962) *Collected Papers I: The Problem of Social Reality*. The Hague: Martinus Nijhof.

Schütz, A. and Luckmann, T. (1973) *The Structures of the Life-World*. Evanston: Northwestern UP.

Schweizer, H. (2008) *On Waiting*. London: Routledge.

Sheller, M. and Urry, J. (2006) "The new mobilities paradigm", *Environment and Planning A*, 38(2): 207–226.

Shove, E., Pantzar, M. and Watson, M. (2012) *The Dynamics of Social Practice: Everyday Life and How It Changes*. London: Sage.

Urry, J. (2000) *Sociology Beyond Societies: Mobilities for the Twenty-First Century*. London: Routledge.

van Doorn, N. (2017) "Platform labour: On the gendered and racialized exploitation of low-income service work in the 'on-demand' economy", *Information, Communication & Society*, 20(6): 898–914.

Veen, A., Barrat, T. and Goods, C. (2020) "Platform-capital's 'app-etite' for control: A labour process analysis of food-delivery work in Australia", *Work, Employment and Society*, 34(3): 388–406.

Wajcman, J. and Rose, E. (2011) "Constant connectivity: Rethinking interruptions at work", *Organization Studies*, 32(7): 941–961.

Watts, L. (2008) "The art and craft of train travel", *Social & Cultural Geography*, 9(6): 711–726.

Webster, J. (2016) "Microworkers of the gig economy: Separate and precarious", *New Labour Forum*, 25(3): 56–64.

13
ON DAY LABORERS' DIGITAL MOBILE MEMORIES

Carlos Jimenez

Introduction

Day laborers stand on street corners, at hardware stores, and sometimes on streets hidden from view in urban areas across the world, waiting for work. Their mobility and sociality stand in stark contrast to the solo travers in Erika Polson's Chapter 10 in this volume who experience an untethered and priviledged type of mobile sociality. Day laborers, who find themselves on the literal fringes of society, experience something more aptly described as a stunted mobile sociality. They are not free to roam a city looking for work in formal realms, but instead are relegated to several nodes where they must wait along with several others day laborers. Yet, these nodes, if they are lucky, operate as a portal when employers shuttle them across the state of Colorado, sometimes several hours into the mountains, to do work very few Americans citizens want to do or that they will do for less money. Day laborers seek out spaces in both their urban environment and on their mobile phones that offer a sense of coherence and opportunity to reconfigure their social, mobile, and economic relations.

Day laborers operate on one-day-at-a-time jobs, which bear some similarity to the "gig economy" we have come to experience in the last decade (Friedman, 2014). However, the current form of the gig economy is dependent on a vast digital infrastructure and its various platforms, which facilitate work agreements so that people do not have to meet face to face – a particular draw in the time of COVID19. Platforms like Uber, DoorDash, and even Next Door provide a sense that the worker has been evaluated and deemed reliable either through employee interviews, background checks, or, in the case of Next Door, word of mouth. Digital technologies have given order and specialization to day labor, which has taken over such industries as taxi driving, food delivery, dog walking, and much more.

Despite the expansive digital network and its platforms, day laborers continue to stand outside on corners waiting for work. One reason is that day laborers have their own informal social organization at outdoor sites that establishes unwritten rules that offers distinction and support, making these sites a welcoming home base (Malpica, 1996). Another reason is that day labor offers a range of benefits not easily accessible elsewhere. These include the fact that a day laborer can earn higher hourly rates than federal or state minimum wages and is exposed to a variety of trades (carpentry, roofing, masonry, landscaping, etc.) from one day to the next that improves their chance of future work in one or more of these trades. Moreover, they have the "extremely powerful and underrated act of turning down a job or negotiating a fair wage for a day's labor" (Valenzuela, 2000: 4). Yet, the reason many wait at day labor sites is that most are undocumented and not fluent in English, which limits the work opportunities they can access in the formal labor market.

Day labor differs from the typical gig economy job, as their situation entails a different type of life world and work ethic that cannot rely on an online profile or mobile phone, but instead on being visible in the public sphere and ready to approach a potential employer's car very early in the morning. For a workforce that is largely unauthorized and that has "low levels of education and a poor command of English," the digital gig economy is still perhaps too rigid and costly (Valenzuela, 2003: 308). Their access to a networked sociality or digital space is further restrained by their precarious technological access and limited familiarity with the capabilities of mobile phones. There are, of course, day labor apps out there, but no one I interviewed in Denver used one. Despite experiencing a sort of stunted mobile sociality, day laborers are using their mobile phones – just not in the same way as food delivery workers or others in the gig economy. Their use operates on a more intimate and personal register.

This chapter seeks to understand how day laborers are using their mobile phones in their everyday life, and also examines what information they store in them. Specifically, I examine a combination of information that draws on conversations with day laborers and the images and videos they capture and keep on their mobile phones, as well as the digital communications they send and consume via text and social media. What does their digital archive say about their lives, needs, mobility, and sociality? While other scholarly work on mobile communication focuses on the global digital network and access to information and to people living near and far, this chapter looks inwards to understand what work mobile phones do at the personal level for low-wage and immigrant workers. Within this chapter, I examine the private mobile phone use and archives of day laborers to conceptualize how mobile socialities contribute to our understanding of how sociality and mobility are being intimately reimagined through day laborers' mobile phones.

Mobile digital memories

The transition from conventional cameras to mobile camera phones has had implications for a range of personal photographic practices and memory, from what we take pictures of to how we share them and what social purpose they serve. In 2007, Lisa Gye analyzed the transition to argue that photos on mobile camera phones reflected a more personal and private experience than the personal photography that preceded it via conventional film cameras. At the time of her writing, mobile camera phones were defined by low resolution and social media photo sharing platforms like Instagram did not yet exist, but already she noted the increasing trend towards the personal and its risks: "By reinforcing the intensely personal, camera phones may also participate in this narrow economy of self, one that becomes almost asocial in its emphasis on the individual and its refusal of the more broadly social" (Gye, 2007: 286). Is being more focused on the self – the personal – in one's photos an outright contradiction to sociality? Gye forms this interpretation in part because her reference point is a mainstream user whose social life may be vibrant and full, making an asocial photographic production concerning and striking. Yet, a turn inwards to the self in mobile phone photos may equally be read in the opposite way as an articulation and move toward a different type of sociality – one that may not at first appear as vibrant or full. For day laborers who experience a precarious type of sociality in their everyday lives, discussed more in the following section, the asociality produced in personal photos may represent a starting point. In another way, the personal turn in mobile phone photos creates a preliminary social stage whereby day laborers are experimenting and trying to cobble together something coherent from a disparate set of social experiences and identities.

Others, like Daniel Palmer, have focused on the way in which image storage has changed in an era with ubiquitous mobile phone ownership. Previously, material photos could be found in the once common household relic of the twentieth century, the family photo album. They contained images of the lives and adventures of a family unit. The collection of photos served to remember loved ones and to cherish milestones of a family's life, yet, according to Palmer, they also "perpetuate myths of coherence and togetherness by favouring happy moments. As such, they are implicated in the naturalization and reproduction of patriarchal structures of the family and leisure" (Palmer 2010: 156). What visual memories we choose to archive – and how – suggests something about our social needs and desires. With mobile phones, the storage of digital images has largely gone online. In Palmer's study of an online photo sharing and storage site, thisMoment, he argues that the site highlights how online photo sharing facilitates the cultivation of one's personal and social identity. In particular, the site allows for a "reflexive project of self" (Giddens, 1991: 218). Martin Hand adds nuance to this by arguing that the mediating capacities of mobile phones is shaped by expectations of online visibility (Hand, 2016). While Palmer and Hand situate self-presentation and reflexive practices in relation to online visibility, this can also be done offline and on one's phone through what individuals choose to photograph, store, or delete. Arguably, it is more personal and authentic since it has not been filtered or reorganized to be viewed by an online public, but is intended for one's self.

Other characteristics that make mobile camera phones different from conventional cameras are the kinds of images they produce and how they are treated. The conventional camera was dusted off and taken out with the intention of capturing a specific moment or event and producing a physical photo to store or display for as long as possible. Comparatively, the mobile phone is always at hand and captures special moments as well as those that are not planned. Here, the minutiae of one's daily experience are captured and result in a highly personal view of a users everyday lives. The ability to take as many photos as they want and to immediately delete has also led scholars to describe the experience of digital mobile memories as impermanent or ephemeral. Very few users have any intention of printing a physical photo; instead, there is a preference for "the communicable and the disposable, at the expense of permanence and durability" (van Dijck, 2007: 110–111). According to Palmer, the value of digital mobile memories is the ability to appreciate them nanoseconds after being taken, to circulate the image on a global online network, in essence, the "experiential immediacy" is up front and center (Palmer, 2010: 158). The study of day laborers points to a different type of social practice with digital mobile memories, one that places significant value on digital images and is not treated as ephemeral. While there certainly can be some experiential immediacy, storing the photos on one's mobile phone gives them a longer social life.

With the ever-expanding availability of digital memory on mobile phones, we must understand what the implications of mobile digital memory are for our everyday lives and sociality, especially among a marginalized population. In 2008, Anna Reading was examining these very questions in the UK. In her article "The Mobile Family Gallery?" Reading is interested in understanding how mobile phones reconfigure memory practices in the digital age (Reading, 2008). She specifically identifies how women in the UK produced domestic images and argues that they constituted gendered memory practices that allowed them to reconstitute their identity and value as mothers on their mobile digital memory. This is in a UK context where women were increasingly joining the workforce. Reading ultimately argues that mobile phones "help bridge or traverse the gendered contradictions of late capitalism" (Reading, 2008: 363), and I extend this argument to highlight how day laborers similarly use mobile phones to address and challenge their social marginalization and contributions as immigrant workers.

It is worth noting that the memories digitally collected, in the case of day laborers, are defined by their mobility. Day laborers zip from one part of Colorado to the next. One day, they may have been working on a roof of a two-story building, and the next they are working in the crawl space of a home built in 1900. Photos capture the exterior, interior, and neighborhood spaces they get to visit. There is no long-term consistency like there is in traditional office work or even construction – the site and employer continually change. Mobility and the range of experiences are an attractive feature for day laborers, but it makes it really hard to articulate a coherent experience if one were to rely just on their memories. The portability of the mobile phone, digital memories, and expansive storage make for a far more coherent experience. This chapter examines the personal content on the mobile phones of day laborers and what it means for their lives and identity. I begin by providing some context for who or what a day laborer is in the United States, and then examine the pictures and videos stored on a day laborer's mobile phone, then finally discuss the long-distance conversations and connections day laborers have through their mobile phones.

Day labor in Denver, Colorado

To understand how day laborers use their mobile phones and what they store in them, it is first important to understand what a day labor is and what their lives are like. The tradition of gathering in public spaces to trade one's labor for pay goes back to medieval times in Europe, and played a vital role in the Japanese economy during the seventeenth century. In the context of the United States, the origins are traced to the 1700s when "common laborers – scavengers, chimney sweepers, wood cutters, stevedores, and dock workers – are said to have sought new jobs each day" (Valenzuela, 2003: 313). Back then, day laborers competed with indentured servitude and slavery for work, and were preferred because day laborers could be laid off during slower times in the season while servants and slaves had to be taken care of (Valenzuela, 2003). The rise of day labor in the last five decades differs, and is the result of immigration, economic restructuring, and the decreasing availability of formal work (Valenzuela, 2000, 2003). The consistent demand for day labor is driven by a need for workers who are both flexible and cost effective (Valenzuela et al., 2006). Day labor represents a unique opportunity for workers to "gain a foothold in the urban economy," especially those who are recent arrivals or do not have a network of support, and for some, it is an alternative to formal employment that is poorly compensated or inaccessible to undocumented workers (Valenzuela, 2000, 2003). The nature of day labor in the United States is temporary, with employment lasting 1–3 days and being primarily relegated to household maintenance (landscaping, painting, roofing, etc.) but can include factory and industrial work (Valenzuela, 1999). The work itself is precarious, provides "no benefits or workplace protections, pay[s] poorly, and [is] characterized by workplace abuses such as instances of nonpayment, lack of regular breaks, and hazardous work" (Valenzuela, 2003: 309).

In a 2019 survey of 411 day laborers in Denver, Colorado, scholars Rebecca Galemba and Randall Kuhn found that day laborers are largely foreign born (88%), Latino (94%), and undocumented (65%). Of the surveyed workers, 23% were homeless and 64% had smartphones, although this percentage increased over the course of survey (Galemba and Kuhn, 2019). While they seek work regularly, many average only about seven hours per day and 17 hours per week, and the inconsistency of the work suggests that many will not break the federal poverty threshold of $15,000 (Galemba and Kuhn, 2019; Valenzuela et al., 2006). In terms of their quality of life, a 2011 study on the well-being of day laborers identified that social isolation was perceived by them as a primary life stressor (Negi). Along these same lines, research has described day laborers as a workforce deeply disconnected to information and resources. In a study on information access of day laborers, Bruce Jensen painted a grim picture:

> It's hard to imagine a more information-impoverished constituency than immigrant day laborers. For a number of economic, linguistic, legal, and cultural reasons, the *jornaleros* . . . are effectively shut out from essential services and information sources that most of us take for granted.
>
> (Jensen, 2002: 288)

Jensen's study was published over a decade ago, and since then, mobile phones have dropped in cost, mobile plans have added benefits, and, as a result, the majority of low-wage and immigrant workers own one (Anderson and Kumar, 2019). Galemba and Kuhn's survey further confirms this, which suggests that day laborers have a new level of access to distant and digital social worlds and to the expanded storage capacity of mobile phones. In a pioneering study on mobile phones and day laborers in Seattle, Washington, Luis Fernando Baron and colleagues found that mobile phones help "immigrant day laborers maintain links with their past and their roots, offer tools to navigate their present needs, and help them build future plans and aspirations" (Baron et al., 2014: 107). Their study also provided evidence that the mobile phone served as a source of connection not only to friends and family, but also to employers and a steady income. Building on Baron and colleagues, this chapter examines more closely the data stored on a worker's mobile phone to understand how mobile storage and data are used to address day laborer's present needs and desires.

For this study, I carried out 28 months (September 2017–January 2020) of ethnographic fieldwork at Centro Humanitario, a day labor center in Denver. I draw on observations, informal conversations and interviews with 15 day laborers and the following closely examines data from six day laborers who were interviewed each between two and four times in sessions that lasted 60 minutes. Each conversation focused on a component of their mobile phone use which included: a) pictures; b) videos; c) contacts, phone logs, and texts; and d) online behavior and applications. The range of stored media on their device comprised what I call a day laborer's digital archive. The next section begins by examining the data and content related to their lives as day laborers, followed by data reflecting the minutiae of their everyday lives.

Day labor data and memories

As one might expect, a prominent type of photo that day laborers take with their mobile phones is related to their work. While day laborers could take photos of every single job they perform, they often do not. One reason is that day laborers regularly get hired to do cleanup or help with moving, which are not tasks that stand out aesthetically or that they find worthy of remembering. In fact, day laborers tend to take photos of work that is unique, complex, or that requires they capture its progression. Ramiro, for example, took photos of a ceiling he painted, but what separated this task from other painting jobs was that the ceiling fan had to be carefully covered so that it would not get paint on it. The height and odd shape of the fan made this task complex, and warranted a photograph.

In a similar vein, day laborers had photos of work they did with patio decks, masonry of cracked bricks, tile work, and roof maintenance. For roof work, Alejandro took photos of a roof that needed a gap filled, but it was not a standard fix. There was an incline between one section of the roof and the other, making the job especially tedious. Another day laborer, Javi, had a photo he said he was very proud of. It was of him standing at the top of a ladder that was set up on a second-floor patio that put him up at a height of well over 20 feet off the ground where he was stripping old paint from the exterior of a house (Figure 13.1). This same day laborer had another set of photos from a job when he was tasked with moving rocks, concrete, and other debris from a demolition job. He and other workers lifted and tossed mid-sized rocks when they came across several black spiders. Intrigued by the critters, he took a picture of one and it looks to measure at just over one inch long. The spider is not your typical scrawny domestic dweller: it is plump, fuzzy, black, and with a few white dots on its very large abdomen. Javi said that after finding more than one, several other day laborers walked off the job.

These examples of workplace photos share many characteristics that capture a completed task, their complexity, and, in some cases, their danger. They represent a type of photographic trophy

Figure 13.1 A friend has taken a photo of Javi standing at the top of a ladder positioned on the patio.

that speaks to their work ethic and skills, but also in some cases captures their bravery and attention to detail. Since most day laborers subsist on one-day-at-a-time verbal contracts, their work would be lost without such photographs and recordings. They do not own the property they work on or fix, and it is unlikely that they will return to the same private space when the work is done, yet they have contributed to its material existence and improvement through their labor. Given this, the day laborer's mobile phone camera is a valuable archiving tool that can be used to document the labor they perform, the specific projects they work on, and the spaces they inhabit. The changes in workplace from day to day means the mobile phone is able to document these spaces and tasks that otherwise would blur together – and along with it, moments of intrigue and pride.

The mobile phone allows them to give each one distinction and to prolong the affective moment. Affect here is drawn from Kindberg and colleagues (2005), and their taxonomy that defines an affective image as one that is used for personal reflection or to capture an experience that can be shared later. While for some, taking a photo involves a desire to remember the affective moment, for other day laborers, it is about having a photographic record of their labor to build credibility and ultimately to get more work. In addition to supporting their need to find continuous work, photos on their mobile phone also serves as an additional layer of protection against criticisms of incomplete work, damage, missing items, or low-quality work.

Some photos are also explicitly produced at the employer's request, especially when an employer is away or pays in advance. It means that a day laborer will take a picture of a finished project like an air conditioner unit installation or a vital structural renovation to show a perfectly balanced beam, which serves as evidence that the work was done and done right. Some day laborers will also take photos of how they are storing the employer's tools or how they protected the floor, which was done when the worker was exposing and cleaning a brick wall with a sandblaster. One day laborer, Bart, took photos of the paperwork that listed the ordered material and a second photo that captured what was delivered. Bart said he did this so if anyone claimed otherwise, he had photographic proof that the order was complete and accurate. Part of this is surveillance to ease the comfort of employers, but there are also cases when it is self-imposed as a form of insurance for the day laborer.

Tangential day labor data and memories

There are several reasons why day laborers stand on corners and wait for work. One study found that many enjoyed the variety of work offered, the power to accept or decline, and the places they get to visit (Valenzuela, 2000). In addition to photos related to their work tasks, day laborers capture objects, moments, or places that stand out from their everyday lived experience. Many of these photos are the result of their day labor where they were able to travel, see, and experience things they might not otherwise. Ramiro took a job that led him to a hangar at the Broomfield Airport where he was hired to help move heavy aviation equipment. Over the course of the day, he learned that the person who hired him had built their own aircraft and while at the hangar, Ramiro was allowed to get into the cockpit of the employer's bright cherry red propeller plane. The employer also took several photos of him and a video as he got inside the plane and smiled at the camera (Figure 13.2).

Another particularly unique place in Denver is the Molly Brown House Museum, where Ramiro was hired to help move materials in the basement of the building. The Molly Brown House Museum honors the memory of Margert Brown, a wealthy philanthropist, socialite, and one of the survivors of the sinking of the Titanic. Ramiro was able to see parts of the home that very few people get to see, but it also exposed him to a piece of Denver history that he had no idea existed. At the end of the workday, Ramiro was able to visit the museum where he took

Figure 13.2 Ramiro getting inside a red propeller plane at a hanger in the Broomfield Airport.

as many pictures as he could. They included images of the Titanic, Molly Brown's suitcase that was recovered, and several pictures of fellow passengers. While not all are as lucky as Ramiro, many day laborers have photos of the scenery that surrounds the home they are working on. Alejandro has photos of a mountain town in Colorado and captured a scenic view that included a log bench overlooking two snowcapped mountains. Photos like these are familiar and often in relation to a vacation or one's leisure time, but here they are explicitly related to work.

Day laborers also had photos of murals and public art that they had encountered near their worksites. Denver hosts a city-wide event called CRUSH Walls that brings artists to create

Figure 13.3 A mural in the RiNo neighborhood highlighting the racial and economic tensions of gentrification.

public art in the city's River North (RiNo) neighborhood. RiNo also happens to be a community that has undergone severe gentrification. It is a process fueled by developers and contractors, many of whom utilize day laborers to cut costs. One image that Ramiro took addresses this issue head-on (Figure 13.3). On the left side there is a small two-story home with a woman standing in front of her house holding a rolling pin. Across from her on the right is a group of people that includes a banker holding money bags, a person on a scooter, a woman doing yoga, and another taking a selfie. Each of them is standing on top of a bright orange rhinoceros with brooding red eyes, which interestingly borrows the image that the RiNo art district uses to represent itself, except for the red eyes. Knowing the role of the day laborer in gentrification, I asked Ramiro to explain the image to me. He started by describing its components and the tension between the two sides. I followed up by asking him where he would put himself in the image, and he responded:

> I would put myself, as a day laborer, to be honest, with my vest, drill, and happy because I made money. I would put myself in the middle, because I'm not on the rhino. In part, I'm not on board with raising prices and not giving an opportunity to people that have been living there for years and want to live modestly.

Ramiro also said, "they are destroying all the old houses in the neighborhood, converting them, and that's what happened. Money and ambition, and not letting people live their life." The day laborers are indeed not represented in the mural, but Ramiro understands his role in the process of gentrification as being neither a driving force nor having the ability to prevent the rhinoceros' destructive potential for long-time and working-class communities of color.

These photos are tangentially connected to their work as day laborers, and are not for employers or meant to bolster their work credentials. Instead, they highlight an affective moment of wonder and fascination for day laborers that departs from the familiar. In their digital archive, these photos of things tangentially related to their work combine with the set of photos that capture only the unique and complex work tasks which they have undertaken. The combination in the archive of these photos creates an entirely different day laborer. Instead of one who sits on corners and gets the typical job to help with moving or cleaning, the images they take cultivate a self-presentation and sociality that is defined by a series of complex and unique work, travels to new and distant places, and has once-in-a-lifetime experiences as a result of their work. The memories that day laborers store on their phones stands in stark contrast to the rigors, risks, and abuses that many face as day laborers. While these moments of wonder may be fleeting, they are one of the core reasons why day laborers stand on street corners.

Everyday life data and memories

Like most mobile camera phone users, day laborers also capture a life beyond their work, which includes mundane moments, materials, and experiences on their phones. Each day laborer with whom I interacted had photos that articulated their own personal identity and preferences that stood out when compared to one another. Ramiro had several images related to his health and hygiene. He had pictures of Centrum vitamins, grooming products, Arm & Hammer foot powder, and a berry trail mix from 7-Eleven. He said that these were items he really enjoyed and took a photo because he wanted to be able to remember so that he could buy them again. These photos are functional in that they serve as a reminder. This is especially the case with Centrum, which comes in a variety of forms for youth, middle-aged, and older individuals, and then breaks down by sex. Certainty, taking a photo is a helpful reminder. When I told him that I noticed this trend of self-care in his photos, he said:

> Yes, it's true. It's a point that you correctly identify, because it is something that I take very seriously; to maintain myself and as a worker. I eat salad, juices, vitamins. Because my dad died of diabetes, my sister got diabetes, and had a procedure to help with her diabetes. I started to think, and I asked my mom, why didn't my dad take care of himself? Apart from his ignorance, he was also very *machista*: "I'm not going to drink that, I'll eat watermelon and menudo." I'm on my way to 50, just two more years.

His digital mobile memory validates and supports the emotional importance of health by remembering the products through photos.

Several day laborers had images of food on their phones, but Bart stood out for the amount of food images he had. His pictures were mostly of Mexican-style food that he cooked at home and that he ate at restaurants. One example of food he cooked himself was carnitas, cooking in large copper *cazuela* until crispy, which is a process that can take a few hours. The image also included friends sitting around, drinking beer, and of one friend making his specialty salsa with a mortar and pestle. The images of restaurant food included menudo, sopa de pescado, and – of course – his accompanying beer. The sopa de pescado is especially mouthwatering with crab legs sticking out on one side and clams peeking out just above the surface on the other side. Bart said that these photos in particular were shared with a small circle of friends and family. The other set of photos he had were of his girlfriend, her niece, and of family gatherings that included a baby shower and birthday party. In his archive, food takes a prominent place not only because of its nourishment, but because of the community and cultural practices that surround it. As with Ramiro, there is a clear desire to document moments of self-care, nourishment, and health.

Since almost one-fourth of day laborers are homeless, many experience a different relationship to their mobile phones and the photos they take. Most research agrees that users largely view digital photos they capture on their mobile phone as ephemeral or more transient (Gye, 2007; Palmer, 2010; Reading, 2008). Digital photos arguably do not carry the same longevity as material photos in a family album or those taken with a conventional camera once did. However, this perspective does not fully capture the experience of those struggling with homelessness. While day laborers have similarly experienced the short life span of mobile phone photos through a lost or upgraded mobile phone, many place significant value on the digital images that they capture. Javi, for example, had images of the homeless shelter that he slept at, but also had a peculiar image of golden trinkets. In the palm of his hand, he holds three small golden trinkets that were shaped like a motorcycle, McDonalds fries, and a Coca-Cola bottle (Figure 13.4). When I asked him why he took this photo he said,

Javi: Because I've never seen those before.

CJ: So you wanted to remember it?

Javi: Yes, I wanted to keep it in mind always, because sometimes – I live on the streets – I get robbed. I lose things; it's very easy. Everything I have, I appreciate it. Many things have come my way and go. And then I remember that I have them, but I don't have the thing anymore. It's to remember that I had it.

Javi's relationship to material goods is much more precarious than for those with reliable housing, income, and storage. As a result, the photographic image on his mobile phone is a precious archive that he can hold on to, at least for as long as he has his phone. Knowing this risk, however, he also uploads his photos to a Facebook account that he uses to stay connected to friends back in Mexico. When we looked at his Facebook account, he had made the photo of the golden trinkets his profile picture. The novelty of these knickknacks appealed to him and not only did he want to memorialize them, but he used them to stand in for himself as his profile picture. He is one of the younger day laborers I interviewed and is more tech savvy that those older than him, but his practice to permanently and digitally store a photo reveals that he values the digital memory as much as the material object.

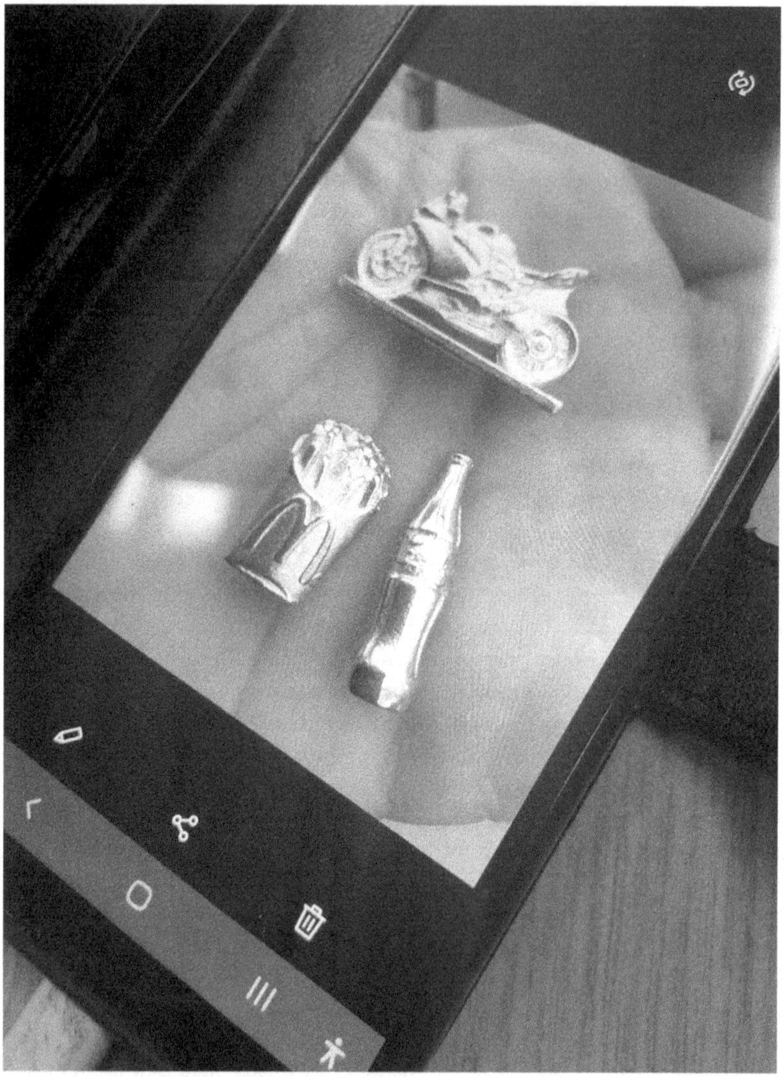

Figure 13.4 Three small golden trinkets given to him by a friend nestled in Javi's hand.

Connecting to places and people

Day laborers also use their mobile phone to connect with employers. As Baron and colleagues discovered, day laborers are able to coordinate and to be rehired later on by sharing their phone number. This enables an additional source of work from employers who were satisfied with their work. More interesting, however, are their connections to distant places and people – whether to hometowns in Mexico or through YouTube video from Indonesia, images of the Great Wall of China, or friends in the Dominican Republic. The photo of the Great Wall of China is a curious one because it was on the phone of a day laborer, Fabian, who captured it by taking a picture of his television screen. It was as if he took the photo through a window of something that sat just on the other side, which is perhaps how undocumented workers may experience distance and mobility, given how their legal status restricts their movement. In an instant, the day laborer watching television added a photo of the Great Wall of China to his archive, to his memory, because he thought it was mesmerizing.

Sitting at home and immobile, Fabian used his phone to save an image from across the world. But why not simply use a search engine and download an image? No, it would not be the same. On the one hand, his practice illustrates the workaround needed to overcome the limits of his technological familiarity, but on the other hand, his action of taking the picture of the television screen also captures his emotions at that very moment. Filled with surprise and awe, the effect is captured in his archive and mimics what someone might feel at the footsteps of the Great Wall of China. It also combines two communication technologies, but shows their differences. Television is a flow of content that is designed for continuous consumption, but the mobile phone is selective and personal, emphasizing its ability to remember and create an affective moment through its stored images.

While one may assume that older day laborers do not use social media, the actual case is that both young and old day laborers were on social media platforms, and primarily Facebook. Several day laborers sought out companionship or someone to talk to. While younger day laborers "friended" those they know in their everyday life, older day laborers' network of "friends" were largely strangers they "friended" based on their appearance or location. Jesse, for example, had several conversations with women in the United States (Arizona and Oklahoma) and in Mexico (Chihuahua and Juarez), but looking closely at their conversations, it was clear that these were one-sided and weak connections. Jesse's messages to the women were often religious in nature, some were chain messages, and others were flirtatious greetings. But taken together, the women very rarely responded, and the chats were largely a string of several messages from him. Jesse said he was motivated by wanting to connect others to religion, but when I asked him whether there was any benefit for him, he skirted the question. In part, Jesse may not fully recognize or acknowledge why he seeks out these female connections through social media.

Other day laborers had similar transnational conversations with women in the Dominican Republic who were also strangers to them. This is similar to Burrell and Anderson's work, where Ghanaians sought out strangers in neither their host nor home country, but beyond both (Burrell and Anderson, 2008). Instead of only looking homeward, they were gaining access to other information flows that presented them with new possibilities, new lifestyles, and new ideas. For the day laborers in Denver, however, close examination of the conversations revealed that they were long-distance and still one-way conversations, as Figure 13.5 illustrates. Despite being weak ties, the conversations offer day laborers a sense of a connection and the opportunity to express kindness and well wishes to their long-distance connection. While many mobile phone owners would not waste their time on these superficial conversations, for day laborers struggling with social isolation, these weak relationships were all they wanted or needed.

Those who were younger and typically more savvy took advantage of Facebook groups to connect locally. Javi, for example, joined over 50 groups, and they were largely related to Colorado. The groups included "buy and sell," several related to available work and community announcements, and others were specific to resale of cars and clothing. These local connections also are not necessarily emotional, but are more practical in terms of finding work and trying to stretch their earnings as far as they can. As younger individuals, their connections to distant lands are not as strong or needed as those who are older.

Figure 13.5 A one-way conversation via Facebook between a day laborer and a weak contact living abroad.

Conclusion: stunted mobile sociality

The digital archive of a day laborer contains images of things they have experienced, touched, fixed, or that they want to remember, and also information of distant worlds and people. This is partly produced by their mobility as they travel across the state of Colorado and into different homes each day through their work with a new task. As a result, a day laborer's digital archive has two primary functions. First, it is pragmatic; day laborers need to find work and showcase their skills and experiences through their digital photos to current or future employers. Photos also help them remember what they have done, and what products they should purchase again. Second, the archive operates on a personal and affective level; what is stored on their phones allows day laborers to continuously reimagine their identities and lives. What they store on their mobile phone rearticulates their mobility and sociality. In some cases, the digital image allows day laborers, especially those who are homeless, to equate the value of a digital photo to the permanence of a material possession or object that they do not have or that might soon be lost or stolen. It defines their work experiences only by those memories that required immense care, skill, and bravery. In an economic and public sphere that places them on the fringes of society and devalues their talents and abilities, the digital archive rearticulates an identity that centralizes their contributions as workers and simultaneously addresses their social and emotional needs.

Day laborers are using their digital memory prosthetic device not only to remember, but to cope. The types of memories they choose and the types of connections they seek work to traverse their struggles with social isolation, economic insecurity, and their value as a workforce. The digital archive allows day laborers to highlight for themselves their contributions to the local economy and, more simply, to a range of people that hire them and their property that needs fixing. As the mural with the rhino illustrated, day laborers are often overlooked as a vital force that improve homes and changes neighborhoods. Their mobile phone is a tool for day laborers to challenge – for themselves – the way society perceives and understands them. A day laborer's digital archive creates a counternarrative. The digital archive filters and stores the day laborer's greatest hits (their flexibility, bravery, self-care, and relationships), but also provides a personalized space of digital accumulation and connection. As conversations with women online and the photographs of worksites multiply, any sense that they do not own the property or that their connections to women are weak fade away through their accumulation of memories. The growing digital archive provides day laborers an outlet to counter the shortcomings they experience materially, emotionally, and/or socially.

The mobile sociality experienced by day laborers may at first appear stunted when compared to more privileged communities, but their digital archives and communications reveal that they have cobbled together both digital memories and weak connections to make sense of their mobility and social experiences. Their sociality and mobility are, no doubt, chaotic and inconsistent. They move from one employer and worksite to the next. If they are lucky, this is a process that repeats itself several times a week, but these repetitive experiences – along with those that are unique or special about a job – quickly fade and blur together. This is the cost of their mobility, and yet their digital memory prosthetic provides a solution by giving longer life to those special moments. With their mobility they sacrifice certain forms of sociality as they struggle with social isolation, homelessness, and a sense of belonging. On their phones, their mobility and sociality are rearticulated. On their devices, they can clearly see the hard labor they do, the material contributions produced by their skill and mastery, and the multiple tethers they have to individuals elsewhere, even if they are weak. Mobile sociality exists in stunted forms whereby the tethers to the digital network are not so strong, but while it is stunted, this does not mean that mobile sociality cannot develop in other ways. In the case of day laborers, mobile sociality is experienced in more intimate and personal ways that makes sense of the chaos and inconsistency they experience.

As a community that struggles with homelessness and legal status, their access to social, cultural, and economic capital is tenuous. The mobile sociality that they cultivate with their mobile phones certainly contains a tether to distant worlds and people over social network sites, but in many cases, that connection is a photo of a television or material objects that will soon go missing, or of weak connections to women living abroad. Yet, for day laborers, the quality of those connections must be put in the context of their lived experience. The value is the mobile sociality that day laborers experience internally through the digital archives they create. Whereas many others can reimagine who they are in the public domain or online through various social media platforms, day laborers are reimagining who they are in their private social world on their phone. The turn inwards is not asocial, as Gye and Palmer suggest, but instead is to rebuild and incubate their sociality. Day laborers create a digital space for themselves to experience wonder, pride, and connectivity – and on their own terms.

References

Anderson, M. and Kumar, M. (2019) "Digital divide persists even as lower-income Americans make gains in tech adoption", *Pew Research Center*. Available at: www.pewresearch.org/fact-tank/2019/05/07/digital-divide-persists-even-as-lower-income-americans-make-gains-in-tech-adoption/ (accessed August 20, 2019).

Baron, L. F., Neils, M. and Gomez, R. (2014) "Crossing new borders: Computers, mobile phones, transportation, and english language among Hispanic day laborers in Seattle, Washington", *Journal of the Association for Information Science and Technology*, 65(1): 98–108. doi:10.1002/asi.22949.

Burrell, J. and Anderson, K. (2008) " 'I have great desires to look beyond my world': Trajectories of information and communication technology use among Ghanaians living abroad", *New Media & Society*, 10(1): 203–224.

Friedman, G. (2014) "Workers without employers: Shadow corporations and the rise of the gig economy", *Review of Keynesian Economics*, 2(2): 171–188. doi:10.4337/roke.2014.02.03.

Galemba, R. and Kuhn, R. (2019) "Wage theft and its victims in Colorado: Research". Available at: www.colorado.gov/pacific/sites/default/files/Rebecca%20Galemba%2008.15.2019.pdf (accessed January 10, 2020).

Giddens, A. (1991) *Modernity and Self-Identity: Self and Society in the Late Modern Age*. Cambridge: Polity.

Gye, L. (2007) "Picture this: The impact of mobile camera phones on personal photographic practices", *Continuum: Journal of Media & Cultural Studies*, 21(2): 279–288. doi:10.1080/10304310701269107.

Hand, M. (2016) "Persistent traces, potential memories: Smartphones and the negotiation of visual, locative, and textual data in personal life", *Convergence*, 22(3): 269–286. doi:10.1177/1354856514546094.

Jensen, B. (2002) "Service to day laborers: A job libraries have left undone", *Reference & User Services Quarterly*, 41: 228–233.

Kindberg, T., Spasojevic, M., Fleck, R. and Sellen, A. (2005) "The ubiquitous camera: An in-depth study of camera phone use", *IEEE Pervasive Computing*, 4(2): 42–50.

Malpica, M. D. (1996) "The social organization of day-laborers in Los Angeles", in R. I. Rochin (ed.) *Immigration and Ethnic Communities: A Focus on Latinos*. East Lansing, MI: Julian Samora Research Institute, Michigan State University.

Negi, N. J. (2011) "Identifying psychosocial stressors of well-being and factors related to substance use among Latino day laborers", *Journal of Immigrant and Minority Health*, 13(4): 748–755. doi:10.1007/s10903-010-9413-x.

Palmer, D. (2010) "Emotional archives: Online photo sharing and the cultivation of the self", *Photographies*, 3(2): 155–171. doi:10.1080/17540763.2010.499623.

Reading, A. (2008) "The mobile family gallery? Gender, memory and the cameraphone", *Trames: Journal of the Humanities and Social Sciences*, 12(3): 355. doi:10.3176/tr.2008.3.10.

Valenzuela, A. (1999) *Day Laborers in Southern California: Preliminary Findings from the Day Labor Survey*. Working Paper 99-04. Los Angeles: Center for Study of Urban Poverty Study Urban Poverty, University of California Press.

Valenzuela, A. (2000) "Working on the margins: Immigrant day labor characteristics and prospects for employment", *Center for Comparative Immigration Studies*. Available at: https://ccis.ucsd.edu/_files/wp22.pdf (accessed January 11, 2020).

Valenzuela, A. (2003) "Day labor work", *Annual Review of Sociology*, 29(1): 307–333. doi:10.1146/annurev.soc.29.010202.100044.

Valenzuela, A., Theodore, N., Meléndez, E. and Gonzalez, A. L. (2006) "On the corner", *Day Labor in the United States, Center for Study or Urban Poverty*. Available at: www.coshnetwork.org/sites/default/files/Day%20Labor%20study%202006.pdf (accessed January 11, 2020).

Van Dijck, J. (2007) *Mediated Memories in the Digital Age*. Stanford: Stanford University Press.

14

FROM SOCIAL MEDIA TO MEDIA SOCIALITIES IN MOBILE WORK

Aspiration in the cases of Australian mining and Everest tourism industries

Jolynna Sinanan

At an elevation of 2,800 metres (9,300 feet), the town of Lukla in Nepal is home to one of the world's highest airports. Commonly known as the gateway to Everest, it is the first point of entrance to the Mount Everest tourism region for treks to Everest Base Camp (EBC) and commercial summit expeditions. It has a main street with globally recognisable establishments bearing regional inflections such as an Irish pub, an Everest Burger restaurant and a Starbucks Coffee Lukla. Dawa Tamang was working as a trekking guide with a Nepali company when I met him in Lukla in 2018. He was 38 years old and had worked as a guide since 2005. He had just returned from taking six people on a 19-day trek. Dawa had constant access to Facebook Messenger throughout the trek as guides and porters had free wi-fi access in all guest houses provided by the company. During the day, wi-fi and phone coverage were patchy due to high altitude. Lukla had the best coverage Dawa had seen for the last few days, so he sat in the dining hall of the guest house, drank some tea and replied to messages while his group checked in. He scrolled down his Facebook feed and saw posts by family and friends, which included clients from previous treks. Many had tagged him in their photos from the trek, thanking him for his role in "achieving their dream" and "ticking EBC off the bucket list". Dawa liked a post from a retired German client in his sixties who had employed him for a private group helicopter tour a couple of years earlier. Dawa had stayed in contact with this client, who wanted to plan another trip in upcoming months and employ Dawa as a guide. Through their conversations in tea houses, this client was also the first person to introduce Dawa to the idea that when Dawa was elderly, if he had earned enough throughout his years working, he could employ someone to provide at home care and might not have to depend on his children to take care of him, an idea that was relatively uncommon for predominantly agricultural and rural populations in Nepal.

Some 8,500 kilometres (5,200 miles) away in the town of Moranbah, in the Bowen Basin coal mining region of Queensland, Australia, Nicole had just finished a 12-hour shift on day three of her seven-days-on, seven-days-off roster. Nicole was 34 and had been working at the same mine for three years. Her children, aged 6 and 4, stayed with their father on the outskirts of Brisbane while she worked on site. Nicole's ex-partner was also a FIFO, a fly-in, fly-out long-distance commuter. They had coordinated their rosters so that one parent was with the children for the week that the other was at work. Nicole returned to the staff changing areas and checked her phone, which she had to keep in her locker during her shift. Her ex-partner had sent her a video clip and two photos of one of their children choosing a book for the other during their bedtime routine. Relaying this to me, Nicole described how she suddenly felt a cluster of guilty thoughts overcoming her. If she had time, she would pick up a couple of new books between the drive to Mackay and the flight back to Brisbane, but she should not keep coming home with gifts every time she returns from work. She thought she would really like to take the children on a short holiday during her week off, but that would mean pulling the kids out of their routine. What can she do for them as their mother who is always away?

These brief portraits have attempted to capture some of the intergenerational and familial aspects of the emotional geographies of mobile work (see Bissell et al., 2020; Holmes and Wilding, 2019; Madianou and Miller, 2013). Dawa and Nicole are both regional commuters who spend weeks at a time away from their families. On the one hand, Dawa might appear to reflect a workforce in the international tourist industry that has been labelled precarious, uncertain and contingent (Çelik and Erkus-Öztürk, 2016; Lee et al., 2015), and Nicole might be representative of more middle-income and Western, privileged mobilities (Baldassar and Merla, 2014; Polson, 2016). However, these seemingly disparate examples are also reflective of what Glick Schiller and Salazar (2013) have termed regimes of mobility. Glick Schiller and Salazar address movement across borders that may facilitate different kinds of economic mobility but also impose measures that affect individual mobility. In particular, and relevant to the themes that I will explore in this chapter, they focus on the role that imaginaries play across physical movement, and how these impact personal aspirations. Regionally commuting workers are less likely than transnational migrants to be separated by cultures or national borders, but the distances between workers, their families and social networks "at home" may heavily impact time spent together, their ability to sustain intimacy at a distance and the everyday negotiations and obligations that accompany maintaining close relationships. Further, within these intra-state (rather than international) regimes of mobility, mobile workers' diverse experiences might impose limitations that frustrate their own aspirations for social and economic mobility.

This chapter contributes to the concept of mobile socialities by situating the role of digital media as communicative affordances of portability, availability, locatability and multimediality (Schrock, 2015) within the socialities created by regional, long-distance commuting work. First, I revisit critical concepts in mobile and social media to establish that although theorisations such as "full time intimate community" (Ito and Okabe, 2005) and "scalable sociality" (Miller et al., 2016) address socialities as social networks, connections and relationships, they only capture aspects of the complex and often contradictory ways that intimacies are imagined and practised in situ and across distance. By critically revisiting these theorisations of mobile and social media, I seek to address the following question: in what ways have digital media impacted the shifting of aspirations within forms of socialities created by mobile livelihoods? To do this, I draw on unfolding, ethnographically informed research on the Australian coal mining industry (from 2017 onwards) and the Mount Everest tourism industry (from 2018 onwards), both of which are sustained by mobile livelihoods. These forms of work entail intimate socialities of temporary relationships and others that are enduring and emotional geographies created by sustained periods of time at home and work. Through these case studies, I also argue that digital media practices within such socialities play a role in shifting the aspirations and imaginations of individuals. These shifts warrant more attention, as they may contribute to further solidifying a collective sense of familyhood or create ambivalences and tensions which challenge collective aspirations in households over time.

Media socialities: rethinking theories of social media

From 2012–2016, I worked on the *Why We Post* project as one of nine researchers across eight countries conducting 15 months of ethnographic fieldwork on uses of social media. Each researcher on the project carried out prolonged and in-depth fieldwork, studying social media with an anthropological commitment to holistic contextualisation and a social sciences approach, more generally, that centres on human experience.[1] In the regionally comparative volume, *How the World Changed Social Media* (Miller et al., 2016), we introduced "scalable sociality" as a theory of social media. We argued that social media created new potentials for sociality by facilitating the degree to which individuals could conduct relationships, from the most private to the most public, and from the smallest group to the largest. Through controlling their presence on social media platforms, individuals can engage in private dyadic conversation at one end of the scale, and communication that is more akin to public broadcasting on the other end of the scale. In the English field site, for example, high school students only used Snapchat with a small group of trusted friends and used Instagram with the intention of attracting as many followers as possible, including strangers outside of their networks who appreciated their images (Miller et al., 2016: 3–5).

Yet, on reflection, the theory of scalable sociality seems rather platform centric. "Scalable" echoes a geographical approach that focuses on delineation and organisation of territory and space (such as home, local, national or global), but scale is also constructed and transitory (Furlong, 2011). The effects of scale encompass relationships, and although we emphasised sociality, privacy and publicness, we did not integrate a consideration of how relationships as intimacies and imaginations are made and unmade through shifting association with different socialities. How do intimacy, emotional connections, individual and collective imaginations play out in the socialities facilitated by Snapchat, for example, versus Instagram? Entire genres of communication migrate between platforms, and culture-bound notions of privacy shape individuals' control of the nature and extent of their intimate associations (Chambers, 2017). Chambers' observation suggests that more attention to locally based socialities can reveal how associations and the meanings associated with them may shift geographically. Social interactions exchanged within differently located spaces of interaction are shaped by respectively located cultural expectations and norms, which become incorporated into conceptions of intimacy more broadly (Farman, 2015; Wilken et al., 2019).

Given that the development and uptake of social media platforms have changed rapidly since the *Why We Post* study was completed, the research did not go into significant depth about social media as a convergence of platforms, locative and mobile media, and the subsequent implications for the material location of individuals engaging in digital practices. With the emergence of a significant body of work on locative media, we now know that place matters and matters of place include ideas of place (Wilken et al., 2019; Schwarz, 2015; see also Morley, 2009). Adoption of locative media responds to intimacies that are situated in cultural practices, norms and expectations.

Mobile socialities as geographies of intimacies and aspirations

While this earlier body of mobile media research tended to focus on youth, Holmes and Wilding (2019) pose the concept of "intimacies at a distance" to explore ideas of intimacy and mobility across different age groups and categories of relationships, and incorporating influencing factors such as gender. Their special issue of *Emotion, Space and Society* brings together research in migration, mobilities and emotions to consider models and practices of intimacy in all mobile lives and the role digital media might play. By considering the broader issue of relationality, contributors consider migrants, relatives of migrants and commuters, and the ways they engage with different cultural contexts as part of their everyday lives. Although this focus moves beyond family members' digital practices within the household, critical to ideas and practices of intimacy at a distance is the idea and ideal of home. In relation to families, Clark (2014: 321) situated uses of mobile media within the "emotional and moral economies of the household". By doing so, she drew attention to the family as an economic, as well as social, structure that entails labour invested both inside and outside the home. In addition to labour, managing and maintaining the household depends on exchanges of emotion between individuals and groups within the family. Considering the materiality of home, Morley (2002: 24) suggests that the idea of home is the feeling of safety and security embodied in the physical structure of a house, such that the idea of home and family become interchangeable. Indeed, Hartmann's (2018) ongoing research on homelessness reveals that within media studies and beyond, the concept of home is taken-for-granted and yet constantly shifting. The household, home and family may be the primary set of relationships where cultural and social values are embodied and reproduced, and yet, the number of families that are characterised by sustained periods of absence is steadily increasing (Baldassar and Merla, 2014). This body of research brings together perspectives on the intersection of place and the creation and maintenance of relationships. Significantly, these authors draw attention the kinds of intimacies that are facilitated by digital and mobile media. In the discussion that follows, time spent away from home and family are typical as an experience of being a mobile worker. Explorations of maintaining familyhood, and the impact of mobile work on family relationships and aspirations, are discussed elsewhere (Sinanan and Baldassar, forthcoming). Here, I focus on socialities as temporary and enduring relationships; relationships situated within emotional geographies facilitated by working away and their intersection with leisure and routines to explore shifts in aspirations.

Mobilities and the social connections and experiences that result are driven by aspirations, to an extent (Polson, 2016). However, although the language of wants and choices leads us to believe that aspirations are highly individual, the desire for a "good life" is part of a system of ideas formed in social life (Appadurai, 2013: 187). While aspirations for life projects (financial improvement, security or further mobility) are shaped by social and cultural relations, they are also facilitated and constrained by opportunity structures. Fischer (2014: 64) describes that opportunity structures that impact individual agency to realise aspirations include legal regulations and market mechanisms that influence labour, but also social formal and informal social norms related to factors of gender, ethnicity and class.

Polson (2016: 9) describes the new global middle class (as a set of cultural practices, rather than empirical fact) as an aspirational social space. Drawing on Bourdieu, she argues that mobility allows for individuals to make up for what they lack in economic capital (due to shifts in labour in a post-Fordist era) by strategically creating and leveraging cultural status acquired through experiences such as travel. The experiences enabled by working mobility creates opportunities, which in turn impacts aspirations. In a similar vein, Dahlgren (2019) examines the mobile working class and how individuals accrue economic capital to become upper middle-class workers in the mining industry in Queensland, Australia. For Dahlgren's mining workers, these opportunities shifted aspirations for life projects based on lifestyle of home ownership or renovations and travel (Dahlgren (2019: 61). Notably, she observes that for the highly masculine workforce, aspirations for life projects are expressed along the lines of the gender divide between workers and their partners in the context of ideal familyhood. These class and gendered aspects of shifting aspirations within mobile workforces resonate with the illustrations in following sections.

The two case studies in this chapter provide snapshots of comparative mobilities within a relatively affluent, "Western" country (Australia) and a country that is more characteristic of what has been described as the "developing world" (Nepal). These sites were chosen to compare how individual aspirations are shaped by relationships situated in differing cultural, social and labour contexts. I draw attention not only to the choices and strategies that individuals employ, but how these choices are structured through political-economic circumstances that transcend the individual (Fischer, 2014). A combined study of comparative populations and demographics can reveal how mobile livelihoods often include circumstances of precarity that tends to be associated with lower-income, less mobile citizens who are typically from the global South. Yet, as the case study in Australia reveals, such circumstances can also be experienced by relatively more affluent populations. Returning to the initial example of theorising media as scalable sociality, mobile media play a key role in facilitating relationships that are created by working across expanded geographical scale.

From mining to mountains: a comparative study of mobile media and mobile livelihoods

In the sections that follow, I bring together these economic, intimate, geographic and social dimensions of digital practices, drawing on research conducted in the Bowen Basin coal mining region in Queensland and in the Khumbu region in northern Nepal as part of two distinct but related studies carried out since 2017. For one study, across four trips from December 2017–March 2019, I spent time in mining workers' accommodation near Moranbah, where workers reside while away from home. Long-distance commuting workers, distinguished as FIFO (fly-in, fly-out), DIDO (drive-in, drive-out) or BIBO (bus-in, bus-out), are typically rostered for shifts of five days on, two days off or seven days on, seven days off. I conducted in-depth interviews with 12 individuals (nine men and three women) between the ages of 22 and 58. Six of the men and two of the women had partners with children in their households. Further to these interviews, I spent time conducting follow-up visits with three men and one woman while they were rostered off in Brisbane, the Gold Coast and Gladstone, and one man who had relocated work places to Central Queensland. These interviews formed part of a project investigating mobile media practices of workers and family members, how they create familyhood, their own sense of well-being across distance and how different stakeholders do (or do not) support them. The interviews fit within a broadly ethnographic methodology including participant observation within the mining workers' accommodation and ethnographic attentiveness to workers' meanings and experiences of digital practices (Pink et al., 2016).

The other study developed out of two preliminary visits I made, in 2017 and 2018, to Kathmandu and the tourist hub of Namche along the EBC trek. These visits provided the groundwork for fieldwork conducted in 2019, with a follow-up period of fieldwork scheduled for the 2020 April–May summit season during the "Visit Nepal Year of Tourism 2020" campaign. This fieldwork was cancelled due to the COVID19 pandemic. With Nepal's peak tourism season cancelled, too, and workers remaining in Kathmandu or in villages in surrounding regions, the future of the tourism workforce in upcoming months and years remains uncertain.

This project in the Khumbu region brings together research in transnational migration, lifestyle mobilities and travel to investigate how mobile media shapes meanings of Everest for tourists, and how this impacts the routine practices of minority workers. Using digital ethnography, participant observation and interviews, I investigate the role of mobile media in the Everest tourist industry for tourists and workers. The material presented in this chapter draws on insights gained from time spent in Namche with 15 tourists (two women and 13 men, aged between 22 and 54), three guides (men, aged between 24 and 38), six porters (men, aged between 19 and 25), two managers of a guest house (a couple: the man aged 37 and the woman aged 35, parents of an 8-year-old daughter) and two managers of a porters' accommodation house (sisters aged 19 and 22). Contrary to the visibility of Khumbu Sherpa, who have been traditionally associated with Everest mountaineering, significant numbers of guides and porters are from the Tamang and Rai ethnic groups – populations from other parts of Nepal who have historically been at the economic margins of Nepali society (Nepal, 2005). All workers who participated in this study are Tamang and Rai who work seasonally in the Khumbu region.

With regard to the labour and movement across these two sites, I apply Sørensen and Olwig's (2002) idea of mobile livelihoods, which takes material and infrastructural conditions of work together with the immaterial conditions: the social relationships, emotional connections and individual and collective imaginations that these livelihoods imply. In *Work and Migration*, Sørensen and Olwig argue that the strong focus on international migration has had the unfortunate result that other kinds of population movement have received less attention (Sørensen and Olwig, 2002: 7). They put forward the notion of mobile livelihoods, which considers the differentiated experiences of making a living over long and short distances, within states and localities and/or across national borders. With this, they move beyond "work" and "economic decision making" to "highlight the ways that social knowledge is produced by affiliations with different relationships in wider spatial contexts" in the movements involved with making a living (Sørensen and Olwig, 2002: 10). The concept of "mobile livelihoods" focuses on the ways that work and earning link up with larger-scale patterns of population movement, the range and variation in mobility that such movements involve, the social institutions and networks facilitating and sustaining mobile livelihoods, and the social and spatial practices of mobile populations. It redirects migration research for a broader investigation of the fluid fields of social, economic and political relations and cultural values that these livelihoods imply (Sørensen and Olwig, 2002: 3). As we go on to see, these fields are enmeshed for commuting workers in Nepal's Khumbu region and in Australia's Bowen Basin.

Bowen Basin coal mining: defying the "golden handcuffs" label

Returning to the story of Nicole, who was introduced earlier, one of her key difficulties maintaining communication with her children was that work routines, and the need to leave phones off while driving or operating machinery, left only small pockets of interstitial time to connect with them. "There's a phone ban on site, but you're always snatching time, snatching moments when you can", Nicole said. "If I have a break, I'll try to FaceTime in the afternoon. I can't really at night, even though I have the time, because it doesn't work with their routine". She explained that night shifts were actually better because when she wakes up at 4.30pm, her children's evening routines of dinner and bedtime had not yet started. She tried to give her children a sense of work also meaning having to live far away, but her toddlers had trouble understanding that the worker's accommodation was not a holiday site. Nicole had sent her children photos of the many kangaroos around the accommodation, thinking they would relate to this, but her daughter became upset that "Mummy went somewhere fun like the zoo and she didn't take me along". Nicole reflected on the difficulty of trying to make small children understand the extra complexities of what "work" is for her. "She [her daughter] calls our home 'my home' and work 'mummy's home'". Nicole's vexations over parenting while away reflect Baldassar's (2015) observations of how guilty feelings play out and create obligation within transnational families, and the role phones might play in alleviating these.

Nicole's partner Ryan, who was in his early 40s, worked for the same company and on the same roster as her. Ryan's 8-year-old daughter stayed with her grandmother while he was on roster. Nicole and Ryan were trying to strategise ways of "being a family" when they were both off roster. Their children had spent a little time together, but they had not yet established how to maintain one household together. And so, their relationship's routines were largely limited to the time they were on roster, when they lived together after hours. Household tasks such as preparing meals were not needed, as the accommodation village has a mess hall serving workers meals around their shifts. Laundry was the most immediate priority, since mining safety standards require workers to wear a fresh set of clothing for each shift. Nicole's and Ryan's laundry routines, and keeping their cabin tidy, were perhaps the most immediate on-roster tasks that resembled housework. They explained that they were used to picking up after one another and taking turns doing their laundry because these were the main routines they shared. When they are at home with their children together, there is far more to do with picking up after children, preparing all meals and integrating the routines that they need. Through their time on roster, Nicole and Ryan established their own sense of making a home together through maintaining household-like routines. They had hoped that these routines would contribute to combining their households, whereby during their time off roster, their children would become integrated into the household they established together as a couple. However, the coordination for their shared time together off roster was largely done by Nicole, who initiated planning around Ryan. The exception was that the couple planned for holidays together that included the children, when the efforts for planning and organisation was more equal between Nicole and Ryan – who expressed more enthusiasm for planning around these key events, rather than during the week to week.

There has been little research to date on the effect of long-distance commuting on intergenerational families at different life stages (Pini and Mayes, 2012). A common term associated with FIFO mining labour in Australia is the trap of the "golden handcuffs", which captures how workers are drawn in to FIFO work by the large salaries on offer, even at entry level. Workers become dependent on these large salaries to the detriment of their own psychological health and the well-being of their families, due to the extensive time spent away from home. While workers in this study discussed some of the psychological burdens that working away for prolonged periods of time can cause, they also emphasised the sense of purpose their work gave them in relation to aspirations for the future, especially for their children.

Steve, for example, was 58 and had worked in the coal mining industry since apprenticing as an electrician after high school. He lived on the Sunshine Coast and his daughter, aged 27, had moved to Brisbane to study at the University of Queensland – "Psychology or criminology, I think, I'm not really sure". He started long-distance commuting during the boom when his children were in their late teens. He says he remembers that when they were in school, they "kind of got used to me being away". Now, they are grown up and he said he "probably talks to them more regularly now, but I'm not very tech savvy". Steve brought his laptop with him onsite, an old Lenovo that he could link to his phone's hotspot for basic browsing and reading, but he did not really use it for entertainment, "just information". As the mine he was working in was due for shutdown in three months, he had started looking for other jobs. In the years following the boom, several workers were contracted to the same mines and resided in the same areas. As a result, they formed social networks based on shared rosters. These networks became a source of information about where other jobs might be coming up and which other companies might be recruiting. More recently however, as FIFO mining work declined, workers became more reliant on digital upskilling to search job websites, and subscribing to mailing lists or Facebook pages. Steve explained, "The problem is, a lot of the [application] deadlines are while I'm away, and you need to complete the application online". The way he was getting around these challenges was by emailing his CV and cover letter to his daughter in Brisbane.

> I'll make a time with her and she'll call or I'll call, and she puts in all the details while I'm on the phone. She'll read it and I'll answer. She's pretty good; she'll go through my letter and CV and she'll fix it or make it sound a bit better, things she can do that I can't do as well because I didn't go to university.

Steve's daughter helped him with three applications in two months and a psychometric test.

> That was great, because she had to log in being me and read out the questions and I'd answer and she'd say "really, dad? Are you sure you want to put that one in?" and we'd have a laugh and I'd ask what does she reckon and we might change it. I know it's not the point of doing it, but if it's due midnight that day, it's not like I can get it done here with an old laptop and a mobile phone. Most of these things don't come up as well on the phone.

Steve's experience speaks to three insights that emerged more generally from the research: intergenerational aspirations, life stage and digital inclusion. First, Steve's experience of entering FIFO work during the mining boom is consistent with that of several workers in Queensland's mining industry in the late 1990s. Significant numbers of men redirected their skills towards long-distance commuting work because industry projections based on demand from China's economic growth expected the boom to last (Dahlgren, 2019). Aspirations of FIFO workers at the time were generally related to property ownership and lifestyle (vehicles, boats and home renovations); however, following the bust, aspirations of mining workers turned towards longer-term security and education for their children. This arc was reinforced by my research participants. Second, Steve's daughter, who had just started high school when he began mining work, is now in her late 20s and is diversifying her career. She was working as a small business owner in their home suburb on the Sunshine Coast, before undertaking her university degree. While parents like Nicole, Ryan and Steve may be dependent on the mining industry and – due to the intensity of their work rosters – may lack opportunities to upskill or diversify, the income afforded by FIFO work may translate to intergenerational achievements, such as their children attaining university education and working in multiple vocations at later life stages. Third, several participants, like Steve, did not consider themselves particularly digitally literate, despite working in technologically complex environments. Many only brought their smartphone with them to the accommodation, while others – mostly those in their fifties – brought a basic laptop for web browsing in their spare time. For men in their late fifties, the attitude of "not being computer savvy" is reflective of wider observations of the limits of digital inclusion, indicating bias against older Australians who are not as confident with navigating digital landscapes, and whose social connections may suffer as a result (Thomas et al., 2019).

This section has presented a brief snapshot of the impacts of long-distance commuting work and the role of digital media in family relationships. The workers involved with this study are seldom associated with employment circumstances that may be called "precarious"; however, workers often emphasised feelings associated with a "regime of precariousness": how precariousness in modes of production bleeds into how social relations are produced and reproduced in different ways (Çelik and Erkus-Öztürk, 2016: 424). These include uncertainty about the future for individuals and their families. These brief illustrations also suggest that women may have to invest more efforts into the longer-term life projects of the family, which is reflected in the emotional burdens expressed by Nicole. The next section presents a comparative example of workers whose livelihoods fit with how precarity of work has been treated in scholarly research. Within the international tourism industry, precarious work has become increasingly characteristic of employees' experiences of labour. Tourism employees are highly mobile and tourism is at the forefront of the emergent collaborative or gig economy, within which long-term employment prospects are unclear (Bertoli et al., 2017; Dredge and Gyimóthy, 2015; Duncan et al., 2013).

The Everest tourism industry: an alternative to state development initiatives

This section highlights the ways in which relationships formed through mobile work can influence aspirations for the future, particularly when work is considered precarious. Gendered dimensions of work, digital media practices and experiences of mobile socialities are expressed along the lines of aspirations for independence and entrepreneurship, particularly for young women working in the Everest tourism industry for whom seasonal work allows time for further education. Porters and managers of porter accommodation lodges generally have less prolonged encounters with trekkers and tourists than guides such as Dawa, who was introduced at the beginning of the chapter and to whom I shall return shortly.

Porters are often from lower-income families from other rural areas of Nepal, yet significant numbers are increasingly from Kathmandu. Those who reside in or come from Kathmandu are required to fly to Lukla, "the gateway to Everest", and from there, villages along the main EBC trek are only accessible by road. Porters who live in the Khumbu valley surrounding Everest typically walk between two and four days to meet their guide and clients in Lukla. Trekking porters can carry up to 40 kilograms (88 pounds) throughout the day to meet groups at their accommodation at the end of the day's itinerary. Depending on the trekking company they work for, some porters stay in separate quarters in the same accommodation as their group. These generally have better facilities than porter houses on the outskirts of towns and villages; usually small and housing up to 50 porters per night, these lodges cater to porters carrying supplies and goods who are not affiliated with trekking companies.

Sisters Etisha, aged 20, and Ditya, aged 18, were managing a porters' accommodation lodge in Namche. They lived in Namche throughout the two peak seasons and off-peak seasons, returning home to their village only for monsoon season between June and August. If they needed to, they would visit home throughout the year for family occasions, but they tried to stay in Namche, where they could earn, for prolonged months. Throughout their two years managing the porter house, they were visited by guides who needed extra accommodation for their porters during peak seasons when tourist accommodation lodges were full. Setting up long-term relationships is vital in vocations of precarious work, and even more so in developing economies where social relations are in many ways a more reliable safety net than other economic activities espoused by development initiatives, including entrepreneurship (Bebbington and Kothari, 2006; Narotzky and Besnier, 2014). While money can be spent as soon as it is earned, personal relationships can be called upon over a longer period of time. Reciprocity in these negotiations included an understanding that the sisters would prioritise porters associated with guides who established good relations with them. Good relations entailed not only good pay and expectations for appropriate behaviour from porters, but also modest gifts of supplies and advice in expanding their business and opportunities.

Through these relations established with guides, especially those attached to international or reputable local trekking companies, Etisha received advice on how she could study online for a certificate in basic bookkeeping from an institution in Kathmandu. She was able to study in slower times of the work day, usually in the early afternoon hours, and she acquired her certificate in six months. She was also connected to a cohort of students who supported her formal learning through informal chat groups over Facebook Messenger. Many porters spend most of their free time visiting one of the many snooker houses in Namche, returning in the evening to spend time on their phones in the porter house Etisha and Ditya managed, which had substantial wi-fi connection. Wi-fi connection is a more valuable amenity to porter houses than an extensive menu, alcohol or cigarettes. Wi-fi on phones is a priority for porters after hours for entertainment "killing time", chatting with friends and exchanging images, videos and music. While wi-fi is a critical resource for Etisha to pursue educational opportunities, for porters, wi-fi might be associated with leisure and entertainment activities independent from work. Rabin, for example, was a porter who was 20 years old. He explained that one of the things he especially liked to do on his phone before he goes to sleep is look up images of sports cars he aspires to own in the future. Rabin was planning to continue working as a porter, hopefully pick up more English and then train to be a trekking guide. He saw his opportunities in his home village as being fairly limited to working in agriculture.

Rabin's experience of working with other young men in the Khumbu resonates with Xinyuan Wang's (2016) ethnography of rural-to-urban migrants in China. In relation to factory workers in industrial towns, Wang argues that for the generation of rural migrants who are younger than 25, the trajectory of rural-to-urban migration has shifted from the "push" factor of economic necessity to the "pull" factor of aspiring for experiences associated with modernity. For young migrant workers, migration means "to see the outside world" and to gain autonomy from one's family. Contrary to much of the transnational families and migration scholarship (see Bacigalupe and Lambe, 2011; Diminescu, 2008; Licoppe, 2004), in this context, digital media is less of a bridge that connects individuals with the family and friends they have left behind in villages, and is more of a projector, illuminating an ideal modern life to which they aspire.

For example, in industrial China, posts of future aspirations around consumption are prominent on social media platforms. Wang argues that this is hardly surprising, given a population of migrant factory workers who see their work as a stage towards obtaining wealth. Creating fantasies of consumption, young men post images of cars, beautiful women and branded clothing, while young women post images of princess-style bedrooms. For youth in particular, posting these images shows a technicolour world that contrasts with lives spent mostly in factory work (Wang, 2016; Miller et al., 2016).

The images posted by young men working as porters were similar to those posted by factory workers Wang describes, to the extent that they portray lifestyle aspirations (through images of cars and motorcycles, which friends like and comment upon). Other genres of images – representing trekking, the Himalayas and Mount Everest – were more reflective of the tourist imaginary of Nepal: a mythologised space of exploration, adventure and remoteness (Salazar, 2012; Norum, 2016). These images posted during treks with groups of clients became advertisements of sorts, and legitimised the identities of young men as being upwardly mobile through tourism industry work.

In this final example, we return to Dawa. Dawa had been working with a young man, Ajay, since Ajay finished high school when he was 18. Ajay worked as a porter for the same international company Dawa was a guiding for, and when Dawa left to join another guide who started his own, locally based trekking company, Ajay joined that company, too. Ajay continued to work as a porter, then as an assistant guide for seven years. He worked closely with Dawa, improved his English from spending time with clients and picked up conversational German and French.

In 2018, Ajay left the company to begin his own trekking business. He wanted to be self-employed and he believed that he had accumulated enough personal contacts to be able to sustain his own business conducting treks and tours for private groups only. His primary clientele are former clients of treks on which Dawa was a guide and Ajay was his assistant. In his first year, Ajay also received groups who came to know of his services through networks of his former clients. He explained:

> I learned so much from Dawa and the international company, but I wanted to own my own business. I wanted to be my own boss. I have a young son now and I did not want to spend so much time away from home. This way, if I can have small private groups, I can be away less but still have enough business for the peak season.

In the conversations that followed, what was striking about how Dawa and Ajay negotiated with their clients was that there seemed to be little sense of financial competition. Ajay sought to be self-employed, while Dawa worked for another company, yet they continued to draw on each other as social and informational resources. Further, Ajay cultivated his business through a Facebook page and an Instagram profile. While curating images and videos taken on treks, he shared these with Dawa so as to update the website of their previous company. They explained that because they were friends and colleagues, if either was impacted by a particularly low season or lack of clients, they might be able to provide the other with work. And although they now worked for different companies, they were confident to share their reputation and refer clients to one another in the case that one of them was unavailable.

Conclusion: mobile media, mobile socialities and future orientations

These insights into mobile work lives in the Australian mining industry and Mount Everest tourism show the intimate digital socialities that accompany working away from home and family for sustained periods of time, and demonstrate the role these digital practices play in shifting individuals' aspirations and imagination for life projects. Through the affordances of locative and mobile media, the placed-based relationships that emerge through mobile work continue to influence how people may imagine their futures and strategise for opportunities towards meeting their individual or collective family aspirations. Such uses of digital platforms occur at the intersection of technological visions and modern ideals, and include the role that new technologies play in managing the possible risks of precarious work (Björkman and Harris, 2018).

More generally, these aspirations are developed in a cultural and social context of what can be made possible (Fischer, 2014). Mobile socialities, navigated by social and mobile media, perhaps collapse geographical borders and the policies that define regimes of mobility, and enable a wider repertoire of cultural and social imaginings. As Nielsen and Pederson (2016: 237) argue, these imaginings might be indispensable in helping people make sense of changing social, political and economic circumstances by straddling the divide between an uncertain present and an unknown future.

Note

1 Detailed findings on topics from political engagement to visual images can be found in each of the regional monographs: www.uclpress.co.uk/collections/series-why-we-post.

References

Appadurai, A. (2013) *The Future as Cultural Fact: Essays on the Global Condition*. London: Verso.

Bacigalupe, G. and Lambe, S. (2011) "Virtualizing intimacy: Information communication technologies and transnational families in therapy", Family Process, 50(1): 12–26.

Baldassar, L. (2015) "Guilty Feelings and the Guilt Trip: Emotions and Motivation in Migration and Transnational Caregiving", *Emotion, Space and Society*, 16: 81–89.

Baldassar, L. and Merla, L. (eds.) (2014) *Transnational Families, Migration and the Circulation of Care: Understanding Mobility and Absence in Family Life*. New York: Routledge.

Bebbington, A. and Kothari, U. (2006) "Transnational development networks", *Environment and Planning A*, 38(5): 849–866.

Bertoli, S., Fernández-Huertas Moraga, J. and Keita, S. (2017) "The elasticity of the migrant labour supply: Evidence from temporary Filipino migrants", *The Journal of Development Studies*, 53(11): 1822–1834.

Bissell, D., Gorman-Murray, A., McNamara, K. and Straughan, E. (2020) "Living apart together: How working away affects individuals, households and wellbeing". Available at: www.workingaway.net/ (accessed August 20, 2020).

Björkman, L. and Harris, A. (2018) "Engineering cities: Mediating materialities, infrastructural imaginaries and shifting regimes of urban expertise", *International Journal of Urban and Regional Research*, 42(2): 244–262.

Çelik, C. and Erkus-Öztürk, H. (2016) "Role of precariousness and space on class formation process: The case of Antalya's tourism workers", *Capital & Class*, 40(3): 419–445.

Chambers, D. (2017) "Networked intimacy: Algorithmic friendship and scalable Sociality", *European Journal of Communication*, 32(1): 26–36.

Clark, L. S. (2014) "Mobile media in the emotional and moral economies of the household", in G. Goggin and L. Hjorth (eds.) *The Routledge Companion to Mobile Media*. London: Routledge, 320–332.

Dahlgren, K. (2019) "Digging deeper: Precarious futures in two australian coal mining towns", Unpublished PhD Dissertation, London School of Economics and Political Science, London.

Diminescu, D. (2008) "The connected migrant: An epistemological manifesto", *Social Science Information*, 47(4): 565–579.

Dredge, D. and Gyimóthy, S. (2015) "The collaborative economy and tourism: Critical perspectives, questionable claims and silenced voices", *Tourism Recreation Research*, 40(3): 286–302.

Duncan, T., Scott, D. G. and Baum, T. (2013) "The mobilities of hospitality work: An exploration of issues and debates", *Annals of Tourism Research*, 41: 1–19.

Farman, J. (2015) "The materiality of locative media: On the invisible infrastructure of mobile networks", in A. Herman, J. Hadlaw and T. Swiss (eds.) *Theories of the Mobile Internet: Materialities and Imaginaries*. London: Routledge, 45–59.

Fischer, E. F. (2014) *The Good Life: Aspiration, Dignity, and the Anthropology of Wellbeing*. Stanford, CA: Stanford University Press.

Furlong, K. (2011) "Small technologies, big change: Rethinking infrastructure through STS and geography", *Progress in Human Geography*, 35(4): 460–482.

Glick Schiller, N. and Salazar, N. B. (2013) "Regimes of mobility across the globe", *Journal of Ethnic and Migration Studies*, 39(2): 183–200.

Hartmann, M. (2018) "Mobilism in translation", in K. Fast, A. Jansson, M. Tesfahuney, L. R. Bengtsson and J. Lindell (eds.) *Geomedia Studies*. London: Routledge, 173–194.

Holmes, M. and Wilding, R. (2019) "Intimacies at a distance: An introduction", *Emotion, Space and Society*, 31: 1–4.

Ito, M. and Okabe, D. (2005) "Intimate visual co-presence". Position Paper for the Seventh International Conference on Ubiquitous Computing, Tokyo.

Lee, D., Hampton, M. and Jeyacheya, J. (2015) "The political economy of precarious work in the tourism industry in small island developing states", *Review of International Political Economy*, 22(1): 194–223.

Licoppe, C. (2004) "'Connected' presence: The emergence of a new repertoire for managing social relationships in a changing communication technoscape", *Environment and Planning D: Society and Space*, 22(1): 135–156.

Madianou, M. and Miller, D. (2013) *Migration and New Media: Transnational Families and Polymedia*. London: Routledge.

Miller, D., Costa, E., Haynes, N., McDonald, T., Nicolescu, R., Sinanan, J., Spyer, J., Venkatraman, S. and Wang, X. (2016) *How the World Changed Social Media*. London: UCL Press.

Morley, D. (2002) *Home Territories: Media, Mobility and Identity*. London: Routledge.
Morley, D. (2009) "For a materialist, non–media-centric media studies", *Television & New Media*, 10(1): 114–116.
Narotzky, S. and Besnier, N. (2014) "Crisis, value, and hope: rethinking the economy: An introduction to supplement 9", *Current Anthropology*, 55(S9): S4–S16.
Nepal, S. (2005) "Tourism and remote mountain settlements: Spatial and temporal development of tourist infrastructure in the Mt Everest region, Nepal", *Tourism Geographies*, 7(2): 205–227.
Nielsen, M. and Pedersen, M. A. (2016) "Infrastructural imaginaries: Collapsed futures in Mozambique and Mongolia", in M. Harris and N. Rapport (eds.) *Reflections on Imagination: Human Capacity and Ethnographic Method*. London: Routledge, 237–262.
Norum, R. (2016) "The hypersocial: Transience, privilege and the neo-colonial imaginary in Expatria, Kathmandu", Unpublished PhD dissertation, University of Oxford, Oxford.
Pini, B. and Mayes, R. (2012) "Gender, emotions and fly-in fly-out work", *Australian Journal of Social Issues*, 47(1): 71–86.
Pink, S., Sinanan, J., Hjorth, L. and Horst, H. (2016) "Tactile digital ethnography: Researching mobile media through the hand", *Mobile Media & Communication*, 4(2): 237–251.
Polson, E. (2016) *Privileged Mobilities: Professional Migration, Geo-Social Media, and a New Global Middle Class*. New York: Peter Lang.
Salazar, N. B. (2012) "Tourism imaginaries: A conceptual approach", *Annals of Tourism Research*, 39(2): 863–882.
Schrock, A. R. (2015) "Communicative affordances of mobile media: Portability, availability, locatability, and multimediality", *International Journal of Communication*, 9: 18.
Schwarz, R. (2015) "Online place attachment: Exploring technological ties to physical spaces", in A. de Souza e Silva and M. Scheller (eds.) *Mobility and Locative Media: Mobile Communication in Hybrid Spaces*. Abingdon: Routledge, 85–100.
Sinanan, J. and Baldassar, L. (forthcoming) "Undervalued infrastructures of care: The role of digital media for mobile and FIFO families", *International Journal of Care and Caring*.
Sørensen, N. and Olwig, K. (eds.) (2002) *Work and Migration: Life and Livelihoods in a Globalizing World*. London: Routledge.
Thomas, J., Barraket, J., Wilson, C., Ewing, S., MacDonald, T., Tucker, J. and Rennie, E., (2019) *Measuring Australia's Digital Divide: The Australian Digital Inclusion Index 2019*. Melbourne: RMIT University, for Telstra.
Wang, X. (2016) *Social Media in Industrial China*. London: UCL Press.
Wilken, R., Goggin, G. and Horst, H. A. (eds.) (2019) *Location Technologies in an International Context*. London: Routledge.

15

WORKAWAY

Cultivating conviviality within mobility, sociality and daily living

Deniz Neriman Duru

Introduction

Workaway.info is a social networking site similar to CouchSurfing; the digital platform is a place, where hosts and guests create a reciprocal barter/exchange of unpaid labour/service in return for free lodging and food. The cost of mobility is shared by the hosts and the guests, whereby the travel costs are paid by the guests and the accommodation and food costs are paid by the host. The guests offer a maximum of five hours of their time for five days a week, in the form of language exchange, babysitting, pet sitting, help in cooking and housework, farming, creating sustainable eco gardens – in short, anything that the hosts are willing to receive in return. The guests vary from being backpackers, who would like to travel in different countries, students in their gap years, people who are looking to settle in a new country and would like to test the labour market options, people, who would like to have a change of career or get experiences abroad to improve their career, to retired people, as well as couples or friends, who travel together. Hosts and guests have to negotiate employee-employer friendship relations, boundaries of private and work life, and where they both share daily experiences, and enter into labour and cultural exchange.

Workaway invites a spectrum of sociability, ranging from the most intensive form – that of conviviality, living together and sharing daily life with the host, and socialising within the host's social networks, such as family, friends and other Workawayers staying with the host – to sociality of meeting new people via different social media platforms, reactivating old ties as well as meeting new people face-to-face. Experiences of sociality and conviviality are shared through personal and shared media use (e.g. WhatsApp picture and video exchanges, watching films and listening together to music), sometimes posted on social media and/or shared privately with family and friends abroad/left behind.

As Workawayers look for new experiences through this digital platform and app, I coin the term "convivial media experiences" and explore the ways in which they live together, and socialise, when they are on the move. Building on Polson's (2018) place-based apps and the travellers' longing for "authentic experiences", Hill's (2019: 34) "roaming audiences" and their media experiences, and my own work on conviviality (Duru, 2015; Duru and Trenz, 2017), in this chapter, I seek to explore in what ways Workaway provides opportunities and constraints for forms of conviviality, sociality, mobility and cultural exchange. The key argument in this chapter is that conviviality within Workaway experiences underscores collective social relations; by using the lens of conviviality, we can analyse collaboration, and in particular trace the ways people collectively manage trust and tensions in labour relations, cultural exchange and daily living.

In the following, I first give a brief account of the socio-economic and cultural context of Workaway and the digital structure of the site. Second, I describe how I think through the meaning of mobile sociality in relation to experiences of conviviality with empirical examples. By doing that, I explore opportunities and constraints of Workaway experiences regarding living with racial, gender, class and religious differences, and managing these differences within a convivial media and home environment.

Researching Workaway

This research builds on ethnography, which is divided into three methodological components: a) unobtrusive methods that analyse the digital content and infrastructures found on Workaway.info via textual analysis of the social networking site (e.g. guidelines for hosts and guests, blogs, videos) and Forum entries on Reddit, where there are critical discussions about Workaway experiences; b) ten video and face-to-face interviews[1] with hosts and guests; and c) 'hosting' as an ethnographic method (19 months between April 2018 and November 2019).

My motivation, both academic and personal, in using Workaway as a host was three-fold. As a new mum, a foreigner –a Turkish woman married to an Italian, living in Denmark and commuting to Sweden for work – with a husband whose work is demanding and with no families to give a hand, it was sometimes difficult to be on top of everything. We thought that any kind of help, for instance to pick up the child from day care when I cannot get back on time, would be very appreciated. We first had a babysitter who helped us, but I needed to keep meeting interesting people, to talk to them, to know about millions of ways of being a human, to talk and feel similar and different at the same time. As an anthropologist and media scholar, in love with diversity, languages, mobility, digital media research and long-term fieldwork, and as a very social person, I was immobile and asocial as a new mum, with a baby attached to my breast, and my mobility was restricted to daily commuting to work. I was sad and concerned for not being able to do long-term fieldwork anymore due to family and work duties. Getting some help, being social and meeting people from diverse backgrounds and wanting to do research let me be a Workaway host and "let the world come to me". I read blogs on the Workaway site, where they post interviews with hosts and guests. A Moroccan host, Mustafa, said "even though I don't travel as much [as] before, the world comes to my door via my guests".[2] Mustafa gets help with housework, cooking and cracking argan fruit, which grows in some parts of Morocco, Brazil and Israel. He is a masseur and makes and uses his own argan oil for massage. He takes his guests to his family and friends, where they eat, talk and through hospitality, they get to know each other. A digital platform and app like Workaway, make immobile people, such as those, who do not have the 'correct' passports or means to travel, to be 'mobile' in the ways in which the world comes to them.

Contextualising Workaway: digital structure, affordances and how to choose guests and hosts

Workaway makes it possible for people from different socio-economic, ethnic and religious backgrounds, who live in different parts of the world to share convivial media experiences, which I refer to in this chapter, as the experiences people share through finding each other through digital platforms, based on their needs and mutual interests, conversing and interacting digitally and offline, entering into any kind of socio-cultural and labour exchange, and sharing everyday life.

The mobilities enhanced by Workaway show variety based on the motivations, length, nature of experiences and the Workawayers' age. Some of my informants/guests did "discovery travel" (Urry, 2007: 10), doing slow tourism (Fullagar et al., 2012), or voluntourism (Lyons et al., 2012; Wearing, 2001), to learn new things. Some wanted to start a new life as a new migrant in a different country, such as in Denmark, before having a family or before having other commitments. Some kept moving from one country to another. Even though there are no statistics available about the demographics of Workaway users, from Workaway blogs, Reddit platforms, my interviewees and informants, I got to know that those, who travel with Workaway come also from less wealthy, less privileged and less educated backgrounds from different parts of the world. This goes in contrast with the fact that geo-social mobility and geo-social media (meet-up groups, expat social networking sites) privileges the global middle class with higher education (Polson, 2016; Jansson, 2018). In voluntourism, the recipient is seen to be inferior; travellers usually come from the Global North and travel to the Global South, to make a difference (Lyons et al., 2012; Sin, 2009). Nonetheless, some Workawayers have low incomes/budgets; hence, many come from the Global South and those who host/receive the help can be highly educated and wealthy, and from the Global North. Hosts and guests can also be equal in terms of their socio-economic and educational backgrounds, or regardless of their socio-economic background, people with similar dreams and ideology can meet as hosts and guests and work for an eco-project. For example, Janka (Central European), who I interviewed, was committed to promoting sustainability, a greener environment, a world that is based on voluntarism and a sharing economy without any money exchange, where one can eat and grow organic food and live a 'natural' life. Janka eats raw vegan food, does dumpster diving, hitchhikes wherever she travels and tries to avoid any money transactions as much as she can. For instance, she did Workaway in a sustainable eco-farm in Denmark, where she could live the way she wanted.

Regarding the digital structure of the site, hosts and guests log in separately. There is already a structural asymmetry due to a high demand of Workawayers but less supply of hosts. While guests pay €39 a year for a single guest account or €49 for a couple account (travel with partner or a friend), it is free for the hosts. The guest fills in his/her skills (e.g. languages spoken, if they hold a driver's licence, babysitting, elderly care, house- and pet-sitting, cleaning, cooking, hotel management) by ticking from the list of skills/help. If they have something special such as teaching yoga, or an instrument or something that is not on the list, they can write it in their profile description. The host fills in the kind of help needed, information on their lifestyle and accommodations offered. Guests and hosts can make a search through skills, help needed and location. Guests can search for hosts without a subscription, but cannot contact hosts without paying the fee. The site is owned by Ven Ltd., an e-commerce company, and the subscription fees cover most of the running costs of this digital platform.

While Workaway provides these opportunities for people from different socio-economic, educational and national backgrounds to live and socialise together, it is still bounded by how open hosts and guests are in accepting each other, by border control regimes and the ways in which the site is structured in terms of providing feedback and reviews for the hosts and guests. Once guests and hosts make a search through skills and location, they contact each other. Usually, guests contact the host first. Some hosts only reply to requests that they are interested in. In order to encourage hosts to reply, Workaway has a rating system based on hosts' replies and feedback received. The more the host replies, the higher his/her ratings become, and goes on top of the host list. Similar to Airbnb, there is a host calendar, where the host marks availability. The guests write in their profile the time period of travel and country of destination. A video talk between hosts and guests is recommended by Workaway prior to choosing, for a better guest-host match, to agree on the dates, on the type of work/exchange, motivations and expectations from both sides.

In order to enhance security, safety and trust, the hosts and the guests rate and leave a feedback to each other and can report any misconduct or uncomfortable situation. Workaway moderators check feedback before they are posted, and negative feedback is not shown publicly. If the host or the guest gets two negative feedbacks or were reported to have engaged in misconduct, their accounts get removed. Thus, both guests and hosts tend to leave mostly positive feedback (the average rating is 4.4/5). If one has an average or mediocre experience, then both hosts and guests tend not to leave any rating/feedback. The lack of reviews signals a poor experience, so guests and hosts check the reply rate and reviews to make a decision; nonetheless, as almost all the feedback is positive, this draws an unrealistic picture of the guest or the host. This grey zone of no reviews or writing/publishing only positive reviews hides reality and lacks transparency. Thus, some Workawayers turn to other platforms such as Reddit to describe uncomfortable situations to get advice on what they should do.

Another non-transparent aspect of choosing is that with such feedback, you can see who is chosen and what kind of experience they had, but you cannot see why one does not choose a particular person. Their choice is affected by the profile descriptions, where they can see dietary concerns, lifestyle and what is meant by the work exchange between the guests and hosts, and by the video calls/messages exchanged. Nonetheless race, gender, ethnicity and religious affiliation, which might affect people's choices, do not need to be mentioned in the profile. For instance, one guest (Central/Eastern European woman in her twenties), Maria, whom I interviewed, did a Workaway in Northern Europe, but she did not want to live with Muslims or Jewish people. A host from Israel approached her to ask whether she would be willing to stay and help with childcare. Maria narrated:

> I really can't – I really can't imagine living with someone who has some any other religion, or something like that. For example, when I was looking for the families, I also got a message from one woman, she lives in Israel, and I was quite worried, like I told her that I am afraid, that I don't want to go to Israel because I am afraid that I should maybe my umm, well [hesitation] that I would adapt myself to their life, or I don't know how to say it, that I think that I am not, I am not religious and I can't imagine, like, living with her two or three months. . . . I don't hate Jews. I have no problem with them, but only for a holiday but not for like for six weeks, living with them. I don't know what to expect, you know. I wanted to be sure that I will be in Europe, that I will live – that I will be living with people with the same thinking as me. I would like to know some Jewish (people), why not, but not Muslims and so on. For six weeks, I did not want to pray every night or morning or do stuff like this. You know it is good to see it, but not to live it every day.

I opened up the conversation about living with ethnic, religious and cultural differences by asking her whether she has friends from different religions or whether she has her views from hearing or readings the news. She then continued:

> No, I don't know about any (any Muslim or Jewish people/friends). I just know that people hate them, the Muslims. All we hear is that they are causing problems in Europe. That is all we hear. We don't hear anything good about them. Maybe they do good things, but all we hear is that they are sexually abusing girls and women, and we know nothing good about them.

As shown in this example, people might not choose a guest or a host due to not being comfortable living with people from a particular religion, cultural, ethnic or national background. This also shows that any kind of prejudice or concerns people might have for one another is also reflected and seen in the ways in which people choose a Workaway host or a guest. The case of domestic labour and childcare requires convivial, social and labour relations, where private life and work life is juxtaposed, where one lives with the host. If the guest (or the host) has some strong prejudices, this might already impact their choice. Maria, in fact, had communicated with the Israeli woman about her concern of living with Jewish people and that the dates suggested did not fit in her calendar. Nonetheless, many people might just not reply back to requests, and the reason of not having a reply seems like a black box, ranging from being busy to reply, to not being available or not liking how the person simply looks like, and so on.

Unlike Grindr, where user profiles are racialised and ethnicised through a drop-down menu (Shield, 2019), there is no categorisation on the Workaway profile entries regarding gender, race, ethnicity, sexual orientation, educational level or religion. This implies that Workaway would like to promote that selection of people should be based on matching hosts and guests with the right skills and type of help/exchange and that gender, race, sexual orientation, education level, religion or ethnicity should not matter or play a negative role in this selection. However, the profile photos and/or video call done prior to choosing reveal the race and physical features, and some Workawayers, who did not get any hosting offers speculate gender and race as the underlying reasons. As there is no space for discussion on Workaway, where the platform tend to hide negativity, potential guests and hosts choose other platforms (e.g. Reddit) to discuss issues of gender, race and physical appearance in being chosen.

Based on the textual analysis of Workaway posts on Reddit, young white females are perceived to have an advantage, especially for hosts, who expect domestic work of cooking, cleaning and childcare. White men complained to be discriminated against compared with women. A few black women posted that hosts were racists in the ways in which they either did not reply to their messages, or replied that they are not available anymore even when they had availability on their calendar. As a reply to these complaints, some black female Workawayers wrote that they were hosted as guests and had very good experiences to justify that race did not have a negative impact. Other Workaway hosts replied that they accepted people from all races, as long as they could see that they are fit for the work/exchange/help expected from the host. One man wrote that as he was tall and overweight, he was rejected to be hosted in an Asian country. The host said that he would not be able to fit the guest in his house. Some hosts in Asia replied back to explain that in Asian countries, houses and furniture are mostly made for small and short people, by stating that the single bed is the same size as a child bed in the United States or Europe, and will not hold the weight of the Workaway applicant, and furthermore the toilet seat would break.

On Reddit, there were also many posts from people who did not get any response from any hosts yet, so it is a common problem for the guests to send even 50 invitations and not hear anything back. Hosts wrote (we had the same experience as well) that many guests send a copy-and-paste message without checking what hosts expect and/or calendar availability, that there is a big surplus of guests in comparison to available hosts (for instance, we could only host eight guests out of 200 requests in 19 months), or that it takes a lot of time to choose and plan for hosting, so that it is not possible to reply to all messages. Hosts gave general advice to those white men and black women to have a complete profile and to write well-thought-out messages to show they fit to the hosts' descriptions and lifestyle, to state that it is not a matter of race or gender but of Workawayers' approach, personality, having the right experience and skills, and matching the hosts' availability.

On Reddit, while almost all the complaints come from the guests who were not chosen, critiquing the power of the host in the selection process, Maria's quote earlier in this section challenges this hierarchy, where guests also make rejection choices. There is an inequality in the ways in which hosts have more power over choosing and accepting guests; nonetheless, if the host is not deemed to be a good host by the guest, they can leave. The last-minute function on the Workaway site is meant for this, and is to be used by both guests and hosts, when one needs to immediately find a host or a guest.

In sum, Workaway creates many opportunities for people from different socio-economic and cultural backgrounds to experience conviviality through labour and cultural exchange. Nonetheless, hosts and guests can have prejudices for people from different racial and religious backgrounds or prefer one gender over another. The digital structure, the rating and reviewing system, the non-transparent negativity and the positive content of the platform make it difficult for guests and hosts to get a 'real' sense of what Workaway experiences can be. After having contextualised the ways in which Workaway works as a digital platform, its structure and affordances, how guests and hosts select and use the platform, in the next section, I conceptualise and illustrate my take on mobile sociality through empirical findings from Workaway.

Conviviality within mobile sociality

I build on Wittel's (2001) network sociality to distinguish the sociality and the conviviality aspects of mobile sociality. The sociality described by Wittel (2001: 71) is individualised, embedded in technology, ephemeral but intense, networked by the integration of play and work and based on getting the necessary information in a rather short amount of time. Network sociality is not based on a common and shared history; hence, people exchange information but do not share experiences (Wittel, 2001: 65). The conviviality within Workaway experiences underscores collective social relations, highlighting collaboration rather than individualism, in particular managing daily tensions, helping each other and responding to the needs of the people living in the home. While people cooperate and engage in labour, and affective and cultural exchange, they build and rely on trust, reciprocity and constructing shared narratives.

In difference to the meaning of 'conviviality' in English as "having a good time in the company of others: i.e. a jovial, festive, companionable state, occasion or atmosphere" (Overing and Passes, 2000: xiii), my use of the term comes from the Spanish *convivencia*, which means living together, having "a joint/shared life" (Overing and Passes, 2000; Erickson, 2011). Conviviality includes not only harmonious times, but also tensions, conflicts and disputes (Overing and Passes, 2000; Karner and Parker, 2011; Duru, 2015, 2016) rooted in the sharing of a communal life, such as fights between flatmates, or arguments between a husband and wife over not cleaning up. I describe conviviality as a lived practice, shared ways of living, a sense of belonging in a place and grassroots' management of diversity, in the ways in which people solve their daily tensions and conflicts, as well as show solidarity and cooperate based on shared needs, lifestyles and interests. Hence, conviviality describes the multicultural/plural society not "as a patchwork of five and ten cultural identities" (Baumann, 1999), neither as "mosaics puts together" (Eriksen, 2001), nor "as billiard balls" (Wolf, 2010), but as intertwined patterns, webs and networks in dynamic relations, constantly forming and reforming clusters based on needs and interests, within a socio-economic and political context.

My approach does not necessarily emphasise differences between ethnic or religious groups, but how people from different backgrounds can form relations through digital media use and share the everyday life. While studies on conviviality emphasised living togetherness of diverse groups in the locality and hence is static (e.g. Duru, 2015, 2016; Duru and Trenz, 2017; Erickson, 2011; Karner and Parker, 2011), mobile sociality through convivial media experiences emphasises transnational mobility and place-making of new mobile citizens. Mobile sociality is located in the sociability of individuals from diverse backgrounds, in their encounters in transnational places, and their networks of interactions and relationships when they move from one place to another where mobility, sociality and conviviality are supported by digital platforms, such as Workaway, without people being dependent on them. The quality and strength of relationships depend on the shared experiences cultivated together during socialisation and conviviality – living and working together.

Sociality and mobility

The Workawayer can travel alone or have a travel buddy (found through the app) who is only an accompanying person; hence, *sociality* takes place during the *movement* and ends at the end of *mobility*. The Workawayer can also travel with a partner, sibling or friend (joint account) and requests the hosts to accommodate the travelling couple. Some further sociality is initiated and/ or reactivated through digital media use. By using social media, location-based apps and digital emplacement technologies (Polson, 2015, 2016, 2018; Duru and Trenz, 2016, 2017) such as meet-ups, Facebook groups and events, and Airbnb's 'experiences' function, travellers make new friends and socialise in the place where they have newly arrived. The Workaway app has a GPS meet-up function, where Workawayers, who are nearby can write messages to each other to meet. Our German Workawayer, Ulrike, met a few other Workawayers through the Workaway app, and they met for coffee and did sightseeing in Copenhagen.

Workawayers also meet friends from childhood, high school, previous university experience or work whom the Workawayer has not seen for some time. They are connected via Facebook and/or Instagram, happen to be in the same city and can reactivate their old ties. Our Workawayers Holly (New Zealander), Carmen and Pilar (Latin American) met up with their high school friends. Mary (American) met her baker colleague from New York, who lived in Copenhagen, for lunch, for a movie night, for Halloween and Christmas parties and for sightseeing. Carmen (Latin America) had two high school friends, who were on their Danish working holiday visa for a year (very common for Latin Americans, who use Workaway), and did part-time cleaning jobs together. Momo (Japanese) went to Meet-up gatherings (mostly attended by foreigners) to practice Danish.

Not all mobile sociality is initiated through digital media. For example, Jason, a New Zealander Workawayer in Latin America whom I interviewed, met his Swedish girlfriend during his Workaway trip, then kept in touch via social media and eventually followed her to Sweden, where he currently lives and works. Satoko and Momo, who we hosted, by coincidence, worked (at different times) in the same Japanese restaurants (part-time) while living with us and they hung out with their work friends, who also came to our house for an evening of cooking, baking, eating and playing. Mary, our American Workawayer, got along very well with her colleagues at the bakery where she volunteered. She wanted to learn how to bake Danish pastries and bread, and as she worked there for free, her voluntary work was very appreciated and she formed friendships. In her relationships with her colleagues, in the interviews I conducted with other Workawayers and in the relationships we had with our Workawayers, the more people give and do for each other and show help and solidarity, the stronger ties of friendship were formed and sociality turns into conviviality.

Conviviality and mobility

The Workawayer (the guest) can travel alone but then joins a host (or host family). For instance, we picked up most of our Workawayers from the airport, train or bus station, welcomed them with a meal, showed the house, gave the wi-fi password so that they could let their family know that they arrived safely, gave our phone numbers to their family members if/whenever they might need to call. We were a host family, and for some, a stepping stone in their Copenhagen life. They stayed with us until they found a stable job, or signed up for a course/education, or while they volunteered in Copenhagen to gather experience for their career. We shared any local knowledge as much as possible – for example, which SIM card to buy for their mobile phone, how to travel, whether to buy tickets or travel cards, where to visit, how to seek a job – as well as loaned our bikes and clothes when theirs did not match Copenhagen's cold weather. So, when they started their Copenhagen life, they were not alone.

Erika Polson (2018: 160) refers to a quote from an Airbnb advertisement clip "Don't go to Paris. Don't tour Paris. And please, don't *do* Paris. Wherever you go, – don't *go* there, LIVE there". She draws attention to the change in tourism and travel experiences. While tourism used to aim to be an escape from the everyday life, the ordinary, and to seek something different, even 'exotic', the trend of the last couple of decades has been to seek 'authentic' experiences that build on 'feeling' like a *local* and experiencing the everyday life, like the locals do (Hoeller, 2015: 1 cited in Polson, 2018: 168). Hence, sharing daily life, and experiencing the life of others through living and working together – conviviality – forms the most intensive form of the mobile sociality experience between hosts and guests. With Workaway, the authentic experience relates to conviviality, in the ways in which people share the intricate, intimate daily moments, and daily joys and tensions. Me running in my nightgown after our toddler son, Antonio, with my hair electricised up in the air, stepping barefoot on Legos, grapes and blueberries thrown away by Antonio, who runs around the kitchen table with a cheeky smile, and me checking on my phone the bus and metro connections to the train is as authentic as it gets for our Workawayer, who tries to catch and block Antonio by holding in her hands Antonio's jumper, jacket and boots in order to dress him to take him to the daycare.

Workaway's aim is to create a win-win situation, where the relationship between hosts and guests is built on reciprocity, where both should gain/receive something from the exchange. The benefit on the one hand is economic, promoting sharing economy, and on the other hand, both learn from each other, share experiences and engage in cultural, affective and labour exchange. While Bianchi (2016) and Laesser et al. (2009) draw attention to the individualism, independence and autonomy of the solo traveller, Workaway users depend on others. Mobile sociality through conviviality constructs almost a symbiotic relationship between guests and hosts, who need to adapt to each other's needs and work together to improve or contribute to each other's lives.

For instance, Emily, a British young woman in her early thirties, had two Workaway experiences in France. In the UK, she teaches English as a second language to adults and wanted some experience to teach English to children. She stayed with a family, where the host mother taught English to French children. When I asked her to reflect on her Workaway experience, she said:

> Workaway is good for solo travellers. Most of my friends go on holiday with other friends. I always go on holiday on my own and it is always hard to book a single room. You turn up at the reception and they say "where is your husband?" "No, I am single and blah blah blah". But doing Workaway, you can be a single traveller and not be on your own. . . . And sometimes you can go on holiday and not to speak to anyone. So it was completely the opposite to suddenly be with this group of people and they want to share their lives with you, so I think that for solo travellers it is fantastic . . . they (hosts) wanted to share their lives with other people, and they were as interested in what I have to tell them and show them about England

> and the films I like, books I read, and you just learn so much about each other. It is not about the work. It is not even about the travel. It is a cultural exchange. It is a really good thing – really good thing.

Emily emphasises the cultural exchange rather than work. She draws attention to being part of someone else's daily life and routine, and how she as the guest, and them as the host entered into a mutual, reciprocal relationship of conviviality and shared everyday experiences. There is no written contract between the host and the guest, or a binding agreement. The guest is free to leave before an agreed date when there is a change of plan. This 'freedom to exit/leave' is also a relief for hosts, who host complete strangers in their house. If the guest cannot meet the needs of the host, or the host is not happy with the guest, they could ask for the guests to leave before the agreed date. This freedom to exit, in a way, creates a better balance of power relationship and diminishes the inequality between hosts and guests, because it encourages both sides to make it work, make an effort every day, and not to take each other for granted. Here in Emily's narrative, the feeling of gratitude for the guest to stay for free and eat nice food, and for the host to be helped for free, learning from each other, listening to people's life stories, form the bonding and contributes to having a pleasant experience. She learnt a lot from the ways in which the host mother taught in class. She enjoyed French cheese and wine, and stayed in a very beautiful room. The cultural exchange she referred to included not only sharing each other's tastes in music, books, film and food, but also learning from that family's daily life and routine. In return, she helped the single host mother to prepare the girls to go to school, to do the shopping and to help in the household cooking and cleaning. When the washing machine broke, or the girls protested having to eat their vegetables, when the mother was stressed in trying to manage work and family life, Emily was there to help as an extra pair of hands. Emily highlighted the conviviality aspect of the experience, where you share the daily life with all its good and challenging ways and how the bad or negative moments play a role in the bonding. In their case, the balance and the reciprocal exchange worked well and turned into an ideal Workaway experience.

Nonetheless, not having a contract and verbal agreements on labour exchange can turn into a negative situation where the host 'uses' Workawayers. When I asked my interviewees how the Workawayers would define Workaway experiences, they used cultural exchange, volunteer job exchange, help or work, or a combination of these terms based on the relationship they had with the host. Every Workaway experience is formed of different combinations of sociality and conviviality. In some Workaways, there might be a limited contact/exchange between the host, and hence the guest and the host do not share conviviality and the exchange is limited to an employee-employer relationship around getting labour done – and sociality and convivial experiences must be looked for somewhere else.

Marko and Eva (whom I interviewed) complained about doing more hours than they were supposed to and with the host not being social and by treating them as "labourers". In their cases, they used the term "work" for the Workaway exchange, because it felt like working rather than a cultural exchange or help, which would have been a more reciprocally shared experience between the host and the Workawayer. Marko, a Southeastern European male in his late twenties, did Workaway in Spain at a hostel reception desk. He had limited contact with the host, so he hung out and socialised with other Workawayers and hostel guests after his reception duty. Eva, a German female Workawayer in her early twenties, did Workaway at a Swedish farm, where she worked much more than the 25 hours maximum per week. The hosting lady had more than 50 animals and was in fact relying on free labour by hosting Workawayers to run a farm without offering any kind of cultural or social exchange with the guests. Eva complained about the fact that the host did not say a proper thank you at the end of the day and put the Workawayers under pressure to do as much work as possible. The host always left positive feedback for the guests in order to receive positive feedback; nonetheless, when Eva checked Workaway again, the host's profile had been taken down. So, some of the Workawayers must have given negative feedback and reported extra free labour. This kind of abuse/misconduct is true also for guests. When I was browsing the Workaway host profiles and reading the feedback given and received by hosts and guests, one host, (who left five stars to all her guests) left three stars for one guest, wrote in her feedback that "I cannot leave a good feedback because X took our car and went away and never came back".[3]

Another Workawayer whom I interviewed, Daisy, an American woman in her late twenties/early thirties, describes Workaway as a "volunteer job exchange and a cultural exchange". She also had a limited relationship with her host in Morocco, with whom she did not share a convivial experience but only labour exchange in return for free lodging. The host needed teachers to teach English at the school, and Daisy needed a place to stay. In fact, the place where she stayed was still being constructed. Hence, she had created her own combination of convivial experience, where daily struggles and challenges created bonding and close relationships between the Workawayers, whom were also in the same situation.

> This Workaway was weird but fun and I have met a bunch of wonderful people. . . . It [the room] was not a regular room and it was a very ugly-shaped room; there was an elevator in it, that did not yet work, so one of the beds were in front of it. At the time, when I got there, there was no kitchen, there was no refrigerator, there was no hotplate, but there was a sink. . . . This Workaway did not provide food. . . . We had a lot of fun. It felt a little bit like, when you are a kid and you are playing survivor and you pretend that you are lost in the woods and to make your own (laughs). We would cut bottles of water in half and use the bottom half as bowls and store stuff in the refrigerator. . . . There was definitely a strong connection with the volunteers, and it was also good fun to get to know the teachers.

We could see here that the host did not provide an acceptable lodging arrangement and treated the guests like employees without engaging in a socio-cultural exchange. Nonetheless, Daisy made her own combination of mobile sociality, through building convivial experiences with other Workawayers, teachers and also the locals. She narrated:

> Morocco is an Islamic country. I had never travelled to one before, but I have close friends who are practisers of Islam, so I knew a little bit, but I definitely learnt more, because the teachers in between classes talked with us; they wanted to know about us, [and] we wanted to know about them. One of them taught me how to tie a *nikab*, if I ever needed to, which I never felt the need, but she taught me how to tie it properly, which was a sharing of cultures that was really lovely and it was an interesting way, different than my other Workaways but still an interesting way to learn about the country that I was in and its practices and its people and I got very used to the call to prayer. We got very used to it. There was a mosque nearby. . . . We would sometimes open the door and we could hear the sounds of the *souk*, which is like an open market down below, that was right next to our building. . . . I got used to bargaining for vegetables and mushrooms and bread, and it was a lot fun. We learnt to cook things; it was a great experience.

In her conversations and during the time she spent with the local teachers, she learned about Moroccan culture and practicing Islam. In her daily explorations in the neighbourhood and the souk, observing and participating in the daily life while she shopped, walked around and conversed with the people she met, she got to learn about religion, culture and daily life in Morocco. At the end, she was proud to have learnt a cultural practice such as bargaining, when she acted/behaved like a Moroccan shopper in the market. Hence, the social relationships she has built, turned into convivial experiences, where she internalised the local culture and practices.

Sharing emotions with convivial media experiences

Similar to Hill's (2019: 34) "roaming audiences", Workawayers are roaming people in the literal sense; most of them have a foreign SIM card and might not buy a local SIM card, and rely on the hosts' or cafés' wi-fi. Workawayers roam in between physical borders and are bounded by their passports, residence and visa permits. For instance, while European and Western travellers have fewer restrictions, non-European travellers face restrictions. Those non-Europeans with the 'good' passports (e.g. Americans, Latin Americans, New Zelanders and Australians, Japanese) can stay up to 90 days in a 180-day period in the Schengen area. They can apply for a working holiday visa, stay up to a year, travel and work in a European country (if they are younger than 30).

Hill's (2019: 34) "roaming audiences" watch and listen to media productions on their phones, TVs and tablets, via social media, at different times and places (at work breaks or home, during commutes), streaming or watching/listening live, or consuming previously downloaded content, being connected to wi-fi or 3G or also offline. Workawayers also roam in between digital boundaries and media affordances. Workawayers' media experiences are provided and enhanced by all sorts of media, yet sometimes they are restricted to how much they can afford during their mobility (how much they can physically carry) and what is available to them – and thus, they roam and negotiate digital and material boundaries. Hill writes: "Roaming signifies the affective, temporal and place-making practices of audiences" (Hill, 2019: 34). In Workawayers' media travels and roaming, they form affective relations with the people with whom they share convivial media experiences. These include material objects, as well as content shared such as books and musical instruments borrowed; wi-fi passwords and internet allowances shared; mobile phone messages, pictures and calls exchanged; films watched on the smart TV or tablets; and songs sung and dances performed together while Spotify, iTunes, or YouTube is on.

We built very strong and emotional bonds with Satoko, our Japanese Workawayer, a kindergarden teacher. She wanted to have experience with small children, was fascinated and curious about the Danish preschool system, and wanted to take art courses to teach art to her pupils. Before taking the art course in Denmark (højskole), she stayed with us for three and a half months. She was like an anthropologist, who wanted to be part of our daily life as much as possible, and we welcomed her anywhere: children's birthday parties, meals, sight-seeing in Denmark and my workplace in Sweden, and she even travelled with us to Turkey and Sicily for one month. We sailed together in Marmaris, where she learnt to swim; she went on a horse in Sicily, met our extended families and we ate fish, pizza and pasta, and learned cooking Turkish food (different types of dolma, grape leaves filled with rice and spices) from my mother. She used to take photos of all the food she ate in our company, whether at home or at restaurants; she took many photos and videos of what we did together, especially of our son and almost every day we exchanged on WhatsApp all the photos and videos we took (she took many more than us). When I asked her what she did with all these photos and videos, whether she posts them on social media (we became friends on Instagram after a couple of months when she started living with us) or sends them to her family and friends, she answered that she took them as memories.

On the day she left, she showed us the video she made from these photos and videos, and we watched them all together. She said she almost cried while making the video, and we all got very emotional while watching it. From time to time, when I miss Satoko and our time together, I watch the video she made for us. In the video, we see how Antonio grew up in these months, the places we visited, activities we did and how much we have grown into each other. The video was a priceless thank you for us. Satoko told us that the Japanese song at the background of the video means "we will see each other". After she moved out to do her art course in Jutland, she kept visiting us – and even now that she is back in Japan, we do video calls; she sends us photos of the Turkish and Sicilian food she cooks there, and we share Antonio's photos on WhatsApp. The emotional and affective relationship we have formed together is still kept alive through the affordances of digital media and all the pictures and videos we share.

Concluding remarks

As seen in the empirical examples from my ethnographic research, interviews and textual analysis on and about Workaway, the sociality and convivial experiences of mobile sociality are very unique, ranging from negative to positive ones, based on the dynamics created between hosts and guests.

Workaway can offer a lot of opportunities for mobile sociality: helping/aiding people to be mobile, to move across spaces, to share their everyday life and build strong, long-lasting relationships. Even those who would not have the means to do so because of financial restrictions, or those who cannot be mobile due to physical constraints or family/work duties, can reach each other through a digital platform, meet and live together.

Nonetheless, Workaway is not and cannot be a remedy for mutual understanding, intercultural communication and tolerance. It is still bounded by how much and in what ways guests and hosts are willing to learn about others, and whether to accept, embrace or reject people's differences. Workaway, reminds me of Derrida's (2000: 75) "pas d'hospitalité" which means at the same time, no hospitality and step of hospitality. If the hosts and guests are willing to take the step of hospitality, then they make their own unique relationship. However, if they are not willing to try out new things and open themselves, then there is no hospitality. It all starts with taking a step.

Notes

1 It was extremely difficult to find interviewees. In my Workaway profile, I had written that I am doing a research on Workaway and hence the first thing that the potential Workawayers see is that the hosting is a part of the research, and in our Skype interview, I got their permission to use our experience as ethnographic data for my research. One interviewee found me through the Workaway profile and also referred me to another Workawayer friend, whom I interviewed. Among the 200 Workawayer users, who got in touch with us and whom we could not accommodate, some of them had written in their initial message that they want to help my research in any way. Nonetheless, when we told them that we could not accommodate them, they stopped replying. I then recruited some interviewees from social media platforms, found some more through snowballing and finally managed to do ten interviews in almost two years.

2 https://workawayblog.com/marrakech-learn-moroccan-culture-build-bridges-create-world-travellers-home/.

3 I have paraphrased, shortened and taken out the details of the quote, so that the host and the guest cannot be traced back on the website, for ethical and anonymity purposes.

References

Baumann, G. (1999) *The Multicultural Riddle: Rethinking National, Ethnic, and Religious Identities*. London: Routledge.

Bianchi, C. (2016) "Solo holiday travellers: Motivators and drivers of satisfaction and dissatisfaction", *International Journal of Tourism Research*, 18(2): 197–208. doi:10.1002/jtr.2049.

Derrida, J. (2000) *Of Hospitality, Cultural Memory in the Present*. Stanford: Stanford University Press.

Duru, D. N. (2015) "From mosaic to Ebru: Conviviality in multi-ethnic, multi-faith Burgazadasi, Istanbul", *South European Society & Politics*, 20(2): 243–263. doi:10.1080/13608746.2015.1047080.

Duru, D. N. (2016) "Memory, conviviality, and coexistence: Negotiating class differences in Burgazadasi, Istanbul", in R. Bryant (ed.) *Post-Ottoman Coexistence: Sharing Space in the Shadow of Conflict*. New York: Berghahn Books, 159–179.

Duru, D. N. and Trenz, H. J. (2016) "Diversity in the virtual sphere: Social media as a platform for transnational encounters", in *Integration, Diversity and the Making of a European Public Sphere*. Cheltenham: Edward Elgar Publishing, 94–115.

Duru, D. N. and Trenz, H. J. (2017) "From diversity to conviviality: Intra-EU mobility and international migration to Denmark in times of economic recession", *Journal of Ethnic & Migration Studies*, 43(4): 613–632. doi:10.1080/1369183X.2016.1249049.

Erickson, B. (2011) "Utopian virtues: Muslim neighbors, ritual sociality, and the politics of convivència", *American Ethnologist*, 38(1): 114–131. doi:10.1111/j.1548-1425.2010.01296.x.

Eriksen, T. H. (2001) "Between universalism and relativism: A critique of the UNESCO concept of culture", in J. K. Cowan, M. B. Dembour and R. A. Wilson (eds.) *Culture and Rights: Anthropological Perspectives*. Cambridge: Cambridge University Press, 127–148.

Fullagar, S., Markwell, K. and Wilson, E. (2012) *Slow Tourism: Experiences and Mobilities,* Vol. 54. Bristol: Channel View Publications.

Hill, A. (2019) *Media Experiences: Engaging with Drama and Reality Television*. London: Routledge.

Hoeller, S. C. (2015) *10 Travel Apps That Will Make You Feel Like a Local*. Business Insider, September. Available at: https://www.businessinsider.com/10-travel-apps-that-will-make-you-feel-like-a-local-2015-8?r=US&IR=T.

Jansson, A. (2018) *Mediatization and Mobile Lives: A Critical Approach*. London: Routledge.

Karner, C. and Parker, D. (2011) "Conviviality and conflict: Pluralism, resilience and hope in inner-city Birmingham", *Journal of Ethnic and Migration Studies*, 37(3): 355–372. doi:10.1080/1369183X.2011.526776.

Laesser, C., Beritelli, P. and Bieger, T. (2009) "Solo travel: Explorative insights from a mature market (Switzerland)", *Journal of Vacation Marketing*, 15(3): 217–227. doi:10.1177/1356766709104268.

Lyons, K., Hanley, J., Wearing, S. and Neil. J. (2012) "Gap year volunteer tourism: Myths of global citizenship?" *Annals of Tourism Research*, 39(1): 361–378. doi:10.1016/j.annals.2011.04.016.

Overing, J. and Passes, A. (2000) *The Anthropology of Love and Anger: The Aesthetics of Conviviality in Native Amazonia*. London: Routledge.

Polson, E. (2015) "A gateway to the global city: Mobile place-making practices by expats", *New Media & Society*, 17(4): 629. doi:10.1177/1461444813510135.

Polson, E. (2016) *Privileged Mobilities: Professional Migration, Geo-Social Media, and a New Global Middle Class*. New York: Peter Lang.

Polson, E. (2018) "'Doing' local: Place-based travel apps and the globally networked self", in Z. Papacharissi (ed.) *A Networked Self and Platforms, Stories, Connections*. London: Routledge, 160–174.

Shield, A. D. J. (2019) "'White is a color, Middle Eastern is not a color': Drop-down menus, racial identification, and the weight of labels", in A. D. J. Shield (ed.) *Immigrants on Grindr: Race, Sexuality and Belonging Online*. Cham: Springer International Publishing, 185–225.

Sin, H. L. (2009) "Volunteer tourism – 'involve me and I will learn'?" *Annals of Tourism Research*, 36(3): 480–501. doi:10.1016/j.annals.2009.03.001.

Urry, J. (2007) *Mobilities*. London: Polity.

Wearing, S. (2001) *Volunteer Tourism: Experiences That Make a Difference*. Wallingford: CABI.

Wittel, A. (2001) "Toward a network sociality", *Theory, Culture & Society*, 18(6): 51. doi:10.1177/026327601018006003.

Wolf, E. R. (2010) *Europe and the People Without History*. Los Angeles: University of California Press.

PART IV

Contrary mobile socialities

INTRODUCTION PART IV

Anna Reading

Part IV on contrary mobile socialities builds on the previous parts of the *Handbook* to analyse and provide empirical examples for ways of thinking about mobile socialities that are contrary in various intersectional ways to both mobility and sociality. These include thinking about how mobility includes stuckness, fixity, blockades and jams or immobilities in various forms. It may also include thinking how socialities also involve what seem to be its antithesis – antisociality or asociality, or that which is antagonistic, hostile, disruptive or threatening to the social order, or unpacking the principles of mobility and sociality on which mainstream society may be constituted or biased towards. We ask here how scholars might conceive of and develop analytical frameworks, approaches and conceptualisations to include these other 'contrary' approaches.

My own contribution, Chapter 16, cycles through the different uses and mobilisations of the slogan "I can't breathe" by the Black Lives Matter movement in 2020 and by climate protesters and how the phrase resonates within the contexts of COVID-19 and the pandemic's lockdowns, involving curtailing freedom of association and movement within a liquidity of data and digital media acceleration. I develop a metaphorical analytical approach out of "I can't breathe" to show how we can surface the tensions between immobility and mobility, sociality and antisociality. I draw on ideas from queer studies and autism studies to broaden ideas of sociality to include productive ideas of antisociality; at the same time, I explore how mobilities feed off and have been supported historically and today through rigid and enforced immobilities.

Maren Hartmann in Chapter 17 provides us with another account that enables us importantly to recognise historical trajectories and historicise mobile socialities. Hartmann draws our attention to the understudied but important to dimension of intangible cultural heritage "the wandering years" ('die Walz' or 'Tippelei' in German) in which craftspeople, after they had completed their apprenticeships, would travel, practising their crafts in different places for three years and a day. This is a tradition that was widely practised across Europe for over 700 years and is still practised, although now confined to German-speaking countries, Denmark and France. One of the key rules, however, of today's practices is that the key tool of mobilities – the mobile phone – is forbidden during the craftperson's travels, which in the 21st century (at least before COVID-19) are global, rather than just within Europe. Hartmann thus focuses on the triangulation of media with socialities and mobilities and the ways in which this ancient form of mobilities generates new kinds of socialities, as well as tensions or 'contrary' elements in the present.

Annette Hill's Chaper 19 takes up Roland's Barthes' key idea of transported mobility to explore more broadly how mobile socialities may be also understood within conditions when mobility of people, data and things are blocked and when media breakdown is a feature of everyday life. In this case, Hill uses the example of the "media blackout" within the Swedish forest which results in the need for inhabitants to use various means to obtain wi-fi connection including 'tethering' through mobile devices. In this way, the chapter points us to thinking about other instances in which there are media blind spots or interruptions and how that then requires us to incorporate such unevenness into work on mobile socialities.

Philipp Seuferling in Chapter 19 explores how changing historical practices around media are an integral element of understanding the broader environment of socialities. The chapter provides an important dimension to the book that draws our attention to how we need to historicise mobile socialities – in this case through empirical work on refugee camps and archives in Germany. The chapter resonates here with Hartmann's chapter that points to how we might historicise mobile socialities more generally and ensure that we do not become fixated on contemporaneous articulations of its technologies, practices and manifestations.

Anna Lisa Ramella in Chapter 20 examines socialities of practice with a particular focus on stuckedness and the ways in which this is in productive tension with mobile imaginaries. Ramella uses the example of Kenyan migrant fisherpeople descendants, drawing on rich ethnographic material produced during fieldwork conducted in the Kenyan Rift Valley during 2019–2020. Through the words of the Luo community on the shores of Lake Baringo in northern Kenya, we see how the concept of mobile socialities lends itself to boundary work that surfaces the tensions between an imagined mobility and present day fixities.

Part IV raises a number of important questions for readers of this collection as to how we think about and develop the right kinds of methodological and analytical approaches for exploring mobile socialities. Is there a preferred methodological or analytical approach that is best suited for work within the area of mobile socialities? Should we use pre-existing methodological and analytical approaches, or should we be innovating and developing our own, since perhaps established tools will simply illuminate established knowledge?

What is fascinating about the chapters in this section is the degree to which they demonstrate that there is no proclivity towards a particular method: they use a range of methods from the metaphorical to the ethnographic, from the infrastructural to the archival. What is evident, too, is that these methods are usually mixed together, blended and extended to create new approaches that bring to light new knowledge: Hill, for example, weaves together autoethnography with topography, upheld through a weft of historical infrastructure. Her analytical approach extends semiotics into the realms of meteorology and the non-human. In so doing, the chapter creates its own method and approach exploring the double meaning and practices of the "tethering" of mobile societies.

Part IV also addresses the broader question of the degree to which a particular discipline might lead the way intellectually and theoretically in relation to mobile socialities. Although all the work is concerned with how mobile socialities are mediated, none draw singularly from media studies – or indeed, from its sister discipline of cultural studies. Rather, each chapter is productively and creatively intellectually eclectic in robustly bringing to the problems they explore a range of disciplines from history to geography, anthropology to queer studies, disability studies to folklore studies. In so doing, the chapters in Part IV – as indeed with the collection as a whole – shake up established ways of both doing and seeing the world around us. We see how through researching mobile socialities, this must necessarily draw theoretical sustenance from a truly interdisciplinary approach, which while being anchored in media and cultural studies, none the less is enriched through the anthropological, geographical, mnemological and sociological.

From this, though, we might also ask what other disciplines might be brought to bear to this enquiry. Are there ways in which scholars might draw on psychology or physics, biology or theology, to understand further layers and complexities of mobile socialities? I am reminded here of the words of Mileva Maric, the brilliant physicist and first wife of Albert Einstein, who brought to their scientific work together the importance of continually going beyond established boundaries. In one of her many letters to Einstein she wrote:

> I don't think the structure of the human skull is to be blamed for man's [sic] inability to understand the concept of infinity. He (sic) would certainly be able to understand it if, when young, and while developing his sense of perception, he [sic] were allowed to venture out into the universe rather than being cooped upon earth.
>
> (Maric, 1992: 3)

What the chapters in Part IV reveal is what happens when, as a researcher, you allow yourself to venture out into the full universe of methods, of concepts and approaches rather than being cooped within the confines of a particular discipline or approach. Part IV is an invitation to do just that.

Reference

Maric, M. (1992) "Maric to Einstein: Heidelberg. 20 October 1897", in M. Maric (ed.) *The Love Letters: Albert Einstein*. Princeton: Princeton University Press.

16
'I CAN'T BREATHE'
Metabolising (im)mobile antisocialities

Anna Reading

Figure 16.1 Children's playground, Brockley, London. June 2020.

Source: Copyright Anna Reading

Introduction

'I can't breathe' were the words repeated multiple times by George Floyd, an African American man who was held down, immobilised, choked and killed by a police officer in the US on 25 May 2020. 'I can't breathe' was the phrase politically socialised through #BlackLivesMatter and mobilised through globalised media coverage of demonstrations that erupted after George Floyd's death during the COVID-19 pandemic and its associated lockdowns (Anon, 2020).

'I can't breathe' were also the last words of Eric Garner, who was also subjected to a choke hold by New York City police in July 2014. He said the phrase 11 times before dying of asphyxiation. At least 50 demonstrations took place in the US in response to Eric Garner's death, mobilizing and socializing the phrase 'I can't breathe'. 'I can't breathe' was the phrase circulated through Black Lives Matter (BLM) with the multiple names of other African American men who had died while placed by the police in chokeholds (Anon, 2020).

When a person is immobilised by being placed in a chokehold, if it is a carotid artery restraint, their brain is deprived of oxygen. They fall unconscious, and then they die. Or, if choked through a restriction to their airways, they become asphyxiated, deprived of oxygen and die. When checkpoints, barriers and vigilantes are used to immobilise and prevent the political socialisation of democratic protests such as BLM, it is an attempt to deprive them of political lifeblood, to kill off protests and protestors. 'I can't breathe' also thus signified the choking of the body politic, of democratic process and representation.

In addition and in parallel to the socialisation and mobilisation of 'I can't breathe' by #BlackLivesMatter, 'I can't breathe' was used as a slogan by climate protestors to capture the asphyxiation caused disproportionately within poorer communities in the Global North and in the Global South by air pollution (Simmons and Ahearn, 2017). 'I can't breathe' thus also became a slogan for this dimension of extractive capitalism, signifying a state of global emergency, drawing attention to pollutants in cities from London to Delhi causing breathing difficulties, health problems and – in some cases – death. As Jessica Rapson shows, racism and the climate emergency are structurally intertwined: the oil industry in the US is mapped onto the sites of slavery that produced sugar and cotton (Rapson, 2018).

Being short of breath, and eventually unable to breathe without the aid of a ventilator, is a key symptom of being infected with the COVID-19 virus which attacks human lungs and airways. National and regional lockdowns, stoppages and confinement to homes during the global pandemic in 2020 and into 2021 were used to prevent its spread, along with the use of masks and shields over mouths and noses to filter the air from viral contamination.

This chapter cycles through some of the intersectional mediations of 'I can't breathe' as a starting point to breathe life into an emergent analytical framework for *im*mobile *anti*socialities, which I argue here is not separate from but part of mobile socialities. Drawing on multidisciplinary approaches beyond media studies to create a conceptual framework, the chapter gives air to elements that are *im*mobile and *anti*social. The chapter asks: how do we understand the experience of interruption and amplified displacement that may include frictions – accidental or deliberate – of locative, temporal and social breakage experienced during stoppages of movement which we might conceptualise as immobile socialities? How do we understand what happens to sociality when the possibility of movement of people, but not data or things, is curtailed, as in the COVID-19 lockdown, as result of protests and blockades or as a result of police or security checkpoints on top of the barriers of racist micro-aggressions and deeply inherited structural barriers which we might call immobile antisocialities?

The chapter's first part works outwards from the slogan 'I can't breathe' as inspiration for a metaphorical method to 'metabolise' some thoughtful vitality for the second part that develops a multi-disciplinary conceptual framework for (im)mobile antisocialities. I use the word metabolise and associated metaphorical language of metabolisation because although we might consider metabolisation to be limited to the transformation of food into energy, biologically it denotes a wider process that includes the act of breathing. It is the breath that supplies oxygen that is then transported through the blood into the cells of the body, where it plays a crucial role in the metabolisation of glucose into energy to fuel the body (politic). The waste products that include carbon dioxide produced during metabolisation are carried back to the lungs by the blood from where they are exhaled in our breath. Humans – and, by extension, human societies (and the body politic, we might add) – need to be able to breathe air to create energy through aerobic metabolisation: Glucose + Oxygen → Carbon dioxide + water + energy.

Thus, the chapter is not seeking to conduct a traditional empirical analysis of the meanings of 'I can't breathe', but rather seeks to work outwards from this multiply layered and powerful phrase to develop insights through 'key metaphors, ideas, concepts, and relations' (Jabareen, 2008: 51) that can help us extend our conceptualisations of mobile socialties through a metasynthesis that suggests a conceptual framework for (im)mobile antisocialities. Two main ideas that are given life from 'I can't breathe' are suggested then for our conceptual framework: the idea of a 'respiratory ontology' to suggest how mobility and immobility, sociality and antisociality give life to each other, and as part of this the idea of radical antisociality. The chapter is deliberately written through the language and structure of the breath. We cycle through the various immobile antisocialities aspirated by 'I can't breathe'; at points we gasp, wheeze, before then catching a breath to see what then conceptually transpires.

'I can't breathe': giving life to immobile antisocialities

The mediation, political socialisation and mobilisations of the phrase 'I can't breathe' became *the* slogan for 2020. It catalyses the individual human everyday act of breathing as a shared human need and experience that is essential to all human mobile socialities. We begin to breathe the earth's atmosphere from the moment we are born. We continue breathing for our entire lives, however long or short. We spend a lifetime breathing in, breathing out, staying alive. When we die, our breath ceases. Thus, 'I can't breathe' oxygenates not just the shared experience of breathing but congests what it means when breathing is difficult or laboured, when we are physically choked, unable to breathe, dying, then taking our last breath. Politically and culturally, it both catabolises (breaks down) the knowledge of the ultimate emergency humans face of biological and social death, of cultural and symbolic annihilation, and yet it also anabolises (synthesises and builds up) hope of restitution and renewal.

'I can't breathe' aspirates and expels the multiple biological, cultural and social impacts of the COVID-19 virus. Like most viruses, the coronavirus thrives on human sociality and flourishes within late capitalist mobilities. It causes infection through droplets of moisture from exhaled human breath in the air and excreted sweat left on surfaces touched or handled by an infected person (Shabir, 2020). COVID-19 then enters the human body through the mouth, nose or eyes and moves within its host to its favoured environment, human lungs, which are transformed into a pathogenic home as it feeds and reproduces (Zu et al., 2020). We humans are its chosen medium and squashy vehicle. And we can see why: it is not only that our lungs provide a perfect home and we are adept at pedestrian movement, but technologically we are supported for hard shell mobile socialities. Planes, trains and automobiles were key to the globalised transportation of the virus and its rapid transmission, creating a pandemic (Zheng et al., 2020). Although the Spanish Flu of 1918 was spread rapidly as a result of troop movements and concentrations (Webel and Freeman, 2020), nowadays 'connected world with air travel and denser populations . . . make the spread of COVID-19 easier and faster' (Terry, 2020) At the same time, human knowledge and understanding of COVID-19, its numbers of deaths, its spread and the tactics of governments to combat its global spread through immobilising human hosts has been mobilised through social and mass media (Costigan and Xin, 2020; Ducharme, 2020). 'I can't breathe' then oxygenates the difficulties of dividing and locking down life from non-life, sociality from antisociality, mobility from immobility. As Povinelli (2016: 42) suggests, 'human lungs are constant reminders that this separation is imaginary. Where is the human body if it is viewed from with the lung?' She cites how the lungs know intimately 'the larger, massive biotic assemblage – plants, bacteria, animals, microbes' (Povinelli, 2016: 42). The lungs are a medium of mobile sociality, yet they are choked by pollution, and they turn us against each other through the incursion of contagious pathogens. The lungs can also be stopped all together through the deliberate racist strangulation and crushing by another human being that intersects with a world running out breath, suffocating through polluted air, the killing of phytoplankton in the world's oceans and deforestation (Keller, 2020). 'I can't breathe' brings attention to how human lungs are at the interstices of global eco-asphysia, and in so doing, analytically metabolises how mobile sociality breathes life and gains energy from immobile antisociality.

'I can't breathe' is also a slogan used in demonstrations and in graffiti by climate activists from around the world (Simmons and Ahearn, 2017): it anabolises the impact on the planet's biosphere of extractive capitalism with its dirty, global travel and greedy hyper connectivity (Chow, 2020). It inhales and exhales the planetary sighs of relief in multiple locations during national and regional pandemic lockdowns: media reports suggested that the pausing of the world economy created a cleaner and easier breathing experience (for some). Some media reports suggested that working from home led to the reconstitution of hyperlocalities that were less frenetic, less driven by consumerism and the tertiary economy of restaurants and shopping centres and in which people relearnt new kinds of connection and being as a result of forced immobility and reduced face-to-face sociality (O'Connor, 2020; Leggatt, 2020). Social media posts bore witness to how cities and people from Mumbai to Venice savoured cleaner air and clearer water as result of a reduction in tourism (Brunton, 2020), the closure of factories and the grounding of the world's airlines (Slater, 2020). Media and social media posts celebrated lockdown rewilding: birds returning to sites they had long abandoned due to human activity; goats descending the Welsh mountains to gather on a village high street (Stuart, 2020); seas, rivers and canals running clear (Capovilla, 2020); the celebration of a quiet sky filled with stars visible in cities from England to India (Bhagat, 2020). At the same time, the media created new supplements, new assemblage production forms and public service initiatives. The UK saw the development of what became dubbed quarantine culture with the BBC developing a new online site dedicated to this and newspapers such as *The Guardian* developing a special on-line section entitled Lockdown Culture. Such media reported on how people missing the face to face social contact of cafés, bars, gyms, concerts, films and arts venues, creatively sought to connect through virtual online quizzes, cook-a-longs and avidly shared and engaged with free online content disseminated by physically closed galleries, museums and theatres, as well as creating semi-professional creative outputs that included videos, songs and street operas.

Yet, 'I can't breathe' simultaneously catabolises – breaks down – and takes us into the traumatic and claustrophobic dissociative disorder of a lockdown immobilising human beings, with states and institutions directing requirements for social distance from others that go against neurotypical modes of sociality. It catalyses lockdown claustrophobia and isolating cruelties for a human species that neurotypically propends to sociality, to physical and spiritual congregation, to our love of being together, face to face and day to day. It reminds us how neurotypical humans are physically demonstrative in friendships, especially in acts of care and rites of passage. It draws our analytical breath to the congestion of lockdown isolation and distance, of state enforced immobile antisocialities: children prevented from visiting elderly parents in care homes; the dying child unable to be attended to by his mother in hospital (Murphy, 2020); mothers giving birth without their partners (Collinson, 2020); those already living alone forced into unbearable loneliness to the point of mental collapse (Shepka and England, 2020); funerals and weddings unable to be conducted except virtually or with limited family members in masks, gloves and gowns; masks and screens thickening sociality of expression and touch; singing together banned because of the viral load in the release of collective breath, the prevention of prayer and worship being corporally conducted together, assembled instead through data connecting distanced human beings; shopping reduced for most to socially distanced queues to essential food shops and expanded for the well off into amplified online delivery; sports activities ceasing all together and then restarting with games played without spectators in closed stadia.

'I can't breathe' also metaphorically inspires and exhales a number of intersectional inequalities that stretches our understanding of (im)mobile (anti)socialites. One report said that the COVID-19 lockdown was enabling some people with disabilities to take part in work, culture and socialising from their own home for the first time (Ryan, 2020). When I complained that I was finding it difficult to interpret the complex feelings and expressions of masked people in shops or disembodied colleagues' faces on screens, my autistic son astutely replied: 'Welcome to my world. For me, it has always been like that'. A friend with multiple sclerosis said lockdown was an equaliser: since no one was going out, she did not feel she was missing out, and social distancing meant she felt less anxious when she did go out since people gave her space. 'I can't breathe' also breathes life into the temporal relativity of the immobile antisocialties of lockdown: two months of quarantine is phenomenologically much longer for the elderly, near the end of life, and the toddler at the start of life.

'I can't breathe' is also an anabolic reminder of the disabling impact of the COVID-19 virus on the organs, and especially the lungs of the most extremely affected patients, many elderly and vulnerable, but also those with underlying health conditions, who then had to be immobilised for weeks or even months through medically induced comas in order to be hooked up to ventilators, machines that do their breathing for them. 'I can't breathe', then, suggests we think about the intersectional inequalities that run through immobilised antisocialities: many of those directly affected and a disproportionate number who lost their lives in the US and UK were those from Black, Asian and minority ethnic communities who were on the front line as care workers, cleaners, nurses and doctors (Keller, 2020; Razaq et al., 2020).

'I can't breathe' gained its most powerful momentum as a slogan mobilised in June 2020 by the political movement BLM catalysing a new generation to antiracism. Despite the lockdowns, and the regulations preventing people coming together outside of their households, BLM protests burst bodily onto the streets of cities and towns not just in the US and then the UK, but globally (Maqbool, 2020). 'I can't breathe' was chanted by angry and grieving crowds to commemorate the dying words of George Floyd, recorded by witnesses as he was killed by a police officer who knelt on his neck for nine minutes, several minutes after he took his last breath (Liubchenkova, 2020). 'I can't breathe', which were his last words, was subsequently inhaled and exhaled by multiple others; the phrase was mediated, mobilised and socialised, and was used on banners and billboards, and during protests. The phrase was given its own entry in Wikipedia, showing its links and origins with the last words of Eric Garner. It brought into public consciousness multiple deaths and police brutality against black people in the US and developed into a wider call against to integrate and understand the global legacy of probably the most shocking mobile antisociality of human history: the transatlantic slave trade.

'I can't breathe' then inhales and exhales how the industrial wealth that allows for present-day mobile socialities was and is accrued from centuries of exploitation and racism. African people experienced forced movement for over 500 years through the European and predominantly British-led traffic in human beings. This forced mobility was mediated through tools and technologies designed by white Europeans to immoblise and desocialise the African people they/we enslaved including the infrastructure of slave ships, prisons, docks and plantations, as well as symbolic culture articulated through advertisements, racist literature and political culture fixed through racist legislation. Integral to the (im)mobile antisociality of the slave trade was the deliberate and cruel severance of family bonds and connections, or what Orlando Patterson (1982) coined as 'natal alienation'. Yet, this legacy which underpins the political, social and economics of mobile sociality is so often dissolved into the 'much more palatable tonic of the nation's role in abolition' (Manjapra, 2018). What 'I can't breathe' also apirates and then expels is a particularly perverse immobile antisociality – the principle of 'compensated emancipation'. Nineteenth-century English law did not allow for moveable chattel to be claimed as property: 'rights to moveable things such as household possessions or tools or livestock – were considered unstable, expendable or ambiguous' (Manjapra, 2018). So, slaves were aligned discursively by slave owners with immoveable land and buildings, and slave owners claimed that with abolition, states should be responsible for paying slave owners compensation for the loss of slaves in the same way they were paying landowners for land taken over for canals, docks, and roads. Hence the UK's Slavery Compensation Act 1837[1] compensated slave owners approximately £20 million in around 40,000 awards – while nothing was paid to those who had been enslaved. The model was copied by the governments of France, Denmark, the Netherlands and Brazil (Manjapra, 2018).

These metabolic pathways of 'I can't breathe' in 2020 during the COVID-19 pandemic and the BLM globalised campaigns takes us next to the toppling of fixed statues and monuments to slave owners and those associated with the slave trade. One example in the UK was when the memorial to slave trader Edward Colston was pulled down by BLM protestors in the city of Bristol and dragged some distance by protestors before being dumped into the docks. This was no coincidence – the monument could have been pushed into other more convenient bodies of water, but the Bristol docks are the site where slave ships traded in black lives (Farrer, 2020) 'I can't breathe' thus metabolises those mediated protests that led to moving the immoveable, to the toppling of multiple statues across Europe, the US and Australia. Sky News in Australia and the US reported this as antisocial cultural vandalism (Statue pull downs cultural vandalism, 2020), rather than acts to topple the antisocial legacy of racism and slavery (See Watts, 2020).[2]

'I can't breathe' metabolises the mobile socialities disrupted during COVID-19; it transfers the oxygen of our thinking to its wider immobile antisocialities: the isolation, disconnections and social destructions caused by extractive capitalism and its associated intersectional racist inequalities. But it also inspires pathways to regeneration and restitution. Cycling through the phrase 'I can't breathe' suggests two points of extension to mobile sociality. The first is a non-dyadic, non-dichotomous or respiratory approach to thinking about the relationships between mobile sociality and immobile antisociality. As Peter Adey (2006) argues, mobility and immobility are relational. To this, we might add that sociality and antisociality are likewise: they require each other; they threaten to choke each other while breathing life into each other. It is not simply that one includes the other, but that the constituent elements – mobility and sociality – break down and synthesise each other. This in turn inspires us to rethink and recentre antisociality not only to take account of its negative, destructive capabilities, but as valuable and generative in its own right.

(Im)mobile antisociality: a respiratory ontology

Much communications work struggles to critique the assumption that there is a division between the self and technology, between the human dimension of communication and mediated technologies we then use to enhance, broadcast, connect or assemble our communications. Mobile socialities, particularly if derived from a Foucauldian or even Deleuzian approach, tend to be inflected with an idea of subject-technology relations that is based on a structural bipartition between the intimate (social) self and exterior mobile technology/media technology. Even if it is understood that the relationship between the two is dialogical or relational (see for example in this volume, Costa [Chapter 8]; Hill [Chapter 18]), there is, nonetheless, an underlying assumption that the technologies that afford mobilities and socialities are in some ways separate from the human. Yet, the discussion of 'I can't breathe' suggests that media technologies – as well as those of capitalism, of slavery, of life-saving ventilators – are intimately connected through, into and out of the human subject. As I have argued elsewhere (Reading, 2016), the dimension of the human self related to memory not only changes through and with technologies, but, building on an approach to subject-technology derived from Pierre Bourdieu's field theory, we might deduce that in the unevenly globalised digitised world of the 21st century, we are born, live and die within a techno-mnemonic ecology termed 'the globital memory field'. Humans are never free from media technologies – they penetrate the body (through the language we speak or through medical technology, for example) and extend well beyond the human subject as we listen to the interior of the planet or the origins of the universe. Miller (2008) suggests that to explain this we draw on the Lacanian idea of extimacy or intimate exteriority which refutes the idea of a clear boundary between inside and outside/ intimate and exterior and hence self and technology. But I would suggest it is in the figural image of the lungs conjured by the phrase 'I can't breathe' that suggests we bring to immobile antisocialities a metabolic understanding of its catabolic and anabolic instability, porosity and changeability, and between the (im)mobilites and (anti)socialities of the self and technologies.

Within the idea of respiratory ontology of immobile antisociality, the mobile and the social includes and is dialogue with its antithesis, in ways that may be characterised as being potentially exploitative. Mobility studies increasingly recognises mobilities interrelationships with immobility, stasis and stillness. As Craig Martin (2011: 192) argues, 'stillness cannot be disentangled from movement – they are always immanent to one another'. Martin draws on Michel Serres's classic essay on the 'spinning top' (2011: 200) which is running and yet stays in the same place, spinning yet not moving, rocking and yet stable.

Some work on immobility suggests the potential of immobility for renewal, for the development of new selves and new actions within stillness: this accords with reports of the pandemic pause and lockdown as periods for some of reflection and renewal (Leggatt, 2020). Coker (2011: 99) suggests that during deliberative and collective stillness at a local festival in Nottingham, UK, people witnessed 'a new social configuration existing alongside or even as an alternative to the more habitual or typical social assemblages operating within the city'. Her suggestion of the 'potential' of active and resistant forms of immobility and stillness were certainly part of what emanated from 'I can't breathe' when footballers and other prominent public figures collectively 'take the knee' to signal support for BLM. We see something similar in the uses of collective stillness by climate activists to stop the circulation and movement of people through physical blockades.

Yet, enforced stillness and immobility, such as that for the migrant, also constitutes chronic waiting. In some circumstances, it leads to death: trafficked migrants in conditions of cramped proximity in shipping containers who then suffocate and die (Martin, 2011: 192). Drawing on Appadurai's work on 'scapes', Martin points to 'disjunctions of contemporary mobilities' (Martin, 2011: 195), arguing that there are different levels and speeds of mobilities: financial capital moves quickly; as does media data or immaterial information. For some, there is 'turbulent stillness' (Martin, 2011: 199) far from respite or calm, which involves acute uncertainty and instability. Indeed, the mediations of 'I can't breathe' take us to the ways in which nation-states during the pandemic used the anxiety of contagion and future uncertainty to lock down the movement of bodies across space and banned congregation and gathering. While the mediascape retained fluidity along with goods, and those identified as key workers were permitted to move, the remaining populations were held still in a state of uncertainty and anxiety while on furlough or disconnected from loved ones.

Mobility scholars have argued that within capitalism moments of controlled immobility are deliberate acts to create more profit. Brett Neilson and Ned Rossiter (2011), in their essay 'Still Waiting, Still Moving', show in their logistical analysis of media data in the world's largest ports there is an absence of both movement and sociality of people. To them, 'stillness is not potentiality. It doesn't necessarily imply the capacity to struggle, subtract or constitute new subjectivities' (Neilson and Rossiter, 2011: 52). Rather, the pause – or what they characterise as 'the grinding halt' – is an occasion 'to discover new and unforeseen frontiers of capital regeneration' (Neilson and Rossiter, 2011: 52). In the case of 'I can't breathe' within the COVID-19 lockdown alongside immobile sociality which the media captured and replayed as citizen-generated opera singing from balconies and virtual cook-a-longs, we also saw the amplification of video call platforms. Every school, workplace, and home was penetrated to the core by cameras with every desktop, laptop, tablet and mobile phone blooming with newly downloaded apps for Zoom, Teams, Slack and Skype. During lockdown, worldwide sales of desktop cameras rocketed and downloading of video calling software boomed: Zoom, for example, reported an additional 169% increase in profits in the three months to 30 April to $328.2 million, with an additional 180,000 customer organisations with more than ten employees and a profit of $27 million in the pandemic lockdown period, which constituted more than the total for the prior financial year (Sherman, 2020) As Neilson and Rossiter (2011: 56) note, 'In software focussed supply chains we see how stillness as delay holds an intimate connection to the replenishment of capital'. At the same time every immobilised and de-socialised work-from-home worker from kitchen showroom designers, games writers, teachers and medical advisers was suddenly deskilled in embodied social communication and was catapulted into rapid proficiency in disembodied assemblage-mediated communication, rewiring our brains and bodies for capitalism's profit-driven next phase.

'I can't breathe' within the context of COVID-19–induced lockdown suggests that the pausing of some dimensions of mobile sociality did not simply mark the capacity to constitute new subjectivities, but was nonetheless an attempt to 'insert asymmetrical modalities' of racism and planetary destruction within this technocratic surge for capital and control. It is also notable that within the media coverage of the pandemic, it became apparent that beyond a certain duration, capital required movement. In March, April and May of 2020, politicians' messages in the UK were to 'Stay home. Save the NHS' within the consequent production of 'lockdown culture' that ranged from books to posters (See Bar, 2020). Then by late June and July 2020, furloughed workers paid for by the UK state, unmoved at home apparently 'doing nothing', sitting around in shorts and jogging bottoms, as the media put it, were suddenly required to get moving and return to productive work. 'Get Britain Moving Again' said the *Daily Mail* (Owen and Cahill, 2020). 'Get UK back to work' reported the *Times* (Parker, 2020). And once people were back at work (in theory, though most were working all along but through their computers), the message then was to 'Get out and shop, PM to tell Britain' (Elliot, 2020), reported the *Times*, and 'Shop for Britain' (Hiscott, 2020) shouted the *Daily Mirror*. As Bissell and Fuller (2011: 8) note, 'to be unmoved, unaffected is merely taken to be a provocation to be corrected and to be made movable, workable and accessible once again'.

Radical antisociality

The metabolic pathways of 'I can't breathe' were also suggestive of 'radical antisociality' as a tenet of departure from mobile sociality. Central to the concept of mobile socialities are ideas of both sociality and, relatedly, the social. The latter etymologically has its roots in the Latin socialis 'of companionships, of allies, united, living with others'. Positive approbation is given to the social and sociality within communication studies and media studies. Ontology from neuroscience to sociology and political science valorises the social arising from a normative assumption that all human beings seek to connect with one another in similar ways and prefer to congregate with each other's interests with studies that seek understanding of forms of social life in terms of discourse (Bakhtin, 1981), things (Appardurai, 2014), memory (Fentress and Wickham, 1992) and political movements (Cefai, 2007; Jasper, 1999; Melucci, 1989).

The specific term sociality, however, originates from the work of anthropologist Marilyn Strathern and colleagues (1990: 8–10), who conceived it as 'the relational matrix which constitutes the life of persons'. The concept shifted anthropological approaches and subsequent social science and humanities disciplines away from analytical frameworks that examined society and the social in terms of a system of rules and structures toward the idea that the social is processorial, with humans operating within 'a dynamic, relational, matrix' (Long and Moore, 2012). Since Strathern et al.'s (1990) originating conceptualisation, developments around the idea of the person and affect theory have in turn destabilised the term, with Nicholas Long and Henrietta Moore (2012: 41) asking 'what is the place of non-human, objects and things?'. They argue that the development of a 'definitional haziness' to the term is a positive factor that provides evidence that the concept of socialities has the 'capacity to take many forms' (Long and Moore, 2012: 41). Given this, I would suggest that within our respiratory ontology, sociality can be stretched to include antisociality in a range of forms.

Certainly, there are precedents for the ways in which mobile sociality can be understood to include dimensions that include exploiting and violating the rights of others. Orlando Patterson (1982), whose work explores the violent uprooting, desocialisation and depersonalisation that encompasses economic slavery of human beings in various historical contexts, does not use the term antisociality, but develops the term "social death" as experienced by those enslaved. Neither does he use the term 'mobility' as discussed within this volume: to Patterson, mobility comes in two forms: intra generational and intergenerational and refers to "better material conditions; mobility up the occupational scale". Relatedly, movement refers to the regional and interregional forced movement of people. His work suggests that the basis for an understanding of the most extreme kind of (im)mobile antisociality is that which arises within the conditions of enslavement and the most extreme forms of domination. From this, it can be seen that the ontological variations of (im)mobile (anti)sociality are metabolised through the asymmetry of power.

What can be added to Patterson's work, however, is an understanding of how the media operate in relation to antisociality. It was Paddy Scannell who developed the idea that at the core of our experience of broadcasting within modern life was the importance of sociability (Scannell, 1996: 23), with studies since recasting this as including dimensions that involve antisociality through the deliberate creation of quarrels and conflict (Richardson et al., 1995: 79). What Robin Lakoff (2004) has termed "new incivility" involves new kinds of mediations of what Michael Higgins and Angela Smith term "belligerent broadcasting" (2016) with rancour and rudeness that is often unscripted. To some extent, then, climate change street protests linked to 'I can't breathe' involving the rapid movement and redeployment of protestors to exercise maximum disruption, using mobile and social technologies to facilitate this, might be characterised in this way. But/and these might also be used to explain antisocial movements on the far right (Eyerman, 2016) or the tensions between sociality and antisociality evident in the PEGIDA movement in Germany (Richardson-Little and Merrill, 2020)

I would suggest, though, that we need a deep breath to rethink and recentre the antisocial within immobile antisocialities in more fundamental and radical ways. We see this approach developed within queer studies, in which the antisocial turn has shifted emphasis deliberately onto examining the productive capacities of the destruction of alliances and an exploration of thinking that includes that which is negative (Edelman, 2004). To Halberstam (2008: 143), such approaches can have the tendency, however, to overvalorise antisociality to the extent that it can lead to a fascist sensibility. Nonetheless, such a refocussing on the antisocial might help us see that it is not protests that are 'antisocial' but rather the efforts to block democratic and transformative processes through antisocial alliances between states, corporations and security services.

Despite variations of approach, however, all of these versions of sociality and antisociality are seen as the lifeblood of being human and are 'inextricably and powerfully linked to certain kinds of language and speech as media for the production of meaning, as well as to certain kinds of mind, perception, intersubjectivity, sources of knowledge and reasons for moral conduct' (Solomon, 2012: 114). Solomon challenges what she sees as the three central tenets of sociality: namely first, that sociality is something only done between humans; second, that it is conducted through language; and third, that is conducted by those who have what is termed a "theory of mind" enabling them to relate to others in normative ways. Importantly, her work on autism and sociality suggests that what is frequently missed within neurotypically biased accounts are the embodied, actional and experiential dimensions of sociality (Solomon, 2012: 112).

At their core, both mobilities and socialities are arguably derived from normative physically abled and neurotypical assumptions that rest on the hierarchical devaluation of both human stasis and the antisocial. As we have seen, while mobility studies have recognised the importance of that which is immobile or arrested, we still give little ontological recognition to the value of different kinds of sociality that to the neurotypical person appear to be asocial or antisocial. The 21st century's rise of 'social media' – and with it, anxieties about what constitutes networks, friends, sociality and reciprocity – is based on the assumption that social communication and sociality are valued highly, and antisociality is of less value and is deemed to be destructive and negative. For neurodivergent people, and particularly autistic people, they/we are defined through diagnostic medical criteria that includes deficits in communication and particularly 'social communication'.[3] Yet, online and offline memoirs by autistic people disrupt what it means both to 'have a voice' and to 'socially communicate', and in turn offer an important critique to some of the hidden assumptions of mobile sociality. The autistic writer and artist Judy Endow, for example, relates the story of two young autistic men who are long-term friends but their carers only learn of this when the two are separated and the men become upset. She argues that sociality that includes authentic relationships and communication can manifest themselves in non-typical ways: sociality for an autistic person may be sitting in silence, physically distant, but knowingly noticing and sharing a sense of the wider world together (Endow, 2013: 113). Or, sociality may be being in a continual conversation with objects, light and water or the universe through what appear to the neurotypical person to be 'meaningless' repetitive movements, or stimming, as the non-verbal autistic writer Mel Baggs articulates in her film *In My Language* (Baggs, 2007).[4]

Connected with this, there are radical insights to bring to the concept immobile antisocialities from indigenous humanities which suggest deeper cycles of sociality and mobility. Indigenous cultural analytics disrupt ontological assumptions of what constitutes life and non-life, and the mobile versus the inert. Social life within Western thought is mapped onto that which is living and animate, with human beings at the top and centre. Yet, indigenous ontologies argue that life and animacy within Western thought is constituted extremely narrowly. It sinecures human activity from the rest of the planet and universe in order to extract and subjugate to capital all that is deemed to be dead and inanimate that might include rocks, minerals, water, mountains, trees, plants and even the sun. Within indigenous knowledge, these animate elements are part of a broad social and cultural ecology of which humans are just a small part.

Just as 'I can't breathe' suggests what Elizabeth Povinelli (2016) in *Geotologies: A Requiem to Late Liberalism* argues is the mobius nature of life and non-life, Western ontologies, she argues, excise life and non-life from each other in an effort to extract value from the earth, while categorising and dividing off indigenous and other people whose analytics of knowledge see them as intimately interconnected. Povinelli argues from this that the distinction between life and non-life is increasingly being called into question: we might in that case take another deep breath and rethink the ways in which the distinctions between mobility/immobility and sociality/antisociality are also mapped onto deeply held assumptions of a distinction between life and non-life, which – if we take a respiratory approach – break down.

Concluding remarks

This chapter used an exploration of the political ecology of the slogan 'I can't breathe', using the idea of the lung, inhaling and exhaling, as an approach to what I have termed as the metabolics of the mediated movements and associated choke points of the phrase. This enabled us to trace the ways in which the slogan catabolised and anabolised, broke down and synthesised different metabolic pathways into and out of (im)mobile antisocialities. This suggested two deep breaths drawing anew from the concept of mobile socialities developed so far in this *Handbook*: the idea of respiratory ontology and new ways of valorising neurodivergent socialities and indigenous cultural analytics.

How then might we use (im)mobile socialities? Perhaps, this might be in terms of analysing media activities, connectivities and content that is deliberately disruptive and destructive; this might be by populist fascist movements intent on division and destruction, or antiracist groups or climate emergency activists seeking to block, change or transform mobile socialities. Or, it might be useful for analysing state militia intervention into democratic processes, the production of corporately organised and state-sponsored mobile antisocialities such as 'troll farms' that use media influencers and fake profiles to deliberately interfere in democratic processes (Pomarentsev, 2019). (Im)mobile antisocialities, though, might also enable us to break down and synthesise the impact of the hidden infrastructures and global supply chains that enable mobile socialities, allowing us to ask how mobile socialities implicate users and consumers within a commodity chain of economic activities which move the antisocial dimensions of capitalism and planetary destruction to elsewhere. While mobile socialities for some may engender sociality, they might also be understood to simultaneously engender neither human connection nor the possibilities to move or escape from environmental toxicity or catastrophe.

The chapter provides some necessary groundwork prior to the subsequent studies in this collection. It links to Hartmann's study (Chapter 17) of how people who move as a result of their work function without mobile media, what it means to live within a digital media blackout (Hill [Chapter 18]), the work of narratives to positively anchor young people in situations of precarity (Lynn Scofield Clark [Chapter 5]), and how people may create socialities in situations of forced immobility such as refugee camps (Seuferling [Chapter 19]). The chapter addressed the contrary realities of mobile media and socialities challenging more broadly our normative assumptions about communication and media, mobility and immobility. Drawing on Choy's (2011) work, which explores the relationships of breathing to the sociality of political spaces, Povinelli (2016) argues that we are witnessing a shift in socio-political discourse from the demand 'Listen to me' to the statement, 'I can't breathe'. By extension, I would suggest that what (im)mobile antisocialities catalyses is an approach to cultural analytics that asks what happens not just when we are breathing, but when we are choking, before it is too late.

Notes

1 The full statute can be found in the Statutes Project 1837: 1 Victoria c.3 Slavery Compensation Act. Available at: http://statutes.org.uk/site/the-statutes/nineteenth-century/1837-1-victoria-c-3-slavery-compensation-act/.

2 When a statue was made and erected several weeks afterwards on the same plinth of BLM protestor Jen Reid, it was quickly taken down on the grounds that it was by a London-based artist and was erected without permission. (Roach, 2020).

3 See *Diagnostic and Statistical Manual of Mental Disorders* (*DSM-5*).

4 Mel Baggs sadly died of respiratory failure in the course of writing this chapter in April 2020. For an obituary on her life and work see: Neil Genzlinger, "Mel Baggs, Blogger on Autism and Disability, Does at 39", *New York Times*, April 28, 2020. Available at: www.nytimes.com/2020/04/28/health/mel-baggs-dead.html.

References

Adey, P. (2006) "If mobility is everything then it is nothing: Towards a relational politics of (im)mobilities", *Mobilities*, 1: 75–95.

Anon (2020) "I can't breathe", *Wikipedia: The Free Encylopedia*. Available at: https://en.wikipedia.org/wiki/I_can%27t_breathe#:~:text=%22I%20can't%20breathe%22,by%20New%20York%20City%20Police.

Appadurai, A. (ed.) (2014) *The Social Life of Things: Commodities in Cultural Perspective*. Cambridge: Cambridge University Press.

"Statue pull downs cultural vandalism" (2020) *You Tube*, June 13. Available at: www.youtube.com/watch?v=Sv1744ZEjPE.

Bakhtin, M. (1981) "Discourse in the novel", in M. Holoquist (eds.) *The Dialogic Imagination: Four Essays by M Bakhtin*. C. Emerson and M. Holquist, Trans. Austin: University of Texas, 259–422.

Baggs, A. (2007) *In My Language, You Tube*. Available at: https://youtu.be/JnylM1hI2jc.

Bar, N. (2020) "'Stay at home, protect the NHS' in 'home, safe home: Artists create inspirational lockdown messages – in pictures'", *The Guardian*, June. Available at: www.theguardian.com/artanddesign/gallery/2020/jun/17/home-safe-home-artist-messages-limited-edition-posters-in-pictures (accessed June 26, 2020).

Bhagat, M. (2020) "Quarantine love: Denizen fix a date with the stars", *Hindustan Times*, May 25. Available at: www.hindustantimes.com/more-lifestyle/quarantine-love-denizens-fix-a-date-with-the-stars/story-1LXYJ2sQStnexTSratmBTL.html.

Bissell, D. and Fuller, G. (2011) "Stillness unbound", in D. Bissell and G. Fuller (eds.) *Stillness in a Mobile World*. Abingdon: Routledge, 1–18.

Brunton, J. (2020) "'Nature is taking back Venice': Wildlife returns to tourist-free city", *The Guardian*, March 20. Available at: www.theguardian.com/environment/2020/mar/20/nature-is-taking-back-venice-wildlife-returns-to-tourist-free-city (accessed July 16, 2020).

Capovilla, M. (2020) "Venezia Pulita", *You Tube*, March 17. Available at: www.youtube.com/watch?v=BfRjsJAW410 (accessed July 20, 2020).

Cefai, D. (2007) *Pourquoi Se Moblise-t-on?* Paris: La Deconverte.

Chow, D. (2020) "Why 'I can't breathe' is resonating with environmental justice activists", *NBC News*, June 10. Available at: www.nbcnews.com/science/environment/why-i-can-t-breathe-resonating-environmental-justice-activists-n1228561 (accessed July 16, 2020).

Choy, T. (2011) *Ecologies of Comparison: An Ethnography of Endangerment in Hong Kong*. Durham, NC: Duke University Press.

Coker, E. (2011) "Performing stillness: Community in waiting", in D. Bissell and G. Fuller (eds.) *Stillness in a Mobile World*. Abingdon: Routledge, 87–106.

Collinson, A. (2020) "Coronavirus: Uncertainty over maternity care causing distress", *BBC News*, April 24. Available at: www.bbc.co.uk/news/health-52356067 (accessed July 16, 2020).

Costigan, J. M. and Xin, X. (2020) "How the coronavirus outbreak played out on China's social media", *The Diplomat*, January 31. Available at: https://thediplomat.com/2020/02/how-the-coronavirus-outbreak-played-out-on-chinas-social-media (accessed July 16, 2020).

Ducharme, J. (2020) "News coverage of coronavirus in 2020 is very different than it was for Ebola in 2018", *Time Magazine*, February 7. Available at: https://time.com/5779872/coronavirus-ebola-news-coverage/ (accessed July 16, 2020).

Edelman, L. (2004) *No Future: Queer Theory and the Death Drive*. Durham, NC: Duke University Press.

Elliot, F. (2020) "Get out and shop, PM to tell Britain", June 26. Available at: www.thetimes.co.uk/article/get-out-and-shop-pm-to-tell-britain-zqv5jshtg (accessed July 20, 2020).

Endow, J. (2013) *Painted Words: Aspects of Autism Translated*. Cambridge: Cambridge Book Review Press.

Eyerman, R. (2016) "Social movements and memory", in A. L. Tota and T. Hagen (eds.) *Routledge International Handbook of Memory Studies*. Abingdon, Oxon: Routledge, 79–83.

Farrer, M. (2020) "Who was Edward Colston and why was his Bristol statue toppled", *The Guardian*, June 8. Available at: www.theguardian.com/uk-news/2020/jun/08/who-was-edward-colston-and-why-was-his-bristol-statue-toppled-slave-trader-black-lives-matter-protests (accessed July 2, 2020).

Fentress, J. and Wickham, C. (1992) *Social Memory: New Perspectives on the Past*. Oxford: Blackwell.

Halberstam, J. (2008) "The anti-social turn in queer studies", *Graduate Journal of Social Science*, 5(2): 140–156.

Higgins, M. and Smith, A. (2016) *Belligerent Broadcasting: Power, Civility and Confrontation on Television*. Abingdon, Oxon: Routledge.

Hiscott, G. (2020) "Shop for Britain! Stores urge brits to spend in stores from Monday to save livelihoods", *The Daily Mirror*, June 14. Available at: www.mirror.co.uk/money/stores-plead-brits-return-high-22190982 (accessed June 26, 2020).

Jabareen, Y. (2008) Building a conceptual framework: Philosophy, definition and procedure. *International Journal of Qualitative Methods*, 8(4): 49–62.

Jasper, J. M. (1999) *The Art of Moral Protest*. Chicago: University of Chicago Press.

Keller, C. (2020) "'I can't breathe': The whole earth echoes the cry for justice", *ABC News*, June 8. Available at: www.abc.net.au/religion/catherine-keller-i-cant-breathe-the-cry-for-justice/12332954 (accessed July 17, 2020).

Lakoff, R. T. (2004) "The new incivility: Threat or promise", in J. Aitchinson and D. Lewis (eds.) *New Media Languages*. London: Routledge, chapter 4.

Leggatt, J. (2020) "Farewell workplace burnout: Will coronavirus slow the manic pace of our modern lives", *The Guardian*, April 19. Available at: www.theguardian.com/commentisfree/2020/apr/20/farewell-workplace-burnout-will-coronavirus-slow-the-manic-pace-of-our-modern-lives (accessed July 16, 2020).

Liubchenkova, N. "In pictures: Black lives matter protests taking on the world", *Euronews*, June 15. Available at: www.euronews.com/2020/06/15/in-pictures-black-lives-matter-protests-taking-on-the-world (accessed July 16, 2020).

Long, N. J. and Moore, H. (2012) "Sociality revisited: Setting a new agenda", *The Cambridge Journal of Anthropology*, 30(1): 40–47, Spring.

Manjapra, K. (2018) "When will Britain face up to its crimes against humanity", *The Guardian*, March 29. Available at: www.theguardian.com/news/2018/mar/29/slavery-abolition-compensation-when-will-britain-face-up-to-its-crimes-against-humanity (accessed July 20, 2020).

Maqbool, A. (2020) "Black lives matter: From social media post to global movement", *BBC News*, July 10. Available at: www.bbc.co.uk/news/world-us-canada-53273381 (accessed July 20, 2020).

Martin, C. (2011) "Turbulant stillness: The politics of uncertainty and the undocumented migrant", in D. Bissell and G. Fuller (eds.) *Stillness in a Mobile World*. Abingdon: Routledge, 192–208.

Melucci, A. (1989) *Nomads of the Present: Social Movements and the Individual Needs in Society*. Philadelphia: Temple University Press.

Miller, J. A. (2008) "Extimacy", *The Symptom*, 9. Available at: www.lacan.com/symptom/extimity.html (accessed May 25, 2020).

Murphy, N. (2020) "Family of boy, 13, who died from coronavirus cannot attend funeral over symptoms", *Daily Mirror*. Available at: www.mirror.co.uk/news/uk-news/family-boy-13-who-died-21805750 (accessed July 16, 2020).

Neilson, B. and Rossiter, N. (2011) "Still waiting, still moving: On labour. Logistics and maritime industries", in D. Bissell and G. Fuller (eds.) *Stillness in a Mobile World*. Abingdon: Routledge, 51–68.

O'Connor, E. (2020) "Yes there is panic: But I'm embracing a slower life", *Washington Post*, March 23. Available at: www.washingtonpost.com/lifestyle/2020/03/23/yes-there-is-panic-im-embracing-slower-life/ (accessed July 16, 2020).

Owen, G. and Cahill, H. (2020) "Get Britain moving again", *Daily Mail*, April 19. Available at: www.pressreader.com/uk/the-mail-on-sunday/20200419/281496458420747 (accessed June 26, 2020).

Parker, G. (2020) "Boris Johnson unveils plan to get UK back to work: Stay at home message ditched in England but most restrictions remain", *Financial Times*, May 11. Available at: www.ft.com/content/a58b1a2b-a1d0-494e-aef8-beab084072a3 (accessed June 26, 2020).

Patterson, O. (1982) *Slavery and Social Death*. Cambridge: Harvard University Press.

Pomarentsev, P. (2019) *This Is not Propaganda: Adventures in the War Against Reality*. London: Faber and Faber.

Povinelli, E. A. (2016) *Geotologies: A Requiem to Late Capitalism*. Durham, NC: Duke University Press.

Rapson, J. K. (2018) "Refining memory: Sugar, oil and plantation tourism on Louisiana's River road", *Memory Studies*, 13(4): 752–766.

Razaq, A., Harrison, D., Kaunanithi, S., Barr, B., Asaria, M., Routen, A. and Khunti, K. (2020) "Bame COVID-19 deaths – what do we know? Rapid data and evidence review", *CEBM*. Available at: www.cebm.net/covid-19/bame-covid-19-deaths-what-do-we-know-rapid-data-evidence-review/ (accessed July 16, 2020).

Reading, A. (2016) *Gender and Memory in the Globital Age*. Basingstoke: Palgrave Macmillan.

Richardson, K., Livingstone, S. and Lunt, P. (1995) "Talk on television: Audience participation and public debate", *Screen*, 36(1): 79–82. doi:10.1093/screen/36.1.79 (accessed July 16, 2020).

Richardson-Little, N. and Merrill, S. (2020) "Who is the volk? PEGIDA and the contested memory of 1989 on social media", in S. Merrill, E. Keightley and P. Daphi (eds.) *Social Movements, Cultural Memory and Digital Media*. Basingstoke: Palgrave Macmillan, 59–84.

Roach, A. (2020) "Bristol mayor defends removing statue of Black lives matter activist Jen Reid: It had to come down", *The Evening Standard*. Available at: www.standard.co.uk/news/uk/bristol-mayor-blm-jen-reid-statue-removal-a4500031.html (accessed July 16, 2020).

Ryan, F. (2020) "COVID lockdown opening up world for people with disabilities", *The Guardian*. Available at: www.theguardian.com/world/2020/apr/20/covid-lockdown-opening-up-world-for-people-with-disabilities (accessed July 16, 2020).

Scannell, P. (1996) *Radio, Television and Modern Life: A Phenomenological Approach*. Oxford: Blackwell.

Shabir, O. (2020) "How is COVID-19 spread?" *News Medical Life Sciences*. Available at: www.news-medical.net/health/How-is-COVID-19-Spread.aspx (accessed July 16, 2020).

Shepka, P. and England, R. (2020) "Coronavirus: Coping with grief during a lockdown", *BBC News*, March 27. Available at: www.bbc.co.uk/news/uk-england-northamptonshire-52048088.

Sherman, N. (2020) "Zoom sees sales boom amidst pandemic", *BBC News*, June 2. Available at: www.bbc.co.uk/news/business-52884782#:~:text=Zoom%20said%20on%20Tuesday%20that,of%20the%20prior%20financial%20year (accessed July 15, 2020).

Simmons, S. A. and Ahearn, A. (2017) "When climate change makes it hard to breathe", *OPN. Org*, December 5. Available at: www.opb.org/news/article/when-climate-change-makes-it-hard-to-breathe-terrestrial/ (accessed July 16, 2020).

Slater, J. (2020) "In India, life under coronavirus brings blue skies and clean air", *The Washington Post*, April 11. Available at: www.washingtonpost.com/world/asia_pacific/india-coronavirus-delhi-clean-air-pollution/2020/04/10/ac23dd1e-783e-11ea-a311-adb1344719a9_story.html (accessed July 16, 2020).

Solomon, O. (2012) "Doing, being and becoming: The sociality of children with autisms in activities with therapy dogs and other people", *Cambridge Anthropology*, 30(1), 109–126, Spring.

Strathern, M., Peel, J., Toreu, C., Spender, J. and Ingold, T. (1990) *The Concept of Society Is Theoretically Obsolete*. Manchester: GDAT.

Stuart, A. (2020) "Goats take over empty Welsh streets during coronavirus lockdown – video", *The Guardian*, March 31. Available at: www.theguardian.com/uk-news/video/2020/mar/31/goats-take-over-empty-welsh-streets-llandudno-coronavirus-lockdown-video (accessed July 16, 2020).

Terry, M. (2020) "Compare: 1918 Spanish influenza pandemic versus COVID-19", *Biospace*, April 2. Available at: www.biospace.com/article/compare-1918-spanish-influenza-pandemic-versus-covid-19/#:~:text=The%20higher%20estimate%20of%2050,million%20figure%20suggest%20about%201%25.&text=On%20the%20other%20hand%2C%20we,COVID%2D19%20easier%20and%20faster.

Watts, J. (2020) "Edward Colston statue toppled: How Bristol came to see the slave trader as a hero and philanthropist", *The Conversation*, June 8. Available at: https://theconversation.com/edward-colston-statue-toppled-how-bristol-came-to-see-the-slave-trader-as-a-hero-and-philanthropist-140271.

Webel, M. and Freeman, C. M. (2020) "Compare the flu pandemic of 1918 and COVID-19 with caution- the past is not a prediction", *The Conversation*, June 4. Available at: https://theconversation.com/compare-the-flu-pandemic-of-1918-and-covid-19-with-caution-the-past-is-not-a-prediction-138895.

Zheng, R., Xu, Y., Wang, W., Nig, G., and Bi, Y. (2020) "Spatial transmission of COVID-19 via public and private transportation in China", *Travel Medicine and Infectious Disease*. March 14–April. Available at: www.ncbi.nlm.nih.gov/pmc/articles/PMC7118651/ (accessed July 16, 2020).

Zu, Z. Y., Jiang, M. D., Xu, P., Chen, W., Ni, Q., Lu, G. M. and Zhang, L. J. (2020) "Coronavirus disease 1019 (COVID 19): A perspective from China", *Radiology: RSNA*, February 21. doi:10.1148/radiol.2020200490.

17

MOBILE FIGURES IN CURRENT TIMES

On the Walz

Maren Hartmann

Introduction[1]

> Move on when the neighbour's dog stops barking.
>
> *(Traditional saying in the context of the Walz)*[2]

> During the Walz you experience a kind of freedom and adventure that is rare to find these days. Earning money is not important during this time. Your possessions consist only of those things that you really need to live and work.[3]

The wandering years (called 'die Walz' or 'Tippelei' in German) were named an 'intangible cultural heritage' by the United Nations Educational, Scientific and Cultural Organization (UNESCO) in 2015. The basic idea was that craftsmen (and -women)[4] – the Gesellen – leave their hometown after they finish their apprenticeship and begin travelling for three years and a day. This hardly studied area of mobile lives has been in existence for at least 700 years. It still consists of many of the original traditions, but the underlying rationale has changed. Originally practised all over Europe, it is nowadays mostly upheld in the German-speaking countries, Denmark and France. The countries travelled, however, have become global, and therefore the current travellers have also been called "jet-set-journeymen" (in Bohnenkamp and Möbus, 1995: 100).

Throughout their three years, the wandering craftspeople are expected to work and thereby learn new aspects of their craft. But they are also meant to get to know themselves and the world. Freedom is the word that gets mentioned repeatedly in this context. On a more mundane level, travelling craftspeople offer their (work) services in exchange for food, shelter, and a negotiable wage. Several unwritten rules accompany this journey; several traditions provide a stable framework for the Walz, an outline for a specific form of sociality. A more diverse range of socialities emerges from the individual enactment of these rules and traditions; hence, divergent practices form sociality into socialities. And many rules have to do with mobilities, while some also deal with questions of media. The most unusual of these is the fact that mobile phones are forbidden during the Walz. This chapter is therefore an account of the exploration of this triangle of mobilities, socialities, and media with an empirical focus on wandering craftspeople.

The base of this triangle is the constant interplay between sociality and mobility, whereby one leads to the other. Media (in all their variety) sit slightly uncomfortably on top of this: they simply portray this interplay, but serve also as an important link to 'the rest of the world.' At the same time, media offer an entry point for those who are thinking of going on the Walz. Once on the Walz, however, media are important exactly through their absence. This creates a tension, as well as the specificity of this mobile (social) figure of the journeyperson.

There are additional tensions within this triangle: the wandering craftsperson's form of mobility is based both in tradition and in hyper-modern times; it is highly mobile, but also highly disconnected. The chapter aims to take a closer look at these tensions. It therefore begins with the theoretical framework of mobile (social) figures, to which it returns at the end. In between, the focus is on the phenomenon of the travelling craftsmen/-women. This figure will be regarded from several angles: his/her historical roots and current forms, habits and clothing, movements, and, last but not least, media use and communication strategies.

Mobile figures in mobile times?

> The ideal of mobility – as distinct from mere physical movement – stresses the social production of movement.
>
> *(Rosenberg and Shannon, 2018: 1)*

As indicated in the Introduction to this book, the mobility paradigm is closely linked to the question of meaning making. As has often been claimed, meaning making is a highly social process – clearly visible in the mobile socialities framework. Within this larger framework, the theoretical emphasis presented here is on *mobile (social) figures*. The social is in parentheses, because wandering craftspeople are defined first of all through their mobility, but second of all through their specific practices of oral communication to organise a strong social network. This allows both entry to and sustenance within this rather specific journey.

While the wandering craftsperson is only one such mobile figure, it offers a special twist: it combines an image of the past with a particular view of the future, especially in relation to the organisation of social networks in times of mobilities and mediation. It is thereby, so the claim is made, a very timely (mobile) social figure. Social figures generally help to characterise society (see e.g. Simmel, 1992). They offer a characterisation of the state of the social at a particular time.[5] They can be descriptive or visionary. Mobility has often been analysed through different (social) figures, as Rosenberg and Shannon (2018: 2) point out. They cite the nomad, the pilgrim, the tourist, and the flâneur (a similar list to that of Bauman, 1996; see also Gebhardt and Hitzler, 2006). While different kinds of movement produce certain places or configurations, these social figures additionally reflect and represent understandings of and embodiments of certain kinds of freedom (and restrictions), and certain kinds of social rules and norms (and again: their upheaval – see Rolshoven and Maierhofer, 2012, for explorations of vagabondage). Rosenberg and Shannon refer to Kesselring (2006) to contrast autonomy and heteronomy, as well as production and adaptation (Rosenberg and Shannon, 2018: 5), as aspects that all play out differently in diverse social figures and to which I will return later on as reference points for my analysis of the wandering craftsperson. In this sense, comparisons to other mobile figures often enrich the description of the specificity of the wandering craftsperson.[6]

According to Moebius and Schroer (2010: 7–11), social figures are not identical with social roles and are therefore not limited to specific spheres. Instead, they typically stem from diverse fields; their actions, however, have developed independently and become generalised. As per

Moebius and Schroer (2010: 8), these have become practices used in the whole of society, such as management, speculation, and consultancy. The generalised practice focused upon here is (im-)mobility (and its centrality in current times), and therefore the use of the label 'mobile figure' (cf. Gebhardt and Hitzler, 2006). These social figures tend to become generalised figures or even styles of thinking (Gebhardt and Hitzler, 2006), and also offer an opportunity for identification. This general framework needs to be filled with individual stories. Similar to Nail, who writes about the migrant, I want to argue that "as a figure" the wandering craftsperson "refers both to empirical" craftspeople "in the world and a more abstract social relation" (Nail, 2015: 16). However, a third practice is important in this context: mediation, understood as creating the social through communication – and which nowadays often finds its expression in the usage of media and communication technologies. The core question is: How far does media use – or its absence – characterise the mobile (social) figure in question? And how does this relate to socialities?

In the long run: a) additional notions might be necessary to better assess any mobile figure (kind and speed of movement, social structures, mediation and media use, etc.); and b) additional analyses of other figures might be needed for a better description of the social in current times. In the following, we begin this analysis with the journeyman/-woman as a rather unknown mobile figure.[7]

The analysis presented here is based on a broad study of the existing literature on the topic. It includes both academic and popular publications (especially autobiographic accounts), as well as a number of newspaper articles and websites on the topic. Other important sources were filmic documentaries. The list is not endless, but nonetheless I eventually reached thematic saturation through the existing material.[8]

Travel and work: the journeyman/-woman

The combination of work and travel is by far not new – indeed, it has always been part of certain work traditions. However, it was always reduced to specific occupations – and also in many cases regarded with suspicion, since the sedentary lifestyle was for long the dominant norm.[9] Nonetheless, some kinds of work remained mobile throughout: the shepherd is one example, the wandering priest another (Winter, 2006). They underline both continuity and change: while the shepherd still exists, he/she does not necessarily travel with the herd anymore, but usually lives on a farm. The wandering priest, however, has ceased to exist in the original form, but missionaries are still quite common, even today, as are other forms of evangelical preaching in mobile forms. The pilgrim is also closely related to the wandering priest, since the boundary between the one who preaches and the one who follows is often porous. None of these kinds of mobile work are as clearly organised and structured as the travelling craftsperson, though. And in contrast to the claim that the Walz is a "German charm that is quickly disappearing" (*Der Spiegel*, 2006), it has instead begun to grow again in recent years. As a rite of passage via an extended period of nomadism, it seems to appeal to a number of young people. This might also be based in its particular mixture of modernity and anti-modernity, where seemingly old traditions (from the clothing via the secret rituals to the three-year framing) meet modern ideas of independence and travel, but also of sustainability, of nature, of slow movement, and of reinvigorating the notion of presence.

This Walz has a long tradition in the German-speaking, but also some other European countries. The starting point dates back to the 15th (some state the 13th) century, when the population began to grow and competition amongst craftspeople became fiercer (see e.g. Bohnenkamp and Möbus, 1995). The solution was simple: send the young people on a longer journey where they could earn money, learn additional skills, get to know the world, and – at the same time – offer fairly cheap labour along the way. This was the beginning of the travelling craftsmen. Women began to join these endeavours only later, but remained a clear minority for a long time. They still represent only about a tenth of the wandering craftspeople (some sources state 15% – e.g. Buchner, 2017), but the increase has been rapid.[10]

At some point, the wandering years turned into a must-do, if one wanted to become a master (Meister) of the craft: "Historically, artisans weren't even considered for entry into their respective guilds before going *auf die Walz*" (*Der Spiegel*, 2006). Not entering the guild meant a huge limitation in terms of where you were able to work, who you could marry, etc. The eventual decline of the guilds let to a decline of these restrictions, as well.[11] Nowadays, the wandering years are an option, not a requirement. Most of those wandering are carpenters, builders, plumbers, cabinet makers, roofers, tile layers, or masons, but dressmakers, gardeners, bakers, and others have also joined the travelling folk. Since the 1970s, the Walz has gained in popularity, partly because new 'Schächte' (see later in this chapter) were founded that allowed women into their ranks and provided an emphasis on travelling rather than working (cf. Buchner, 2017: 161ff.). There are no official statistics, but some estimates state a doubling of the numbers in the last ten years: there were supposedly only 250 journeymen/-women travelling around the world in 2005 (*Welt am Sonntag*, as cited in *Der Spiegel*, 2006); now the number has risen to 500 yearly (FNP, 2018) or even 800 (Walbröhl, 2015). Next to the new organisations, other reasons for the increase included the growth of social movements, the general increase in travel amongst young people, the fall of the Berlin Wall and the unification of Germany (since wandering craftsmanship was forbidden in the German Democratic Republic, the former East Germany).

The rules

By and by, the journey began to be shaped by rituals that later turned into traditions (cf. Hobsbawm and Ranger, 1992). One of them is the outfit. It adheres to very clear rules, which partly depend on the particular craft, and has to be worn throughout the whole journey (see later in this chapter). The length of three years is another rule that was eventually implemented: this is the minimum amount of time on the road for most crafts (for some crafts, it is only two years). Next to the temporal component, there is a space component: after leaving his/her hometown, the journeyman/-woman is 'banned' for the next three years from that particular place and is not allowed to enter a 50-kilometre radius around their home (see Figure 17.2). They are not allowed to enter this 'no-go' zone unless an immediate family member dies or some other emergency occurs. The new craftsperson is accompanied away from home by a group of travelling craftspeople and is equally accompanied home after the three years by a similar social group. As part of these events, they also have to perform smaller rituals, such as climbing the town-sign into different directions each time, held by the accompanying craftspeople when falling down. Another important aspect is the nailing at the beginning of the journey, where a nail is hammered through the earlobe, allowing an earring to be worn and symbolically nailing one to keep one's promises.

There are other rituals that are not communicated to the outside world. These are important entry points into the world of free food and shelter, into work, etc. Sometimes they are simply particular formulae that need to be uttered in order to identify oneself as a genuine travelling craftsperson and gain entry. If these formulae were not secret, these kinds of offers could be abused. Other rituals serve to underline the membership in a particular brotherhood (see later in this chapter).

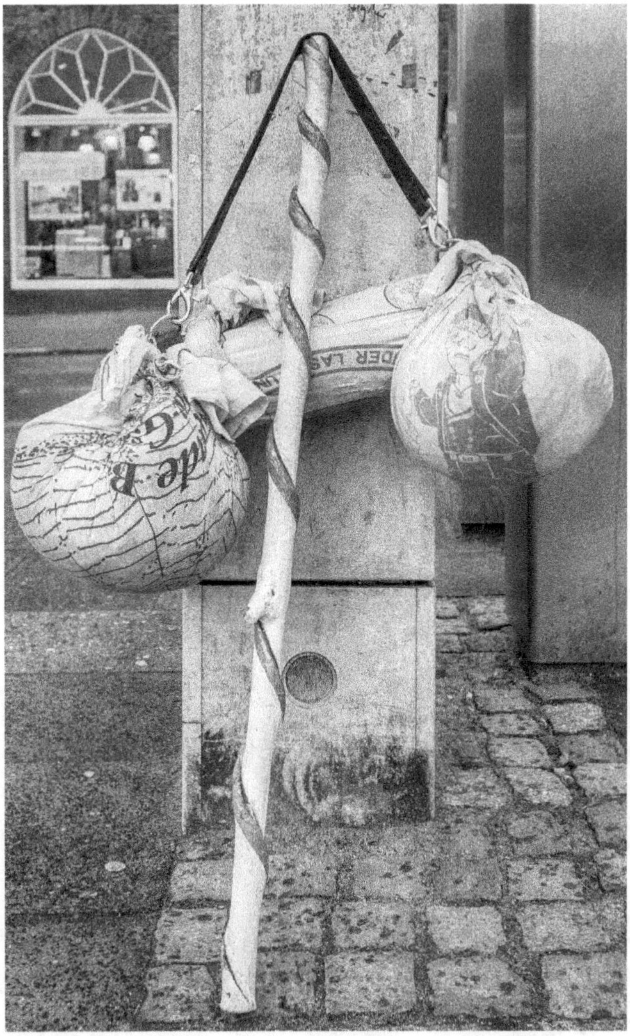

Figure 17.1 The Stenz (walking stick) and the Charlottenburger (linen cloth).

Source: Copyright Ichwarsnur (Marvin Radke) / CC BY-SA https://creativecommons.org/licenses/by-sa/4.0; https://upload.wikimedia.org/wikipedia/commons/5/55/Handwerker_Wanderjahre_Gep%C3%A4ck.jpg

The clothes

> Dressed in a black slouch hat, a double-breasted vest with huge white buttons and black corduroy bell-bottom pants.
>
> *(Der Spiegel, 2006)*

The journeyperson is clearly detectable, because he/she wears a certain 'uniform' known as the 'Kluft.' It consists of trousers often with large flares, a vest, a jacket, and a white collarless shirt, as well as a hat and black boots.[12] This is protective work clothing on the one hand, but also provides a certain visibility for outsiders.

There are also less obvious signs of tradition attached to the Kluft. The jacket, for example, has to have six buttons, which represent the six(!) working days of the week. On each sleeve one finds three buttons, representing the three years of the apprenticeship – as well as the three wandering years. The vest has eight buttons that represent the eight hours of work a day. Particular crafts also have particular colours: the most common colour – black – is used by carpenters and other woodcrafts, while red is used for cloth-related crafts, beige or grey are used for crafts related to sand or stone, blue for metal. There is also a short tie in the colour of the particular organisation (brotherhood) the wanderer is subscribed to, not to be worn by those who travel without such affiliation. It represents the honourability of the craftsperson and is often regarded as the most valuable piece that the (organised) wandering craftsperson carries.

Additionally, the journeyman/-woman carries an individually crafted walking stick (the Stenz; see Figure 17.1). It is one of those objects that the aspiring traveller has to find and handcraft in his/her early days of the journey. A wandering craftsperson was not meant to carry jewellery apart from the earring, which again shows the particular craft that they belong to.[13] This rule, however, does not seem to apply anymore: many can be seen with quite a bit of jewellery. The Kluft is always to be worn throughout the three years – in fact, part of the introductory ritual to the journey can include the burning of the clothes worn beforehand. On top of this, the journeyman is allowed to carry (in yet again a particular set of linen cloths known

Figure 17.2 Linda, a goldsmith, with her book, which includes 5 Euros and shows her 50-kilometre 'no-go' zone.
Source: Copyright LokalPlus/W. Schneider

as the Charlottenburger; see Figure 17.1) a set of shirts and undergarments, but also a sleeping bag, tools, and a particular book, as well as other small things and the previously mentioned trademark walking stick.

The book in particular is a must – and something that cannot be stolen or lost. It is the 'Wanderbuch,' the travel diary, in which the whole journey is documented. This takes place through writings by the Geselle, through stamps that are offered by those companies or individuals he/she works for and from the places he/she visits, as well as other materials (again: these are not to be shown to outsiders, so therefore the exact contents are difficult to assess).[14] It used to be a legal obligation to carry this book, because it served as a form of passport in case the authorities approached the traveller. The book therefore used to be issued by the authorities (the police).[15]

> This institutionalization can be explained by the fact that the privilege to move freely from place to place was historically reserved for the upper class, and traveling common people were viewed as vagrants. For this reason, efforts were made to officially control them. . . . However, due to this institutionalization, the dominant classes had a relatively favorable view of tramping.
>
> (Phillips, 2019: 46–47)

This passport-character of the wanderer's book has been replaced by a slightly less official meaning. One current version is issued by the CCEG, the 'Confédération Compagnonnages Européens/Europäische Gesellenzünfte,' a European umbrella organisation for several local organisations of the different brotherhoods. Other books are organised individually (see Figure 17.2).

Not everyone can become a travelling craftsman: apart from having to have completed the apprenticeship, the journeymen/-women should not have debts or have been in conflict with the law; they cannot be married, i.e. they should be single and childless; and they need to be under the age of 30. Another signifying feature is language: the travelling craftspeople use a jargon called 'Rotwelsch,' which stems from subcultures such as beggars or criminals in earlier times (see e.g. the glossary in Amrehn, 2016). It encompasses several terms that signify a group coherence and allow the travelling craftspeople to bond.

The journey

> We want to travel the cosmopolitan way and thereby get to know others and ourselves. We are just as interested in distant countries and cultures as we are in the different mentalities of our homeland. As strangers we want to live and work with strangers for a while, to know and understand more about the world than those who stay at home. We are proud of our skills and are committed to tradition. The craft and the common experience of the 'Tippelei' keep our community together. The passing on of living customs is just as important to us as is the practice of our craft today and in the future.
>
> *(Website of the Rechtschaffen-Fremde-Schacht)*[16]

When the wandering journeymen/-women start the journey, one part of their initial 'equipment' are 5 Euros, not more (see Figure 17.2). One of the great original rules around the Walz was that they were not allowed to pay for food or shelter – but also not for their travels. This is why they (still) tend to walk, tramp, or, if needed, work on boats that carry them. Nowadays, the rule is partly loosened: if the money has been earned on the Walz, it can also be spent on a flight ticket. As Franz, a locksmith, outlined, he was not meant to leave the German-speaking countries in his first year, was allowed to travel in Europe in his second and the world in the third year (Zschornack, 2015: 125–126). He also described that he was not meant to stay longer than three months in any one place, which can be up to six months in other crafts. But while the rules differ slightly, some of them apply to everyone. While on the road, for example, journeymen/-women make a living by taking on day jobs or other temporary work. They tend to work for master craftspeople, who (nowadays) pay them at least a small, often the standard, wage.[17] These masters often provide them with a place to stay for a short period of time, as well. Additionally, the wandering craftspeople can accept donations to continue their trip.

For the first period (up to three months, often less), the journeyperson travels with someone else – someone who has been on the road for a while and serves as a teacher. This is part of the highly ritualised beginning of the journey. The organisation of these aspects of the journey is made easier if one joins a 'Schacht,' a traditional organisation or brotherhood. At the moment, there are eight of these organisations in Germany.[18] They offer not only a virtual organisation, but a meeting point and orientation. The Heidberg Frei-Zunft, for example, meets once a month in a pub, in a separate room, to discuss matters and uphold traditions, such as singing certain songs and performing other ritual social interactions (such as beer drinking rituals). The meetings also serve to organise the support for the travelling craftsmen/-women, both materially and immaterially. Plus, the Schacht has rules that need to be followed.

Additionally, journeymen/-women have increasingly begun to travel 'free.' But even the free travellers tend to be organised (a contradiction in terms, which therefore has become a point of discussion) and live according to rather similar rules and regulations as those who are organised. A third way is to travel 'wild' (or 'colourful'), which connotes travelling freely and without rules (Amrehn, 2016: 80 – footnote). These do not wear the clothes and do not even call themselves 'Wandergesellen,' since they want to connote their differences. However, the majority of the wandering craftspeople are organised in one form or another.

Journeyman socialities

The whole idea of travelling craftspeople is based on sociality. While many travel alone (or as two), there are many meeting points, rituals and other aspects that are built on the idea of a close-knit community, which takes one on board, teaches one the trade, helps wherever and whenever necessary, and generally guides the process (see e.g. Figure 17.3). There are also phases when working or travelling is taking place as a group, as in the send-off. These travels are based on the idea of both following traditions but also re-inventing them in the present. As such, the social construction, the practices, etc., are highly social. Many of the traditions are not written down, but are re-enacted in and through the practice, whereby they cannot continue without communication around them.

> Journeymen travel in groups or on their own, depending on their trades and their routes, often finding one another by sight. "Have you seen people who look like me?" Mathias Müller, a carpenter, asked people in Tübingen after arriving there to meet up with friends.
>
> (Eddy, 2017)

Many of the previously mentioned structures help in constructing the necessary social networks: one can recognise other wanderers through the Kluft; the brotherhood offers introductions to

Figure 17.3 Two wandering craftsmen in 2006.

Source: A.stemmer/CC BY-SA; http://creativecommons.org/licenses/by-sa/3.0/; https://commons.wikimedia.org/wiki/File:Zwei_wandernde_Gesellen,_2006.jpg

both people and rituals, and it offers material support and regular meetings; the tradition of having a fellow traveller in the beginning connects the new traveller to the existing social network. Even before the actual journey starts, the future wanderer is taken to meetings, is slowly introduced to certain traditions, is shown what to wear, etc. Sometimes they 'share' a building site, as for example a summer building site, which is set up for a good cause and where several craftspeople are needed.

The journeyman/-woman sociality is characterised by these traditions, forms, and customs that everyone has to learn and later pass on to the next. This social network is one that cannot simply be entered, but instead entry needs to be 'earned': a) finish the apprenticeship; b) express an interest; c) meet people and have your interest checked; d) utter interest more precisely; e) find supporters (and the initial guide) who will introduce you officially; f) make it known; and g) begin the rituals that officially let you become part of this community. Even after entry, it is important to stay in touch with your community, to regularly attend the meetings, to perform certain rituals that you will be taught, and to generally adhere to what is expected of you. At the same time, the daily doings and the travels themselves are characterised by great freedom. This is an aspect that most journeypeople emphasise. These are socialities expressed in practice – enacted every day. It is also, crucially, a community based in oral communication. While there is a growing number of accounts of these travels and the related social structures (e.g. Amrehn, 2016; Buchner, 2017; Wadauer, 2005; Zschornack, 2015, as well as the aforementioned articles and videos), it is still a secret sociality. In the books, the rituals are never described in detail; nearly every website has a closed section; even the filmic accounts do not give away too much (sometimes the door is closed in front of the cameras). Once a member, one is introduced to these secrets and literally has to learn and uphold them. The ritual re-enactment allows the sociality to continue and thrive.

One of the major changes in recent times is the underlying rationale behind the travels. As mentioned previously, freedom and adventure are mentioned repeatedly nowadays as a driving force for becoming a travelling craftsperson. This includes, as we will see, the freedom from digital communication patterns. When the idea of the Walz was initially developed, it was for economic reasons – and therefore not necessarily a choice. Later, it officially became an obligation before becoming a choice again in more recent times. Even if it afforded the act of travelling to people who usually would not get to travel, it was still closely linked to work. And while the latter still holds true today, the 'six days a week' does not necessarily apply and periods of leisure are clearly common nowadays (e.g. Zschornack, 2015). Nonetheless, the travelling craftsperson is predominantly working class, white and male (approximately 10–15% are female); some additionally claim there might be a rebellious streak (Zschornack, 2015: 214; Lemke, 2002). The latter is expressed in the hat that is never taken off – for no one. It is an expression of freedom, not having to bow to anyone. The Walz can be described as a rite of passage, which paves the way not only from the apprentice to the master, but also from the young (wo)man to the adult. This passage is expressed through an extended period of nomadism.

For the outside world, the fascination with this rather secluded world has not ceased. Representation is not lacking: it is thriving. This does include mis-representations, for which both sides are responsible. The journeymen/-women and the brotherhoods necessarily thrive on secrecy, while the media tend to romanticise the travelling years, which is often a source of frustration for the travelling craftspeople (see Buchner, 2017). Especially the concealment and the old traditions gain attention, but also the lack of digital media use.

The journeyman and (mobile) media use

> In an adaptation of the old rules to modern times, journeymen do not carry devices like cellphones that allow them to be found. They carry digital cameras, if they like, and write emails from public computers.
>
> *(Eddy, 2017)*

> Furthermore, it can be said that journeymen carry no new media on them, such as mobile phones, laptops or similar devices.
>
> *(Buchner, 2017: 85; translation by author)*

This is one of the rules around the Walz: no phones (mobile phones, smartphones, or similar devices) are allowed. It is clearly fascinating for the media reports about the Walz, since this rule seems – on first sight – anachronistic. It has clear implications, though: the organisation of the social network and of work, etc., needs to take place through other means. Hence, there is a clear focus on oral communication during the Walz. Family and friends, who were left behind in the hometown (or elsewhere), cannot easily be contacted. As one wandering woman reported, she gathers bits of loose change every couple of weeks and uses a phone booth to be in touch with her mother (CKnet Videos, 2016). Connectivity of this kind is possible; it is just difficult. If necessary or desired, the journeyperson can also enter one of the remaining Internet cafés or a public library and use these terminals to send emails or look things up, or he/she can borrow someone else's phone. All that is allowed.

But it is not only public perception that makes the non-use of mobile phones a central aspect of the journey. When asked about what he fears most about the beginning of the Walz, Sönke, a carpenter, states that he will probably miss his phone and 'Dr. Google' most (NDR Doku, 2019). Just as another journeywoman (a seamstress/dressmaker), when asked about the most unusual aspects of going on the Walz, immediately answers that leaving behind the smartphone (in times like these) is most unusual about her journey (CKnet Video, 2016). In the same video (in which three journeywomen of different crafts are interviewed about their experiences), another woman emphasises that not having a smartphone is crucial, but that there are other ways of contacting the family, such as using the aforementioned phone booth. One of these women goes as far as stating that it is extremely liberating not to have a mobile phone (CKnet Videos, 2016). While she started out with "phantom vibrations," still feeling the phone she did not actually have any more, it soon became a memorable experience. For her, after starting her journey only two and a half months earlier, the lack of a phone allows memories to return and allows her talk to people more, to actually look at them. She claims to hear and see more. Away from the screen, the world is visible again; the present moment (and whoever is present) moves to the forefront. Pushed a bit by the interviewer, she continues to proclaim that she recommends everyone to put away the phone for a while. Others state similar experiences:

> In the beginning, it was actually a liberation – because you don't have to be reachable and you actually are not reachable. And therefore you don't have the feeling of missing out on anything. You are simply where you are; you have the conversation that are you are currently having; you drink the beer that is in front of you; you eat the food that's there – and you are just there, in that moment. And that, I think, is what distinguishes a journeyman from a normal person, I would say.
>
> (Kilian, 27, just returned from his Walz – in Bayerischer Rundfunk, 2020; translation by author)[19]

This trope is reflected in most of the documentaries, diaries, and reports about the Walz that have appeared in recent years: to get rid of the mobile phone, to cut the ties to constant connectivity is a liberation, but also a source of fascination for others. It might be scary to begin with, and it might take some getting used to, but it lets one 'live in the moment' much more than before and therefore care less about past or future, about missing out on something, about needing to rush. While this belief in 'offline' cultures is also increasingly proclaimed in other contexts (and detox retreats, apps, etc. have become popular in recent years; see also Syvertsen and Enli, 2019), there is one important difference from these other contexts: the travelling years are not a short-term retreat in order to recharge the batteries and return to normal – this is a three-year-minimum kind of move, as well as a change in attitude and/or lifestyle. The explicit nailing of the mobile phone to the wall or tree in the beginning of the journey is therefore part of the transition to the new life as a wandering craftsperson (e.g. Daschner, 2009) and has become part of the initiation rituals. It is also a transition to a lifestyle that seems very 'out of time' or anti-modern. At the same time, this is part of the attraction and leads to a great fascination: most articles about the wandering years mention the fact or even begin with it.

A crucial question is whether this important difference to mainstream society and its socialities is likely to continue. As another (re-settled) journeywoman proclaims:

> I claim that the mobile phone will win. AT THE MOMENT it is still the case that it is not well-received in many journeyman brotherhoods (*Schächte*), while it is explicitly forbidden in only a few. . . . However, I can let you know that most journeymen/-women have an email address, but how and when you check your mails is entirely up to you. So one at least does not have this constant availability that one has with a mobile phone.
>
> (Journeywoman, settled – quoted in Buchner, 2017: 139; translation by author)

With experiences as the one quoted earlier, however, where the liberation from current connectivity norm is hailed, it might not be the brotherhoods as much as the passage of knowledge and experience from a senior journeyman/-woman to the junior one that lets this rule continue. In fact, it might become one of the defining characteristics of the Walz, whereby both sociality and mobility are possible without the form of mediation now common. In a world where digital detox is either bought as a temporary escape from quotidian perils or ordered from above in larger companies, the idea of a three-year-long period of freedom might become a trademark.

It is important to note, however, that oral communication is not the only kind of communication used by these wandering craftspeople. The coins for the telephone were mentioned previously, i.e. using a phone booth or a phone call from work is not forbidden – nor are mobile phones that are not owned by the wanderer him-/herself. As Linda, a goldsmith, explains: "Once in a while, I can use the mobile phone of the master for whom I work. Or otherwise, I go to an Internet café, when I am planning my route and am looking for a new workplace."[20] Hence, temporarily accessing other people's communication devices is one solution; another is Internet access via other means. Linda is not the only one who additionally uses the Internet to find work and plan ahead. This has become a common trend in the last few years. Another wanderer, Theresa, differentiates between the traditional ways of finding work – going straight to the local carpenter, baker, or whatever your trade is and asking for work or following other travelling craftspeople's hints – and the new version:

> The wandering craftsperson 2.0 accesses the Internet, when he/she is looking for work. In many message boards for 'Wandergesellen,' one can exchange information about work and being on the road and one can also find hints about employers who take on wandering craftspeople.
>
> (Amrehn, 2016: 92; translation by author)

This applies to most – but as Amrehn herself also outlines, by far not all – of the wandering craftspeople. If, for example, there are only very few of the particular craft on the road (as in Amrehn's case – she is a church painter), one needs to use other existing networks to find work. In her case, this consists of contacting people she went to crafts school with and using the landline phone. Nonetheless, the trend seems clear: while constant connectivity is both forbidden and seen as not desirable, the wandering craftspeople are not anti-digital. Instead, they use up-to-date technologies to interact, communicate, and organise – but they do so much less frequently and much more consciously.

A contextualisation

> It remains controversial, which kind of mobility is legitimate and desirable, necessary or useful for whom.
>
> *(Wadauer, 2005: 22; translation by author)*

By far not all kinds of movements and/or all kinds of occupation are regarded (and treated) as equal. Mistrust and recognition are both reserved for the wandering person: the flâneur in Berlin 100 years ago suffered the same fate, when he was characterised as a thief because he was moving slowly and without an outwardly visible purpose through the city (Hessel, 1929). One instead needed to have a purpose or aim to be socially recognised. Journeypeople achieve this through their clothing, which underlines their line of work. Earlier on, I noted the distinction between autonomy and heteronomy, as well as production and adaptation (Rosenberg and Shannon, 2018: 5) as possible framings. When we now adapt this to journeypeople, a first declaration is easy: the craftsperson's journey is generally defined as useful, socially acceptable, and even honourable (see Wadauer, 2005). Other kinds of journeys could also be used to distinguish the travelling crafts: while also young and on the road, the gap year or semester abroad student, for example, has a different purpose and standing: his/her journey is less work- than experience-oriented, as well as less independent. Another way to map these movements would be as official or unofficial, as fleeting or permanent; one can ask whether the journey is officially registered or simply condoned (see Wadauer, 2005).

The journeyperson, in contrast, has a purpose. He/she is meant to improve his/her skills, to understand the world, to offer a temporary relief to the labour market while nonetheless enriching it. Phillips, when reviewing travels and their societal understanding over the centuries, underlines that travel was not common in earlier times and tended to be conducted only when necessary or when 'fate' was involved. Nonetheless, she emphasises, even these early accounts describe travels as transformative (Phillips, 2019: 43–44). This applies even more to those kinds of travels that are conducted 'for travel's sake'; i.e. those that have no other purpose than travelling itself. As Phillips emphasises, these continue in yet another tradition: that of pilgrimage, the travelling for enlightenment's sake. The mobile (social) figure of the pilgrim is often contrasted with other adventurers at the time: those who conquered the colonial worlds. These travels are critically described as only indirectly transformative; i.e. they changed the world (not always in a good way), but did not reflect enough about themselves and the travels (cf. Phillips, 2019).

The Grand Tour of the aristocracy, on the other hand, i.e. young men travelling the cultural centres over a period of three years, educating themselves and strengthening the existing social networks (see Phillips, 2019: 44–45), is characterised as a direct predecessor of the Wandergesellen today, although the latter actually reach further back.[21] While the Grand Tour helped to establish the idea of learning through travels more widely, it was still the privilege of the upper class.[22] In contrast, the journeyman represented a democratisation of (non-leisurely) movement across different classes, as a form of institutionalised travel that became increasingly socially accepted. While the transitory phase was still characterised by the need to show the wanderer's book to justify one's movement, tramping subsequently came to be "seen as a kind of education and often romanticized as the workingman's way to experience some of the values of the aristocracy" (Phillips, 2019: 47).

The Grand Tour was made possible by an elite network that consisted mainly of nobility who were ready to accept other nobility's travelling sons as guests. A crucial element was therefore a social network of trusted others (as also brilliantly described in Fermor, 2013). On top of this, knowledge of Latin or French as an elite lingua franca was an important resource to carry out the tour.[23] Hence, the second crucial element is communication and related forms of knowledge. This combination of well-established networks and specific kinds of communication established the trust necessary to move on, or rather to move on and find work and shelter.

If one compares this to other forms of youthful travels, they are either movement without work (such as interrailing) or work, but limited movement, such as au pair jobs. The closest comparisons are work and travel programmes or voluntary work abroad. Nonetheless, they do not carry the same weight of rules and regulations as well as risk, of e.g. being without a roof over one's head. Neither do they allow the traveller to build up further knowledge and reputation in their skilled job.

Looking elsewhere, the journeyman also shares certain traits with the figure of the tramp or vagabond (see Rolshoven and Maierhofer, 2012).[24] Nonetheless, this figure was (and is) perceived much more ambivalently. As he/she travelled on the same roads as the journeyman and was also looking for temporary work and shelter, there was no great difference on first sight. The journeyman's clothing (the Kluft), however, and the eventual social decline and urbanisation of the vagabond, have made the distinction much easier. The journeyman still offers a sense of lawlessness, but this is nowadays framed as 'freedom.' This freedom, however, is accompanied by unwritten – but clearly reinforced – rules. These are based in sociality and traditions; they are reinforced in the secret rules of the brotherhoods.

The journeypeople manage to organise their socialities primarily offline. They eventually become part of a fairly close-knit – albeit mobile and shifting – community with a strong network structure and culture. This community constantly provides the necessary information via oral communication. However, the link to older social networks (family, friends) is kept to a minimum. During the wandering years, work, food, and shelter are organised primarily via face-to-face interactions. The public presentation of the trade, on the other hand, takes place in media representations (films, videos, newspaper articles, etc.) and on official websites organised by those who have settled down.[25]

Another crucial aspect of the journeymen's travels is that they have a time limit to them. This is set by tradition: three years and one day are the minimum; some people carry on for another year or two. Indeed, setting a return date is often a difficult task and one that many tend to avoid. This time delineation, however, provides a useful framework. Time can be counted down; it can serve as a source of reassurance during difficult times. Spatially, the journeyman has a no-go-zone and is otherwise free to go wherever his/her will and desire or chance take him/her (possible limitations have already been mentioned). Nowadays, many journeymen/-women leave the German-speaking countries or even Europe, working in South America, the USA, sometimes Africa or even Australia. The freedom of movement (albeit with limited resources) is used extensively. Crucially, however, journeypeople have a place to return to. Their homecoming is an important part of the ritual. This differentiates the wandering craftsperson from most other people on the road, who often tend to be socially expelled (as Nail [2015] shows for the figure of the migrant). The travelling craftsperson is a figure defined by movement, but a movement that ultimately comes full circle.[26]

The travelling craftsperson: a mobile figure of (non-)mediation?

As has hopefully become clear, the travelling craftsperson is a highly mobile social figure. While limited in their mobility by some rules (no payment for travels, need to work, the no-go-zone, etc.), one main characteristic of the travelling craftsperson is his/her movement. While it is not daily travel, he/she is expected to move on after either a few days or weeks, sometimes months. The movement is also somewhat visible to the outside world, because tramping is not an extremely common feature on German roads nowadays. It contributes to sociality in the sense that it often affords a chance to talk about the tradition of the Walz, to explain the craft – to get to know people (see e.g. Amrehn, 2016: 122 ff.).

If, as Hannam et al. (2006) argue, mobilities are enabled by moorings; the question is whether these authors have defined these moorings broadly enough when they speak of "spatial, infrastructural and institutional" moorings. What about social network kinds of moorings, i.e. socialities, which can emerge offline as well as online? The socialities presented here and the related mobile figure are based on a dependence on the social. Located in work commitments, but also in the extension of skills and experiences, the journeys by the journeypeople could be described as working-pilgrimages. These journeys originate somewhere and also return to their anchor. Even the longing in between, the homesickness, can provide a positive force, since there is something to look forward to. At the same time, these travels allow an extension of the self, based on secure bounds of the social. Media are not necessary to secure these, as long as the traditional communication forms are practised and updated. Instead, it is the more or less loose framework of rules and traditions that is nowadays coupled with an equally loose use of existing communication technologies (not so much mobile phones or smartphones, but Internet facilities) that allows these socialities to thrive.

Mediation is interesting when applied to *non-use* of dominant media. If "mediation is a practice in which producers, subjects and audiences take part, and take part together" (Silverstone, 2007: 38) and in which 'listening to each other' becomes the normative aim, then non-use of digital media seems on first sight a problematic way of pulling out of the larger conversation, since nowadays this seems to be where participation – to a large extent – takes place. However, the communication and interaction rituals presented here promise a more complex picture. Put rather simply, it is exactly the non-use that creates a basis for 'taking part together.'

Mobile phones – especially smartphones and their associated applications — mediate immediacy. Their non-use, on the other hand, mediates non-immediacy and something different than the smartphone kind of immediacy. The immediacy can be found in individual encounters, often with strangers, which – thanks to their visible set of clothing – take place frequently; while non-immediacy characterises the journeypeople's 'inside' sociality, which often takes place in groups. Sociality of the latter kind, but also information-seeking and some forms of entertainment, have to rely on slower forms of communication, which require more effort than those forms of communication that are currently widespread. These range from networked face-to-face communication, in which information is passed on through other people, to finding a phone booth or emailing when in an Internet café. One consequence of this mediation of non-immediacy is a more conscious enactment of communication. The effort needs to be made.

Communication has been key to this form of mobile sociality, but it seems to become even more crucial now, when mediated communication is becoming more prominent in the rest of the world. The Walz is affected by this, too – more attention is given to the Walz on websites, online fora, etc.; more information about the Walz can be easily accessed on similar sites, but also via networked communication, and work can be organised more easily. But the emphasis on the organisation of the social networks via oral networks enacts a rather specific form of sociality – and an exclusive one that is only accessible to insiders, while it is also highly visible to outsiders. It is an exclusionary practice, based in mobility, which creates a layer of security and stability in this otherwise rather unstable social world. This emphasis on the social is the reason why I call it a mobile (social) figure.

It is, however, not non-mediation, but mediation in the sense of Roger Silverstone (as outlined in Hartmann et al., 2019: 5–10), who analyses everyday life as a moral and social space – a description that easily fits the framework of the wandering years. Silverstone furthermore "presumes that it is in the everyday, and above all in the detail of the relationships that are made with others and which constitute everyday life's possibility, that our common humanity is created and sustained" (Silverstone, 2002: 761) and describes an "ethics of care and responsibility." For him, however, everyday life is very much about continuity, about ontological security. This is less clear in the wanderer's life: he/she repeatedly needs to create this trust – both to be trusted, but also to trust. This takes place through rituals and through communication (and partly through ritualistic communication), but also through the existing network structures. I here follow Silverstone's claim that "no ethics of, and from, the everyday is conceivable without communication, and that all communication involves mediation, mediation as a transformative process in which the meaningfulness and value of things are constructed" (Silverstone, 2002: 761). Additionally, it is important to reiterate that the mediation in our case study is very much non-media-centric;[27] i.e. it takes place mostly outside of technologically mediated spheres.[28] It is this special combination of an outward backwardness coupled with a subtle up-to-date-ness (the wandering craftsperson 2.0), plus the knowledge of what non-use enables, that make this mobile (social) figure specific and future-oriented. It is the partially non–media-centric mediation, coupled with mobilities, that makes up the socialities of the wandering craftsperson. Non–media-centric translates very concretely into a non-use of media, but also emphasises that the aforementioned mediation in the sense of 'listening to each other' is taking place more intensely than elsewhere. Indeed, the Walz could be described as an extended period of intensive listening to oneself and the peer group.

But it is also the particular combination of past and present, modernity and anti-modernity, which makes up the journeyman/-woman. It underlines a form of sociality that allows a strong individual, but one that is clearly embedded in and adheres to specific rules of the collective, whereby the freedom is clearly based on a set of constraints. Nowadays, people choose such a life – and seem to (mostly) enjoy this clear-cut framework of sociality. The anti-modernity aspect is expressed best in the ambivalences of mediation. It is the most visible anti-act, and thus serves to clearly differentiate this group from others. It is this combination which might become a model for other socialities in the 21st century.[29]

Notes

1 I would like to thank the Mobile Socialities network (especially the participants of the workshop in Lund in autumn 2019), as well as my reviewers, David Morley and Orvar Löfgren, for their valuable comments on earlier draft(s). All omissions are clearly mine and some thoughts still need to be developed further. The process, however, is already very inspiring.

2 Available at: www.spiegel.de/karriere/wandergesellen-letztes-grosses-abenteuer-auf-der-walz-a-1054643.html (accessed February 18, 2020).

3 Available at: www.z-wie-zimmerer.de/karriere/die-walz/ (accessed February 18, 2020); translation by author.

4 As will become clear throughout the chapter, there are journeymen and journeywoman alike, and they will appear as such in the text. However, whenever the generic is mentioned, I will use the term 'journeyman' or 'craftsman.' This has to do with the flow of reading, but mostly with the dominant tradition, which originally allowed only men to pursue this kind of travel. This subsequently shaped many of the rituals and rules. Where and how these apply to women will be part of the discussion provided here.

5 Simmel's writings, for example, include chapters on the poor and the stranger, as well as the nobleman/-woman (Simmel, 1992).

6 I tend to speak of craftspeople or craftsperson when I am addressing the general level, while I speak of the craftsmen or -women when I am addressing particular individuals and their stories.

7 Elsewhere, I write about the homeless – another important (im)mobile figure.

8 Further questions that arose through the analysis would best be answered in an empirical research project. An application is currently being prepared.

9 An example of this, as Magnus Andersson rightly pointed out to me, is the low esteem in which the sailor – another globetrotter – was generally held.

10 An interesting question here is in what form 'becoming a servant' served as a differently gendered form of 'mobile work' away from home, but in a housed environment (again, I am grateful for Magnus Andersson for pointing this out).

11 The decline of the guilds also let to a huge decline (nearly a demise) of the number of people on the Walz.

12 For a good overview of the pieces of clothing, see e.g. www.reidl.de/zunftbekleidung-und-zubehoer.

13 If someone ended their journey prematurely (before the three years and one day were over), he/she lost their honourability and the earring was ripped out (and left a mark) – this is the basis for the German expression 'Schlitzohr' (literally a ripped or cut ear), which translates as chiseller.

14 For a particularly nice (but not very typical) version of a current Wanderbuch, see www.typowalz.de/wanderbuch/ (accessed March 3, 2020).

15 Available at: www.rechtschaffene-zimmerer.de/informationen-zur-wanderschaft.html (accessed March 2, 2020).

16 „Wir wollen weltoffen reisen und Andere und uns selbst kennen lernen. Ferne Länder und Kulturen interessieren uns genauso wie die verschiedenen Mentalitäten unserer Heimat. Als Fremde wollen wir eine Zeit lang mit Fremden leben und arbeiten, um mehr von der Welt zu kennen und verstehen als ein Stubenhocker. Wir sind stolz auf unser Können und fühlen uns der Tradition verpflichtet. Das Handwerk und die gemeinsame Erfahrung der Tippelei halten unsere Gemeinschaft zusammen. Die Weitergabe des lebendigen Brauchtums ist uns genauso ein Anliegen, wie die Ausübung unseres Handwerks heute und in Zukunft." (Website of the Rechtschaffen-Fremde-Schacht; Available at: www.rechtschaffen-fremde.de/seite/wanderschaft-heute [accessed March 4, 2020]).

17 The recently introduced minimum wage was originally discussed as a potential threat for wandering craftspeople, but usually they are paid more than the minimum wage, anyway. They also have a special health insurance throughout their journey. Since it is skilled work, they are usually not in direct competition with day labourers or other unskilled workers looking for jobs.

18 These eight organisations/brotherhoods are: Gesellschaft der rechtschaffenen Fremden (actually two closely related organisations), Rolandschacht, Freie Vogtländer Deutschlands, Fremder Freiheitsschacht, Freier Begegnungsschacht, Bruderschaft zur Rose Quedlinburg und Döllnitz Gewandhausgesellen Leipzig, and Heidberg Frei-Zunft Güstrow und Moratneustetten.

19 „Es war eigentlich am Anfang eine Befreiung – weil man einfach nicht erreichbar sein muss und auch nicht erreichbar ist. Und deswegen hat man auch nicht das Gefühl, irgendetwas zu verpassen. Du bist einfach da, wo Du bist; hast das Gespräch, was Du gerade führst; trinkst das Bier, was Du gerade vor Dir stehen hast; isst das Essen, was da ist – und bist einfach da. Und das unterscheidet, finde ich, einen Wandergesellen immens zu einem normalen Menschen, würde ich sagen."

20 Available at: www.lokalplus.nrw/nachrichten/verschiedenes-olpe/goldfisch-aus-dem-norden-seit-dreieinhalb-jahren-auf-wanderschaft-34355 (accessed March 4, 2020); translation by author.

21 Phillips places the beginning of the travelling/wandering craftsmen (the journeyman) in the mid-19th century only. This contradicts most other accounts of their history.

22 In contrast to something like the 'Grand Tour,' quotidian mobilities were more widespread amongst the working class. This is also one of the reasons why the question of female flânerie (and flânerie in general) is so clearly a question of class.

23 I am grateful to Orvar Löfgren for pointing this out.

24 The ragpicker (see e.g. Highmore, 2002: 63ff.) is another such figure.

25 See, for example, www.facebook.com/permalink.php?story_fbid=1225824450765529&id=160949777253007 (accessed March 9, 2020).

26 Another difficulty that presents itself through the academic study of such practices is that, while we tend to listen to the individual story, our main interest lies in its abstraction. Or, as one of the wandering craftsman states in an interview: "Suddenly being of interest to people not as a person, but as a figure – because you represent something that fascinates or interests people or whatever. Yes, and as a result, most of the encounters with people are actually quite cliché" (original: „So auf einmal für die Leute nicht als Person, sondern als Figur von Interesse zu sein – weil man irgendetwas repräsentiert, was die Leute fasziniert oder interessiert oder sonst was. Ja, und daraus resultierend sind die allermeisten Begegnungen mit den Leuten eigentlich ziemlich klischeeartig") (quoted in Bohnenkamp and Möbus, 1995: 95).

27 One interesting twist here, though, is that the media seems to be quite fascinated with the lack of media in use.

28 Additionally, the study tries to follow a non–media-centric path. It does not put the question of media and technologies first, but concentrates on broader issues around the everyday life of a wandering craftsperson.

29 One could instead end with a caveat, since the tendency to romanticize the travelling craftsperson looms large – for any academic, as much as for everyone else. This is partly due to a tendency by the craftspeople themselves to slightly colour their travelling time positively or to at least describe it as a positive rite of passage. As Steffen, a travelling carpenter, stated: "I did imagine it all a bit more poetically; I built a few castles in the air" (quoted in Bohnenkamp and Möbus, 1995: 110, translation by author). However, the power of imaginary castles should not be underestimated, either, especially in times like these.

Bibliography

Amrehn, T. (2016) *Königin der Landstraße: Meine Jahre auf der Walz*. München: Piper.

Bauman, Z. (1996) "From pilgrim to tourist – a short history of identity", in S. Hall and P. Du Gay (eds.) *Questions of Cultural Identity*. London: Sage, 18–37.

Bayerischer Rundfunk (2020) "Was junge Handwerker auf der Walz fürs Leben lernen". Available at: www.youtube.com/watch?v=RAP_PhkENxE (accessed March 9, 2020).

Bohnenkamp, A. and Möbus, F. (1995) *Mit Gunst und Verlaub! Wandernde Handwerker: Tradition und Alternative*. 3rd edition. Göttingen: Wallstein-Verlag.

Buchner, L. (2017) *Über das Leben von Handwerksgesellen auf der "Walz": Eine empirische Analyse*. Münster: LIT-Verlag.

CKnet Videos (2016) "Begegnung in Attendorn – Handwerks-Gesellinnen auf der Walz". Available at: www.youtube.com/watch?v=483sr0KXgkk (accessed March 4, 2020).

Daschner, J. (2009) "Auf der Walz", *Documentary Film*. Available at: www.alleskino.de/empfehlungen/auf-der-walz-2009-stream.html (accessed March 9, 2020).

Der Spiegel (2006) "Craftsmen awandering". Available at: www.spiegel.de/international/funny-looking-hobos-craftsmen-awandering-a-416564.html (accessed March 4, 2020).

Eddy, M. (2017) "Journeymen in an age-old rite", *New York Times, International Edition*, 2. August 9.

Fermor, P. L. (2013) *A Time of Gifts*. London: John Murray.

FNP (Frankfurter Neue Presse) (2018) "Zahl der Wandergesellen hat zugenommen – Drei Jahre unterwegs. Available at: www.fnp.de/hessen/zahl-wandergesellen-zugenommen-drei-jahre-unterwegs-10402184.html (accessed March 4, 2020).

Gebhardt, W. and Hitzler, R. (eds.) (2006) *Nomaden, Flaneure, Vagabunden. Wissensformen und Denkstile der Gegenwart*. Wiesbaden: VS Verlag.

Hannam, K., Sheller, M. and Urry, J. (2006) "Editorial: Mobilities, immobilities and moorings", *Mobilities*, 1(1): 1–22. doi:10.1080/17450100500489189.

Hartmann, M., Prommer, E., Deckner, K. and Görland, Stephan O. (2019) "Mediated Time", in M. Hartmann, E. Prommer, K. Deckner and S. O. Görland (eds.) *Mediated Time. Perspectives on Time in a Digital Age*. Cham: Palgrave Macmillan, 1–21.

Hessel, F. (1929) *Spazieren in Berlin*. Leipzig and Wien: Verlag Dr. Hans Epstein.

Highmore, B. (2002) *Everyday Life and Cultural Theory: An Introduction*. London: Routledge.

Hobsbawm, E. and Ranger, T. (1992) *The Invention of Tradition*. Cambridge: Cambridge University Press.

Kesselring, S. (2006) "Pioneering mobilities: New patterns of movement and motility in a mobile world", *Environment and Planning, A* 38: 269–279.

Lemke, G. (2002) *Wir waren hier, wir waren dort: Zur Kulturgeschichte des modernen Gesellenwanders*. Köln: PapyRossa.

Moebius, S. and Schroer, M. (eds.) (2010) *Diven, Hacker, Spekulanten: Sozialfiguren der Gegenwart*. Berlin: Suhrkamp.

Nail, T. (2015) *The Figure of the Migrant*. Stanford: Stanford University Press.

NDR Doku (2019) "Die nordstory: Auf der Walz: Heute hier, morgen dort", *Documentary Film*, March 22. Available at: www.ndr.de/fernsehen/sendungen/die_nordstory/Heute-hier-morgen-dort-Auf-der-Walz,sendung880980.html (accessed March 9, 2020).

Phillips, B. (2019) *Learning by Going: Transformative Learning Through Long-Term Independent Travel*. Wiesbaden: Springer.

Rolshoven, J. and Maierhofer, M. (eds.) (2012) *Das Figurativ der Vagabondage: Kulturanalysen mobiler Lebensformen*. Bielefeld: Transcript.

Rosenberg, E. and Shannon, K. (2018) "Im/mobility: Connecting disciplines – an editorial introduction", *Geography Research Forum*, 38: 1–15.

Silverstone, R. (2002) "Complicity and collusion in the mediation of everyday life", *New Literary History*, 33(4): 761–780.

Silverstone, R. (2007) *Media and Morality: On the Rise of the Mediapolis*. Cambridge, MA: Polity.

Simmel, G. (1992) *Soziologie: Untersuchungen über die Formen der Vergesellschaftung, Gesamtausgabe Band II*. Frankfurt: Suhrkamp.

Syvertsen, T. and Enli, G. (2019) "Digital detox: Media resistance and the promise of authenticity", *Convergence: The International Journal of Research into New Media Technologies*, 1–15. doi:10.1177/1354856519847325.

Wadauer, S. (2005) *Die Tour der Gesellen: Mobilität und Biographie im Handwerk vom 18. bis zum 20. Jahrhundert*. Frankfurt: Campus Verlag.

Walbröhl, T. (2015) "Wandergesellen und das Problem mit dem Mindestlohn", *Tagesspiegel*. Available at: www.tagesspiegel.de/wirtschaft/handwerker-walz-bedroht-wandergesellen-und-das-problem-mit-dem-mindestlohn/11309432-all.html (accessed March 9, 2020).

Winter, C. (2006) *Die Medienkulturgeschichte des christlichen Predigers von den Anfängen bis Heute. Entstehung und Wandel eines Mediums in kommunikativ-kulturellen Vermittlungsprozessen*. Wien: Nausner & Nausner Verlag.

Zschornack, F. (2015) *Franz im Glück: Meine Wanderjahre auf der Walz*. Köln: Bastei Lübbe.

18
TRANSPORTED IMMOBILITY

Annette Hill

Introduction

This chapter uses Roland Barthes' concept of transported immobility to understand the contrary experiences of living in a digital media blackout zone. By using Barthes' protocol of attention to contingency in everyday life, I explore the wi-fi tethering experience in the Swedish forest, in which I am tethered to particular indoor and outdoor places, and dependent on intermittent digital media connection (see Figure 18.1). Tethering is a useful term to consider for the contrary experience of living in a digital media blackout zone. Such tethering fixes me to a place in the rural forest that enables wi-fi, provides access to and a flow of communication during in that particular moment of connection, and is also a source of frustration as rural infrastructures and inclement weather routinely affect the experience. The chapter explores a double meaning of tethering. First, in the media context, tethering is the act of staying fixed in one place where telecommunications devices can enable effective data movement; this creates a fixity and flow in the media environment. Second, in the mobility context, the tethering of a human, non-human, or object, where there is the action of

Figure 18.1 Tethering: where the wi-fi is in rural Sweden.

Source: Photograph by Annette Hill

restricting or disabling movement, creates a regulation and constraint in everyday life. This tethering experience brings into relief assumptions we often make about normative digital media experiences. In this chapter, I offer reflections that challenge general assumptions of the always already connected digital society. What does rural disconnection mean to these assumptions, with a reality that is defined by media breakdown, an experience that is contrary, where fixity and flow are constitutive of the media landscape? These other stories of the media and its social and technical affordances compel us to review dominant thinking about digital society. As a rural experience at the margins of socio-technical visions and dominant media theories, such an experience signals the mirage of a digital society for rural citizens in specific regions where there are topographical and infrastructure challenges to communication and transportation.

The concept of transported immobility

The inspiration for this chapter arises from a chance encounter with a leftover book *The Eiffel Tower and Other Mythologies* by Roland Barthes (1979). Left behind by a friend when they moved out of their office, the book lay on the bottom shelf, part of a collection of the author's work that included more recognisable titles: *Mythologies, The Pleasures of The Text*, and *Camera Lucida*. This particular collection of thirty essays presents a "sociopathology of everyday life", drawing on Barthes' writings from the mid-1950s and originally published in *Les Lettres Nouvelles* (Barthes, 1979: back cover). One of the stories concerns itself with the experience of eating in a railroad dining car, an episode described by Barthes as "transported immobility". In this chapter, I want to take Barthes' concept of transported immobility and allow it to travel from the dining car of a French train to the forests of Sweden and to my own experience of living in a digital media blackout zone.

In "Dining Car", Roland Barthes reflects on the contrary experience of eating an elaborate meal in an upscale restaurant carriage. A galley kitchen produces gourmet food: dishes are given complicated names, the menu extends across five courses, occupying the traveller on their journey. Barthes meticulously notes thirteen waves in the dining car experience, following a certain order from being seated at the table, through the various courses, to the digestif and paying of the bill. Everywhere he observes "an implacable materiality" (Barthes, 1979: 143); there is silver cutlery on the linen tablecloths, napkins create a "cloud of napery" at the table, even the breadbasket is wrapped in a napkin. This entire mise-en-scene Barthes' describes as a "mirage of solidity" (Barthes, 1979: 143), referring to the material and symbolic construction of security within the contingency of the traveller's experience; the mobility of travellers is transformed into the illusion of sitting in a French restaurant on the streets of Paris, and as the shipping stations slip by, they each participate in the experience of transported immobility.

The thirteen waves of the dining car experience "conceal by a protocol of attention its very contingency" (Barthes, 1979: 141). Time and space are not in short supply: there is time to eat five courses, space to occupy over a long period of time, stomachs to fill en route. For Barthes, the protocol of attention surrounding the elaborate dining car experience is a mythic substitute, a "spectacle of stability" (Barthes, 1979: 144). Such an experience is not uncommon when travelling; think of the dining rooms on cruise ships and the endless entertainment on offer, all designed to create a mirage of solidity and space whilst travelling in cramped cabins across the oceans in all weather. The heavy materiality that Barthes observes signifies an "illusion of immobility" (Barthes, 1979: 144). He writes: "each constraint seems to produce its contrary freedom, each gesture is a denial of its original limits" (Barthes, 1979: 141).

We can extend this idea of transported immobility from its original story of the railroad dining car to other kinds of contrary experiences. Here, the juxtaposition of mobility and immobility is suggestive of how movement and transportation are often accompanied by constraints on mobility in a variety of contexts; for example, the railroad dining car as a mode of transportation for a certain type of passenger who can afford to spend time and money in the dining car. It is also suggestive of how objects and embodied experiences are connected to subjectivities, stories, and myths. What Barthes describes in the idea of transported immobility is how some experiences have a contrary mixture of freedom and constraint, of spectacle and mundane reality.

Barthes' thirteen waves in the dining car experience can be linked to Georges Perec's writings on the infra-ordinary (Perec, 2008). His emphasis on the micro-scale in everyday life and technology is useful in analysing the ordinary and habitual within the idea of transported immobility. For Bassett (2017: 124), "what Perec was looking for" through his observations and writing process was the value of micro-description in re-defining ordinary spaces, paying attention to the mixing of the old and the new in everyday life. For example, in *Species of Spaces*, Perec muses on the bedroom as a space ("I like ceilings"; Perec, 2008: 18), which inspires him to invent a typology of bedrooms: there are nine in total, including number five on the list "makeshift beds: settee, moquette plus cushions, carpet, chaise longue, etc.", and number nine "unusual conditions: nights on a train, on a plane, in a car" (Perec, 2008: 23). Perec reflects on the "resurrected memories of these ephemeral bedrooms", asking abstract questions ("what does it mean to live in a room?"), and making concrete observations, such as "placid small thought number 1" on cats finding favourable places to sleep in bedrooms (Perec, 2008: 24–25). As Morley (2007: 95) notes, "the re-construction of the taken-for-granted, unspoken assumptions that frame our everyday actions and statements" is significant to a sociology of everyday life. Perec's methodology offers an exhaustive attention to the micro-universes we inhabit, allowing for comparisons within and across routines and spaces; inventories of actions, objects, people, and places; and enumeration of the things we encounter (Morley, 2007: 94–95).

Barthes' "protocol of attention" (Barthes 1979: 141) offers precisely this contribution to a sociology of the everyday. Three types of linen, which I outline in what follows, enumerate an upscale restaurant experience. For example, Barthes compares linen napkins to paper ones and ready-made meals. He notes the habitual action of snatching a paper napkin on the go, and the ritual unfolding of the linen napkin in the dining car. To hold a napkin, to be draped by linen, means understanding how that napkin is made from cotton, washed, starched, ironed, and carefully folded into fancy shapes, such as a fan or bishop's hat. Barthes' attention to the linen napkin is suggestive of the taken-for-granted old world restaurant experience, the mirage of solidity. In fact, the dining car has three kinds of napkins to underscore fine dining: the linen tablecloth sets the scene, the linen napkin enfolds the traveller, and the wrapped bread is an invitation to start the meal. The "cloud of napery" (Barthes 1979: 141) embeds the traveller in a mirage of solidity, that "implacable materiality" which Barthes (1979: 143) notes is so significant to transported immobility.

In this chapter, I take inspiration from Barthes' protocol of attention and Perec's micro-description to reflect on the small scale experience of rural life and enumerate the details of the wi-fi tethering experience in the Swedish forest, where I am tethered to particular indoor and outdoor places, and dependent on intermittent digital media connection. Such tethering fixes me to a particular place in the rural forest that enables wi-fi in that moment of connection, while also being a source of frustration as rural infrastructures and inclement weather routinely affect the experience.

Living in a digital media blackout

In this section, I reflect on tethering through a detailed account of my experience in the forests and waterways of Ydre (see Figure 18.2). This experience is particular and unusual, in that it is linked to personal choices to live off the grid (see Vannini and Taggart, 2014) of communications infrastructure in the Swedish forest; nevertheless, the experience serves as a starting point for a broader critical reflection on dominant thinking about digital society – this is a rural experience at the margins of such a technological positioning of the country of Sweden as a place of digital connection. The micro-details of this experience serve as "a protocol of attention" (Barthes, 1979: 141), to the mirage of a digital society for particular rural citizens in the kinds of topographical regions such as Ydre, where energy and communications are a limited resource and subject to breakdown on a daily basis.

There is something contradictory about living in a rural area without robust digital connection. Ydre lies in the Swedish county of Östergötland, a region rich in forest and lakes, with a population of three and a half thousand people. The topography means there are challenges – twists and turns in the roads, elevated areas, small valleys: a geography where wi-fi technologies and mobiles die and come to life again in odd spots in the landscape.

Our house was the one-time home of the local forest and game warden, appointed by the Swedish crown to manage the upkeep of the land and hunting licences. The property was originally built in 1864 by a private forest company with a licence from the Swedish state to conduct forestry; as the forest industry developed in the 1800s, the house and land were bought by the state for the appointed forest and game warden, whose main job it was to oversee the forests and lakes in the area, ensure the replanting of the trees (according to laws dating back to the 1700s), and manage the hunting of elk, deer, and other small game. The house was built to signify authority, with a gravel pathway to the front porch that invited forest workers and local inhabitants to stand and admire the hand-crafted wooden latticework while knocking on double doors – the entrance to the main office with three large windows overlooking a small lake and the Silver Creek which runs through the region.

Figure 18.2 The waterways of Ydre's forest lakes and rivers.

Source: Photograph by Annette Hill

The warden was an important manager of the state interests; he and his family lived in the house and the state ensured they had access to resources, including robust communications. In the late 1800s, daily communication would have been achieved through the state postal service. This public communications infrastructure was manned by post farmers ("postbonden" in Swedish), a long-established postal system of riders, horses, and post houses, dating back to 1663 and the establishment of efficient postal services during wartime. The state postal service modernised in the 1900s to include bicycles, boats, and cars for the postal workers. In the 1920s and 1930s, further layers of infrastructure were added; electricity and a telephone line were installed in the house. These significant layers of infrastructure would have transformed the house into a local hub of energy and communication. The telephone was adopted in the 1880s in Sweden, across which homes would use the phones of Lars Magnus Ericsson. Our house may have accommodated the imposing Bords Telefon by LM Ericsson (BC310), a grand piece of equipment that would adorn an office desk, or possibly a more modest bakelite Eriksson phone (DBK1101). There are still traces of the original posts in the forest, and on the side of the house, one still finds ceramic telecommunications connectors.

We can trace the decline of the state's interest in managing the local forest and game – and by extension, the rural infrastructure for communications and energy – by two key events. The house was rented out to local persons during the 1950s, no longer the dwelling for the forest and game warden. At some point during this period, lightning struck the house, travelling along the telephone line and into the building, destroying one wall in the process. The wall was rebuilt, but the telephone line was left unrepaired, thus marking the end of telephone communications infrastructure in the local area. In the early 1990s, the state dismantled the entire forest and game system, opting for centralised and dispersed nature and forest specialists. The house was then sold to private persons, retaining the distinctive three crowns on the porch by the double doors, the house name of Kronojagarbostaden ("crown hunters house"), and the defunct telephone points and obsolete wires which comprised the last public communications to be installed and maintained by the Swedish state.

The history of the home signals the ebb and flow of layers of infrastructure for energy and communications. The house and grounds were on the grid (connected to regional infrastructure) during the years of the state interest in forest and game, which included delivery of energy (the water mill, wood, and electricity), communications (post and telephone), and water (underground well). Second, the house and grounds were off the grid from the standpoint of privatisation during the 1990s of the forestry industry, energy supply, and communications and water works, all of which were previously managed by the state. The significance of this on-the-grid, off-the-grid status of the house and local area in relation to local politics and privatisation of infrastructures and resources offers a useful context for the analysis in this chapter.

We were the third private owners of the home, arriving in the mid 2000s. We chose to live in Ydre because of the natural landscape, rich rural history, and personal connections with the area and local residents. Alongside installing running water from an underground source, repairing the woodburning stoves and drilling for geothermal heat, we took on the task of a communications upgrade. This involved locating, through trial and error, a mobile communications hotspot nearby in a blueberry patch that proved to be rich both in forest fruits and a network connection. The network was maintained by Telia 2, and with a subscription service to this provider, we could access it from the hotspot. In the summer, I would take the dogs and let them bound around the forest while I caught up with emails or made phone calls. In the winter, we rationed digital communication to accommodate the scarce resources of signal strength in the cold climate; winter and early spring offered a stable 2G signal, but shorter days and freezing temperatures restricted digital media activities due to the darkness and cold weather. In the summer and early autumn, there were longer days and more temperate weather, but a weaker signal once trees came into leaf. I became tethered to this 2G spot.

We invested in a range of routers and signal boosters, installed by a father-and-son team from Stockholm who ran their own business troubleshooting digital media blackout zones. Thanks to this technical intervention, the house became 4G-enabled with a router that offered a reliable data and voice connection in one small area by the living room window. In this location, older models of mobiles, laptops, and computers seemed to work best. An Eriksson Shark Phone, complete with its iconic fin, was a lifeline. This R310 phone, made by the same manufacturer of the LM Eriksson phones that may have occupied the house in the 1930s, was released in 2000 and designed for rugged lifestyles, with the colour of burnt orange; its distinctive black fin contained a rubberised antenna that was robust enough to maintain a stable network signal.

My regular routine involved remaining tethered to this indoor spot for phone calls and emails. Heavy digital work involved taking the car to a layby or the local supermarket. There, I would join other inhabitants also taking advantage of an area with a strong signal, a shared hotspot for multiple people, and reliable network connections. Local inhabitants maximise these hotspots. The supermarket car park is not an upscale digital marketplace, or a social street of the kind described by Polson (2018) in Havana. During the summer months, we sat alone in our parked cars, Skyping with family or downloading files, a living example of Raymond Williams' (1974) notion of mobile privatisation. In the winter, the laptops lit up the dark car park. Now another form of tethering took place – the cars stationary but the people both mobile and immobile in their social relations.

When at home, everyday media routines hark back to the old ways. Sitting in the same place for phone calls, or sending and receiving emails; restricting data traffic rather than always online. If our neighbours are home, it's a bit like sharing a party line; we negotiate data usage in a neighbourly way, mornings for work, evenings for streaming, but not all at the same time as wi-fi is a limited resource. We often stream drama series after dinner of an evening, knowing that our neighbours, when at home, like to watch films on their DVD player. When our neighbours needed a robust wi-fi signal for working away, they parked themselves outside our house for some intense activities, negotiating for sole access to the data traffic for a block period. Here, we can observe a neighbourly exchange of resources that highlights the significance of equitable social relations. We maintain the digital communications infrastructure and share the resources together; and we maintain civility in our social interactions, observing and respecting varying work and entertainment practices that may rely on a wi-fi connection. In such a setting, infrastructure and social relations can and do break down, and it is vital to know how to repair mobile communications and social interactions.

This example of a digital media blackout illustrates Barthes' concept of transported immobility. It's that contrary experience of going out on a forest walk to then remain still in a hotspot, or driving in the car to find a place to park and then go online. We see how vital old media can be – broadcast television and radio all work fine; it's the Shark Phone that connects to the mobile network, the old MacBook that never crashes when endeavouring to update itself. These tried and trusted old media devices become much loved objects – literally lifelines.

As noted by David Edgerton (2006) in *The Shock of the Old*, a user-based perspective on technologies in specific places at particular times opens up our understanding of how old and new media mix together in our everyday experiences. Edgerton points out that a user perspective challenges the agenda set by new media promoters: "when we think about information technology we forget about postal systems, the telegraph, the radio and television" (Edgerton, 2006: xvi). In avoiding an account that focuses on the most modern, the fastest internet connection, or the newest mobile device, my reflection on the traces of old media technologies – such as the telephone connectors – and the more contemporary devices – the Shark Phone – signal that innovation and speed are not the most valued technological terms in this place. Of most value is reliable connection: technologies of trust rather than technologies of acceleration or innovation.

The tethering experience

To take a cue from Barthes' thirteen waves, what are the waves that give momentum to the journey from home to the hotspot in the blueberry patch? I count nine waves in the experience, waves which have temporal and geographical momentum, from exiting the home, crossing over the grounds of our property into the forest, the moment of digital connection, and the return journey. The intention of noting these nine waves is motivated by my engagement with the experience as contrary, signalling a playoff between being connected and disconnected, the personal affordances and structural (dis)affordances enmeshed in the experience (Mannell, 2019). The details in the nine waves are there to energise other ways of understanding transported immobility and the experience of tethering (see Figure 18.3).

First, there is the gathering of materials for the data transfer, phone calls, and emails, including charging devices so the technology is operating at full capacity. In this preparation phase, there is a certain amount of forward planning, such as emails drafted in advance, and preparation for the weather conditions, avoiding rainy days or wearing fingerless gloves. The last thing to do before going out the door is to gather the dogs for the short walk to the forest hotspot. Thus, this preparation phase includes three waves – data and device preparation, weather and clothing preparation, and human and non-human attachments.

Next comes the walk on the lawn, the crunch of gravel underfoot, unlatching the metal gates and crossing the private road and creek that runs under and alongside it. The Silver Creek is an ancient waterway protected in a nature reserve; it's home to otters, crabfish, and a rare kind of beetle. At one time, the creek was a waterway for forest timber, with a mill on the river

Figure 18.3 Tethering and play.

Source: Photograph by Annette Hill

bank; now, all that remains are some foundations to the building and an old millstone which you can spot as you cross the road. This double crossing has visible and auditory marks, signs of both a natural waterway and manmade industry. Once in the forest, the nature reserve sign, there to mark the Silver Creek, serves as the wi-fi spot. Through a process of search and locate, we found this hotspot in the first year of living at the house. This journey phase also has three distinct waves – exiting the boundary of the home, crossing the waterway and road, and finding the signpost for nature and digital connection.

To stand in the blueberry patch is to feel a physical positioning of the body and the device tethered to the signal. There is a small boulder that serves as a makeshift seat or low table. This tethering of the body to the place fixes you in the natural landscape, at the same time as the technology takes you places. After a while, the dogs circle back, rolling in the blueberry, nipping at the berries when in season. It's a reminder this place-making practice has another natural meaning for animals. In the winter, you can spot deer tracks or small signs of hares searching for food; herons – birds so associated with this region that a local town is named after the species – call to each other. The experience is marked by nature, literally with the sign and in the feelings it evokes as you return to home, the digital work done, and the nature work always in motion. This tethering phase has three waves that are variations on place making, from physical marker, to technological communication, and environmental interaction.

Tethering is a technical designation for linking multiple mobile devices to electronic spaces and physical locations of signal strength. Common examples are tethering at airports, train, and bus stations, and at town squares; telecommunications and transportation hubs that maximise data access and connection for mobile media and social interaction. Tethering also means fixing something or someone in one place, restricting movement. And tethering can also suggest emotional states, such as being at the end of one's tether, frustrated to breaking point. Tether as a noun is the actual tie, or rope, and tether as a verb is the action of restricting movement. Here, then, we have a paradoxical meaning of tethering: in the context of media, there is the action of staying fixed in one place so mobile media devices can enable data efficiency and flow; and in the context of mobility, there is the tethering of a human, non-human, or object in order to halt and restrict movement.

The tethering practices for rural people in digital media blackout zones such as Ydre are connected to freedom and constraint; there is the communicative freedom that comes from hotspots at the layby or the living room where tethering enables mobile communications and mediated social interaction, and there are the constraints that arise from infrastructure issues. There is the tethering of people to the home or locality as a means of coercion, restricting their freedom to communicate with others through digital media, or restricting their ability to mobilise themselves.

Another meaning for tethering is the mobile phone or laptop as a tether that binds people to the internet, social media, and streaming services. This speaks to a round-the-clock tether that maximises media time but reduces time spent on other things. The mobile does not bind in the same way when living in a blackout zone, reducing the risk of the "crisis of the summons" (Licoppe, 2012). This tether is one you choose to use in one spot in the home, or for particular activities when out and about. Thus, the constraints of the communications infrastructure enable another kind of freedom from the pressures of mobile and mediated interaction. There are even services that disable digital media in urban environments, offering a digital detox for the busy business executive, a momentary de-tethering from the mobile phone, signalling the purchase of freedom from the media for entrepreneurs and privileged mobilities (Polson, 2015).

The transported immobility of this experience modulates one's sense of place and time. Media time is parcelled into packages: wait, activate, and move on (see Scannell, 1996; Hartmann et al., 2019). Cyclical time is heightened as the seasons become part of how people plan routines for media and everyday life, such as sending a batch of emails in the morning, and taking a walk in the forest afterwards. Lifetime is enriched with nature, silence, and productive time for work, play, and family and friends. A sense of place also changes. That lake is great for swimming and surfing the internet. You tend to view the natural landscape and see through it to the possibilities of digital connection. There is a promontory that offers a sweeping view of the forest – and it doubles as a place to make phone calls. I once saw our builder sitting astride the roof making calls to clients. It is a creation of places and temporal experiences that are contiguous to each other, and the contrast of the natural place and communication practices complicate what we mean by place and time in these contexts.

The tethering experience brings into relief assumptions we make about normative digital media experiences. Me, myself, and my devices are not the most significant source of activity in this place. Place making becomes enmeshed with communication practices; and in these experiences of tethering to specific places, your sense of time changes so that the natural environment has precedence. The growth of the pine and birch trees outlasts my own lifetime, the blueberry bushes bed down for winter, conserving energy, then perennially coming alive each spring ready for summer fruits, a constant source of renewal.

Rural media topography

Ydre has been rated among the top three best places to live in Sweden for the elderly. The citizens of Ydre are a bonus for the area (generations of rural traditions) and a problem (unattractiveness for younger generations). The main work is farming, forestry, and small businesses in building or tourism (there are 650 local firms). There are some urban-to-rural migrants, and a small number of migrants from other countries – there are fewer than 500 non-Swedish citizens in a population of 3,500 and most of these are made up of Danish and German migrants (see https://kommunsiffror.scb.se). Seasonal visitors may revel in midsummer festivities and take valuable time to unwind from urban life, to digitally detox in the wilderness. But locals have to make do, vulnerable to the seasons, weather, technological breakdown, and local community and national politics. Some local communities reactivated a tradition of resident associations and have banded together to lobby for infrastructure, such as fibre broadband installed in the local village.

The drive for privatisation in Sweden over the past few decades has impacted transportation (trains and buses are privately run by companies) and communications (mobiles, wi-fi, and antennas are privately owned and maintained). Customers pay high monthly contracts for mobile and wi-fi that works intermittently. Every contract with telecommunications provider Telia, for example, is affectively shaped by seething frustration, as a long wait in the telephone customer service queue can be suddenly cut off by a drop in the signal. A typical monthly contract is around 400 Swedish Kroner (40 Euros) for a mobile service and 400–500 Swedish Kroner for mobile broadband connection in the home. Unlike public service media organisations SVT and SR, these companies are under little pressure to ensure round-the-clock services; furthermore, many mobile telephone providers typically share local transmitters, reducing differentation in cost and delivery, and leaving customers with little real choice.

In her book *In the Meantime: Temporality and Cultural Politics*, Sara Sharma (2015) develops the concept of power chronography. She explains how we have a shared imaginary of time that constructs it as 9-to-5, a Monday-to-Friday working week. She conducted ethnographic research with people whose job it is to make time for those of us caught in the imaginary of the 9-to-5, from taxi drivers to yoga instructors, or frequent flyers for business class. In these studies, she found that people worried about time, there was not enough time in their week, and they felt the need to speed up and slow down time, depending on the professional labour they carried out. She found there are "differential relationships to time", and these relationships "organize and perpetuate inequalities" (Sharma, 2015: 137). Sharma draws on Doreen Massey's concept of power geometry, which sought to highlight the existing differential relationships between people and space, arguing for a sense of space that can accommodate the differentiated mobility of people in global contexts (Massey, 1994).

There are differential relationships to the power geometry in this region which impact rural inhabitants in various ways, depending on their access to media and transportation, knowledge of different technological devices, and their daily needs, from talking to a friend, listening to the radio, or using a mobile banking app. As Green (2010) noted in her research on inhabitants in the Australian outback, their top priorities were radio signal, recharging batteries for devices, and speaking on the telephone. Basic communication rights – the ability to rely on a radio signal for up-to-date news, or make a phone call to a friend – are central to a power geometry in this area.

For novelist Jane Harper (2018) in her work *The Lost Man*, the extreme conditions of the Australian outback are hauntingly evoked in her story of domestic abuse. A crime is committed, a body found, a police investigation underway. Set against this harsh environment are families trying to survive not only the elements and high temperatures, or the economic pressures of farming, but violence in the home, where access to cars and phones, and maintenance of local antennas, are vital human resources. It is no coincidence that the murder victim is found in the outback, off road, without a car, and close to the local antenna. Communication and transportation become a means of controlling women and migrant workers, a spiral of violence in this crime genre setting. Here, the transported immobility that Barthes describes takes on other dimensions with this case of domestic abuse; immobility becomes a form of coercion. Such an example explored by Harper in her novel suggests the darker dimensions of transported immobility for gender and disability issues, or for the elderly and reduced mobility. Neighbours, family, and friends are both social resources and pose risks, a type of relationship management that underscores both the pragmatic aspects of robust communications and the power relations within such rural communities.

The contours of power in its spatial and temporal forms are visible in the natural and digital landscape of Ydre. The romantic myth of rural Sweden, a place to escape to nature from the stresses of urban living, long established in the Swedish social imaginary (Taylor, 2004) works with – and against – the digital media blackout zone. Previously, local inhabitants were included in the state infrastructure for water, power, and communications, as we saw in the historical context to the house and local area. In a contemporary setting, local inhabitants live here with the expectation that they need to maintain and pay for digital media connection. There is a transition from public service, state-backed, infrastructure to private, neo-liberal services. Those with time and money, health and mobility, cars and bicycles, and a variety of old and new media devices, can switch from one social imaginary of Swedish forest life, to another popular Nordic media imaginary of digital connection. But for those people without such resources, they can experience constraints on their mobility, media, and sociality, by being denied reliable access to public services such as the telephone and digital communication.

Reflections on the vicissitudes of media

In Brian Larkin's work on Nigerian radio and film, he offers a useful counterargument to Western media theory and assumptions about technological plenty and progression. In *Signal and*

Noise, Larkin (2008) underscores the materiality of technologies in context; in Nigeria, media technologies do not always work in the way developers planned; for example, radio signals can be intermittent or distorted, or there can be power failures and unstable networks. Such a perspective serves as a reminder that there are limits to overarching theories regarding the smooth efficiency of media technology in urban spaces. He alerts us to be attentive to other stories of the meaning of media technology and its social uses (Larkin, 2008: 3).

Media breakdown is a regular feature of rural life in Ydre. The natural topography produces and defines everyday media experience. The landscape blocks smooth delivery of wi-fi and mobile signals, producing regular media breakdowns and defining the stop/start media experience. It is an inevitable consequence of living here, but it is not an inevitable political outcome that local inhabitants have to privately source and maintain communications infrastructure. As we saw in the short history of the local area, the state ensured the house was on the grid until it was sold off and became part of the private market, alongside other services. Quite why the state turned away from basic provision of communications is difficult to rationalise, given that large parts of Sweden are rural and it is inevitable that a number of citizens will experience a media topography of the kind described here. Sweden has vast regional variation, with climates that include the snow-covered mountains of the north and the warmer flatlands of the south. The socio-technical vision of a highly modernised and progressive society is predicated on urban spaces and reliable wi-fi provision; this rough and unready rural media experience is at odds with such a vision.

Mark Deuze (2011, 2012) argues in *Media Life* that people co-produce themselves through media and by this process of continuous identity work they may not even recognise the power of media in their lives because they are a constitutive part of the media landscape (Deuze, 2011: 138). Influenced by David Harvey's work on the changing nature of the human experience of space and time as fast and precarious, a kind of intensified living, media become a playground for co-creating identities through "a whole series of simulacra as milieux of escape, fantasy, and distraction" (Harvey, 1990: 302). In a more abstract sense, the contemporary human condition is "a socio-technical experience of reality – a reality that seems to submit itself (potentially) to the affordances . . . of media: a reality that could be cut, pasted, edited, remixed and forwarded" (Deuze, 2011: 139). What Deuze proposes is a media-life–based ontology, theorising media as "the invisible interlocutor of people's lives" (Deuze, 2011: 139). For Deuze, "the governing principle of media life is completely mediated self-creation in the context of always-available global connectivity" (Deuze, 2011: 145).

What happens if we consider people and their experience of living with intermittent digital connectivity in a rural area? This is a media experience of another character, both visible and invisible. Indeed, the breakdown of the communications and transportations infrastructure serves to make the media visible not invisible. Where is that hotspot exactly in the layby by the lake? It's not visible to the naked eye, but the mobile device will tether itself to that place of signal strength, and this moment of digital connection is physical, embodied, and lived. You have to stand still, tethering your body to that precise spot. The need to repair technological breakdown on a routine basis underscores the media's material presence in everyday life; to find a replacement battery for the 2G flip phone, or search for an engineer who understands the workings of the Shark Phone, signals the "implacable materiality" (Barthes, 1979: 143) of the media. And all the small details of what media device, what exact location, what route to take, build into a "protocol of attention" that makes the media visible and invisible in everyday life (Barthes, 1979: 141). It projects a mirage of solidity, and when digital media breaks down in a rural environment, we see a reliance on old and new ways of living within the media, an "oddness of life" and the myriad ways "people get on under challenging circumstances" (Moss, 2019).

Barthes notes an odd combination of spectacle and substance to the dining car experience – what lies behind the illusion, so to speak. He sees this in other myths he recounts in the collection of essays; for example, a spectacle of labour is apparent in the way acrobats entertain a live audience – their effortless movements and choreographed routines of physical theatre offer a sublimation of human work as all the hours of training, the build-up of skills, and the accidents and repairs, come together in this one live entertainment spectacle. Similarly, Barthes' discussion of transported immobility is suggestive of a spectacle of solidity to media presence, and behind the illusion, we find that there are all sorts of things people do, day in and day out, to make media work as regular communication, to find a connection, to repair media that has broken down, offering through their practices a sublimation of human and digital labour that comes together in a spectacle of media presence. Indeed, that heavy materiality and attention to the smallest of details underscores the particular, situated, and contrary reality of living with and without digital media, the vicissitude of media, and its place and power in our lives.

References

Barthes, R. (1979) *The Eiffel Tower and Other Mythologies*. New York: Hill and Wang.

Bassett, C. (2017) "What Perec was looking for", in *The Afterlives of Georges Perec*. Edinburgh: Edinburgh University Press, 120–135.

Deuze, M. (2011) "Media life: A commentary", *Media, Culture and Society*, 33(1): 137–148.

Deuze, M. (2012) *Media Life*. London: Polity.

Edgerton, D. (2006) *The Shock of the Old: Technology and Global History Since 1900*. London: Profile Books.

Green, L. (2010) "Imagining rural audiences in remote Western Australia", *Culture Unbound*, 2: 131–152. Available at: www.cultureunbound.ep.liu.se.

Harper, J. (2018) *The Lost Man*. London: Little, Brown.

Hartmann, M., Prommer, E., Deckner, K. and Görland, S. (eds.) (2019) *Mediated Time: Perspectives on Time in a Mediated Age*. London: Palgrave Macmillan.

Harvey, D. (1990) *The Condition of Postmodernity*. Malden, MA: Blackwell.

Larkin, B. (2008) *Signal and Noise: Media, Infrastructure and Urban Culture in Nigeria*. Chapel Hill, NC: Duke University Press.

Licoppe, C. (2012) "Understanding mediated appearances and their proliferation: The case of the phone rings and the 'crisis of the summons'", *New Media & Society*, 14(7): 1073–1091.

Mannell, K. (2019) "A typology of mobile messaging's disconnective affordances", *Mobile Media & Communication*, 7(1): 76–93.

Massey, D. (1994) *Space, Place and Gender*. Cambridge: Polity Press.

Morley, D. (2007) *Media, Modernity and Technology*. London: Routledge.

Moss, S. (2019) "Martin Parr's 40 years in the life of Wales", *The Guardian*, October 14. Available at: www.theguardian.com/artanddesign/2019/oct/14/martin-parr-photos-wales-national-museum-cardiff.

Perec, G. (2008) *The Species of Spaces and Other Pieces*. London: Penguin, original translation 1974.

Polson, E. (2015) "A gateway to the global city", *New Media & Society*, 17(4): 629–645.

Polson, E. (2018) "Information supercalle: The social internet of Havana's Wi-Fi streets", in R. Wilken et al. (eds.) *Location Technologies in International Context*. London: Routledge.

Scannell, P. (1996) *Radio, Broadcasting and Modern Life*. London: Wiley Blackwell.

Sharma, S. (2014) *In the Meantime: Temporality and Cultural Politics*. Durham, NC: Duke University Press.

Taylor, C. (2004) *Modern Social Imaginaries*, Durham and London: Duke University Press.

Vannini, P. and Taggart, P. (2014) *Off the Grid: Re-assembling Domestic Life*. New York: Routledge.

Williams, R. (1974) *Television: Technology and Cultural Form*. London: Routledge.

19

IMMOBILE SOCIALITIES?

Historicising media practices in refugee camps

Philipp Seuferling

Stuck on the move

> A village at the edge of the Thuringian forest with clean, red-roofed farmhouses. A river cuts through the valley. Cows on green meadows. Forest-filled hills. A train station: *Friedland*. . . . Before the train stops at the station, the traveller sees countless rows of barracks, which disturb the peace of the landscape, with their round corrugated tin roofs and poverty, close to a state of neglect. . . . Two women with three children between three and eight years old take the same way as me. They must have missed the sign, as suddenly the older woman turns to me asking: "Is this the way to the transition camp?" The woman is pale and obviously malnourished. She could be 60, maybe she is only 30. She carries the youngest child on her arm, in her free hand she holds a bundle. Also the older children, happily walking, carry bundles. Illegal border crossers: They have been on the move since dusk. Now it's 3pm. Over there they had a home. Now they own nothing but a small bundle. "No one should notice that we wanted to leave", the woman explains. "And why did you want to leave?", I ask. The answer is always the same. Fear! Fear of deportation. "Now just before the elections it's worse. My husband died in the war. My sister's husband got deported to Siberia". "What will you do in the West?" – They don't know.[1]
>
> *(Metzner, 2016: 17–18)*

In this on-the-ground story, photojournalist Ré Soupault[2] leads us into the world of refugees in 1950s Germany. In September of 1950, she took a tour through German refugee camps, which resulted in an impressive journalistic travel report under the title "West Germany's expellee and refugee problem" (republished by Metzner, 2016). Immediate post-war Europe was characterised by immense forced migration, with Germany accommodating millions of displaced individuals in thousands of camp structures. In the introduction to her article about the camp Friedland (until today the largest still operating refugee camp structure in Germany, founded in 1945), Soupault introduces us to a group of East German refugees going west, ending up in Friedland. She met them on the move, herself travelling by train, the refugees arriving on foot, after having crossed the "green border" through forests and wilderness. They exchange a few words and establish a social connection, the refugees sharing their fate with the journalist, who is interested in their story. This reported scene is a snapshot of a sociality on the move, a mobile sociality of 1950: refugees trying to safely navigate their forceful mobilisation towards the next safe haven – the refugee camp.

In this chapter, I delve into the context of refugee camps in post-war Germany, in order to explore them as historical sites and spaces of mobile socialities and their mediation. The transition camp Friedland was and is the first stop for many forced migrants arriving to (West) Germany. In administrative procedures of registration, documentation, hearings and redistribution to further accommodation, camps and shelters like Friedland institutionalise the first arrival at supposed safety and a new home. In doing so, camps condition mobilities and socialities of the subjects finding shelter. Designed as humanitarian tools of the modern Western refugee regime (Gatrell, 2013), refugee camps are inherently institutionalised spaces. By temporarily immobilising its inhabitants, the camp materialises border management and negotiates the refugee subject's social position – as an outcast of the nation-state system, fallen out of territorial imaginaries of belonging and citizenship into the naked existence of exception (Agamben, 2005; Arendt, 1956 [2017]). The act of escape, characterised by fear and hope to find safety, comes to an abrupt end in the camp, waiting for onward migration, a residence permit or dreading deportation. Until today, refugee camps are home to millions of people. In the Global South, but also in the Global North, such as on Greek islands or in the "Calais Jungle", camps and shelters have become protracted makeshift homes with little to no prospect of fulfilling the alleged humanitarian promise of being a temporary solution before arrival. Originating in post-war camps such as Friedland, the institution of the refugee camp (including all shared institutionalised shelters, such as asylum-seeker centres) has become a default tool of Western asylum regimes, separating and fixating individuals through hollow promises of further migration and eventual arrival. It is the place where the asylum regime unleashes its force, enacted through material infrastructures of separation and social practices of control. Thus, the camp becomes the ultimate mash-up of mobility and immobility: refugees are stuck on the move.

The goal of this chapter is to historically theorise and explore mobile socialities and their mediation by uncovering these paradoxical workings of mobilisation and immobilisation in forced migration processes. Within the empirical context of refugee camps in post-war Germany (roughly 1945–1960), this chapter focuses on media practices as an analytical prism, through which I argue we can historically observe how practices around media shape an environment of both mobile and immobile socialities, emerging in the paradoxical, layered experience of forced mobility and immobility.

Mobility, sociality and media from the margins

> Having arrived at the camp, I go to the police shack. Where "unclear" cases are investigated. When I enter, I hear the hard voice of a German policeman: "You're lying! This photography depicts yourself. Just younger. And it's weird that you have a birth certificate in your bag, but no other papers". . . . On the bench outside countless "unclear" cases are waiting. Their downcast faces reveal their fear. "Will the police believe me? Will they register me?" They fled fear and now encounter the same spectre: fear. . . . A "registered" enjoys strong advantages [compared] to the "illegal". He can register with the housing authority and receives a room. Worst case a sleeping lot in a camp or a bunker.
>
> *(Metzner, 2016: 18–20)*

Refugee camps turn mobile socialities on their head, as well as take them to the extreme. Forcefully *mobilised* individuals are forcefully *immobilised*, and subjects become dependent on socialities both to authorities and peers, as well as forced to socialise, deprived of privacy. This chapter outlines a perspective on mobility and sociality from the margins, looking at "contrary mobile socialities" seemingly at the peripheries of societies, yet at their centres. Ré Soupault describes how the refugee camp physically cuts through and disturbs the landscape and village, almost like a scar. Camps are upside-down spaces of society, negotiating normality and exception, materialising belonging and disbelonging, and an uneasy moment of disturbance in nationalised imaginaries of territorialist sovereign states (cf. Bromley, 2012: 344–345). Simultaneously, such marginal counter-spaces are always already reflections of social centres, projections of utopian imaginaries, enactments of discourses and practices. By zooming in on such margins, we can arguably nuance our insights on mobility and sociality.

This also concerns the fundamental role of mediation and communication. The meaning of media as technologies and practices in the context of forced migration and mobility has been widely acknowledged, and this research field provides a hook for the explorations of this chapter. The refugee's photo in the story related by Soupault, as a medium to prove his asylum claims, is a historical predecessor of the smartphone as a "pocket archive" (Leurs, 2017) for refugees. Today, authorities can even seize smartphones for checking on asylum-seekers' stories (Biselli and Beckman, 2020). Media as technologies and practices are hence inherently tied to being (made) a refugee and navigating this experience (Leurs and Smets, 2018; Morley, 2017; Smets et al., 2020). As a subject of forced mobility, the "connected migrant" (Diminescu, 2008) is intrinsically dependent on mediations of her (im)mobility, in the sense that both "mediated communication and representation are shaped by and implicated in social processes of mobility", while simultaneously, "practices and experiences of mobility are mediated" (Keightley and Reading, 2014: 286). But also of immobility. Camp-based refugees are specifically subjected to regimes of immobility and their mediation. This chapter explores how historically grown asylum and mobility regimes in conjunction with changing media-technological environments form parameters for socialities and experiences of camp-based refugee subjects.

Chapter outline

The objective of this chapter is threefold: a) to exploratively chart the concept of "(im)mobile socialities" within the realm of the refugee camp; b) to map out how (im)mobile socialities in camps are entangled with media technologies and practices; and c) to outline a historicising approach to mediated (im)mobile socialities. To do so, the chapter applies three lenses – theoretically, empirically and methodologically. First, I draw on media practice theory (Couldry, 2004) as a prism for grasping mobile socialities in the camp context, overcoming a functionalist perspective on media for forced migrants and instead being able to capture from below what involved actors, such as refugees and administration, *do* around media and how such media practices create, negotiate and represent (im)mobilities and socialities. Second, this approach is situated in the empirical context of Germany's post-war refugee camps, roughly 1945–1960.[3] Archival records from state and non-governmental organisation (NGO) archives serve as empirical material for an historical analysis. Third, this leads to a methodological perspective on historicising mobile socialities, by extracting, tracing and interpreting media practices from archive material. Focusing on a non-digital media environment, a historical approach can reveal roots, trajectories and genealogies of how media technologies, mobility regimes and embedded socialities have interacted. Drawing on examples, vignettes and contextual descriptions of refugee camps in post-war Germany, the spectrum of mobility and immobility – and emerging socialities – in the confined camp space are explored through the lens of media practices. Here, three dimensions of mediated mobile socialities are identified: socio-material immobilisation through material spaces, media-technological environments in camps and mobilising communication practices of resistance, activism and protest. This leads in a concluding section to a discussion of how we can see spectrums of *mobile and immobile socialities and anti-socialities* unfolding in media practices in refugee camps.

Media practices and mobile socialities

Narratives of mobility in Western modernity, and immobility as its binary opposite, are torn. David Morley (2017: 59–75) distinguishes between a "sedentarist" and "nomadic" imaginary and construction of belonging, movement and territoriality, whereby sedentarism as the rooted stability of a community is traditionally valued, while those who are forced to move are outcasts, vagabond nomads. On the other hand, mobility can be seen as a goal to be achieved, in a progressivist narrative of modern technologies of communication and transport, which enable and shape a globalised and individualised culture, whereby the immobile subject falls short of participating in hypermobile connected civilisations (Urry, 2007; Keightley and Reading, 2014). Within the figure of the refugee and her relation to media technologies, this contradictory understanding of mediated mobility becomes evident. When sedentarism reflects the idealist construction of territory-based nation-states, refugees are outcasts. At the same time, media and migration research underlines refugees' vital dependence and eager adoption of media technologies as a "lifeline" (Alencar et al., 2019) or "logistics from below" (Arnold and Görland, 2019), providing essential connectedness in the experiences of fleeing and arriving. Hence, the figure of the refugee embodies contradictory narratives and experiences of mediated mobility – a fact that could be observed during the so-called European "refugee crisis" in 2015, where media comments doubtfully questioned refugees' necessity to carry allegedly expensive smartphones. Accordingly, the emerging field of "digital migration studies" (Leurs and Smets, 2018) is charting the impacts of digital media technologies: either for creating connectivity and enabling social and corporeal mobilisation (Bishop, 2017), or as sources of immobilisation and encapsulation, through lack or denial of access to media which creates "information precarity" (Wall et al., 2017), datafied surveillance (Metcalfe and Dencik, 2019) or hindering visibility and recognition of refugees through practices of "digital bordering" (Georgiou, 2018). The interlockings of mobility, media and (forced) migration have often been analysed as technology-based processes, which enable and disable mobility (social and bodily) and thus foster inclusion and exclusion in territorialist societies: media fulfil specific functions for migrants. Here, I suggest nuancing this understanding of mediated refugee mobilities through the lens of mobile socialities, by re-embedding the role of media technologies into a less functionalist narrative of forced migrants' media experiences. This arguably enables us to see how practices around media are part of an environment of socialities emerging in the experience of forced mobility and immobility – instead of perpetuating utilitarian views on technology, which are apt for humanitarian and integrationist discourses but fail to question actual desirability of continuous and free connectivity by refugees (see Awad and Tossell, 2019).

To do so, I draw on the concept of "media practices". It was proposed most eminently by Nick Couldry (2004, 2010), but stems from longer interest in holistically scrutinising people's media habits through practice theory and ethnography, with considerable roots in Latin American (Stephansen and Treré, 2020; Martín-Barbero, 1993) and Scandinavian media and communication studies (Drotner, 1993; Bolin, 1998). Keightley and Reading (2014: 292) call for a conceptualisation of mediated mobility, which centrally takes into account processes of *mediation*, instead of starting from the observation of mobility (such as forced migration) and then looking for the functions of media therein. Such an approach, they argue, would need to overcome the traditional distinctions of media studies – audience, reception and production. This is exactly what Couldry (2010) outlines as the potential of a media practice approach. Drawing on sociological practice theory, media practices capture, in a text-decentred way, what people *do* around and oriented towards media. Practices to be understood as ritualised actions consist of various elements, including materialities (infrastructures, tools), skills (know-how, tacit knowledge) and motivation (purpose, aim) (Gießmann, 2018; Shove et al., 2012). In this sense, media practices are open sets of social practices around media, which are always embedded in the totality of other practices. This perspective enables us to overcome a functionalist, utilitarian over-emphasis of media as it

> distances us from the normal media studies assumption that what audiences do ("audiencing") is a distinctive set of practices rather than an artificially chosen "slice" through daily life that cuts across how they actually understand the practices in which they are engaged.
>
> (Couldry, 2004: 121)

Paying attention to media practices thus enables us to see more holistically what people say and do about media, which hierarchies and orders they enact – ultimately, how media (as technologies for storing and disseminating information, as well as institutions) are (made) part of socialities. I argue that media practices offer a helpful theoretical lens to observe mobile socialities. Asking what diverse actors in the camp, like residents and administration, *do* opens up a way of describing and analysing the role of media(tion) in relation to mobility and immobility and emerging socialities in the space of the camp. Scrutinising how forced migrants and administration employ media practices unpacks how the institutional space of the refugee camp within the asylum regime is realised, experienced and enacted. By examining how camp-based media practices by these actors are assembled out of materialities (media technologies), skills, permissions, access, motivations and purposes, we can thus gain a better understanding of how mobility and immobility are mediated and how socialities take shape in and through mediated communication in the camp space.

Media practices are embedded in both the social and the material construction of the refugee camp: as a space realised through relational practices (Massey, 2005), as well as socio-material configurations. Media are constitutive elements of such shelters and their experiences, and thus their socialities (see also the works by Saskia Witteborn [2014]). Refugee camps work by controlling and surveilling space and time of their inhabitants, following legislations of nation-states' asylum regimes. In the case of post-war Germany, beds and rooms were allotted, camp IDs controlled access, food was handed out and rationed, access to welfare services was controlled, the asylum process was administrated and documented. This process decided how long the waiting time of individual refugees would be. Finally, the camp space itself is designed as a separated collection of shacks, buildings and rooms with different functions, all of which shape spaces of community and privacy. Thereby, spatial and temporal parameters are set for socialities – for how communities and networks, stability and permanence, can be established through social ties, communication and media practices (see Wittel [2001] for the connection of socialities, networking and media). A wide notion of media, from spatial design to communication technologies and practices of mediation and communication, leaves room for situating socialities that appear in and through camp-based media practices. Practice theory locates the social in practice itself, so that practices in relation to media become sites and expressions of socialities. The goal of this chapter is tracing how such media-oriented socialities – mobile and/or immobile – could emerge.

Historicising mobile socialities in archives

The outlined media practice lens on mobile socialities has methodological consequences, especially when combining it with a historical approach. Taking up the call of Keightley and Reading (2014) to refine our methodical toolbox for mediations of mobility, I here draw on classical historical methods – archiving and the interpretation of archival documents – and relate them to mobile socialities, through conceptualising them as *historical media practices*. In order to gain empirical access to media practices in the past, the archive acts as a medium: records collected in the media infrastructure of archives, systematised remains and residues of the past, materialised in paper. As such, I grasp the archival documents as *traces* of media practices, allowing for an interpretative, historiographical reconstruction of gone, once ephemeral media practices. Thereby, I draw on approaches to media history from the user's practices' end, inspired by the work of e.g. Kaun (2016) and Marvin (1990).

The records have been collected from state and NGO archives. In Germany's federal system, refugee camps and shelters have been (and are) administrated on state and municipal level; therefore, several local state archives (Lower Saxony, Hamburg, Bavaria), as well as the national archive (Bundesarchiv Koblenz) have been scouted for records related to refugee camp structures. Key word clusters denotating the different terms for "camp"[4] have been applied to the respective search catalogues and crossed with key words around "media".[5] Thereby, a corpus of files on media in refugee camps started emerging. The same has been applied to the Germany Mission file in the United Nations Archives in New York City (collection on the activities of the United Nations Relief and Rehabilitation Administration [UNRRA][6]), as well as the Red Cross Archive (Berlin). Also included are several other smaller holdings, e.g. from the museum of camp Friedland. The documents consist of diverse genres: photographs, administrative files (e.g. reports, bills, registration documents, furniture lists), newspaper clippings, letters, and other random paper notes.

The post-modern archive researcher finds herself in a complicated positionality vis-à-vis archive documents, being subjected to records, which are preserved (and lost) both randomly and as according to archives' rules and systems of preservation. From today's perspective, documents are mediated memories of refugee camp histories, revealing narratives, trajectories and meanings of mobile socialities and media practices through the archive's power–knowledge nexus. Investigating historical media practices through the archive craves an analytical step of reconstruction, a moment of interpreting historical material out of the present perspective. Documents, as a genre, have a "know-show function" (Gitelman, 2014), simultaneously evidence and constructor of reality. As such, they are traces and relics of media practices: they are either media practices themselves, in their materiality as media and documents, or they are reports of media practices. For example, a photograph is a preserved media practice in itself (of a photographer storing and circulating an experience and moment in time), or the photo's content can report of a media practice, for instance in depicting refugees reading newspapers. Applying a media practice lens to this material, hence, enables a historicising exploration of mobility, sociality and their relation to media, taking into account the role of media as technologies and materiality, as well as social processes of asylum regimes and the refugee camp space at large, which condition, affect and shape media practices. This is what the following section will outline, through questioning how historical media practices in refugee camps in post-war Germany shape mobile socialities. Drawing on examples from the archive material – vignette-like stories, which the archives offer – media practices are examined in relation to how they affect (im)mobilities and socialities in construction and subjugation of inhabitants to the space of the camp.

Stories of mobile socialities in refugee camps

The refugee camp is a space of contradictions. At the margins of society, yet often at centres of cities; temporary, yet often permanent; a space of control, oppression and separation, yet often a hub of resistance, agency and resilience. The field of media and migration studies has long been occupied with situating the role of media technologies in these contradictive experiences of migrancy (Athique, 2017), describing the mutual shaping of social contexts and technologies and their affordances as "polymedia" (Madianou and Miller, 2012) or migratory "media repertoires" (Hepp et al., 2011). By applying the lens of media practices as a prism on socialities in the refugee camp context, I want to map out uses of media technologies beyond such dichotomies. In what follows, I carve out three dimensions of mobile socialities, which are contextual components and levels of mobilities and socialities in the refugee camp. These comprise: a) practices of constructing the physical, socio-material space of the refugee camp, whereby (im)mobility, communication and socialities become materialised in the set-up of the camp and its operations; b) media practices and socialities embedded in the media-technological repertoires which existed in the camps; and c) media practices of symbolic mobilisation, activism and protest, of communication beyond the immobilised space of the refugee camp.

Socio-material and embodied (Im)mobilisation

After having arrived in Friedland, the refugee women Ré Soupault introduced us to entered the administrational system of an emerging refugee regime in post-war Germany. Following documentational procedures, their asylum claim was verified, accommodation allotted, probably in shared barracks, benefits of food and social services distributed, maybe a workplace offered. Friedland was a transition camp, registering new arrivals into (West) Germany and distributing them across the country, into other camps or local accommodation (Schießl, 2019). Thereby, mobility, temporary immobility and remobilisation – as well as structures, boundaries and conditions of sociality – fell under the control of the refugee camp and its institutions. This story happened millionfold in post-war Germany.[7] The decade after 1945 saw around 12 million German refugees and expellees from ceded areas in Eastern and Central Europe, as well as eight million Displaced Persons (DPs) liberated by the Allies from concentration camps and finally, with the division of Germany and Europe, millions of East Germans going West, as well as foreign asylum-seekers, e.g. from Hungary in 1956, claiming asylum in West Germany. Different actors, like local governments and the Allies in the form of UNRRA, and later International Refugee Organization and United Nations High Commissioner for Refugees (UNHCR), as well as the Red Cross, developed a legal and institutional system of "handling" and "managing" these individuals, the camp becoming a basic solution for accommodation and administration: "refugee camps", "war returnee camps", "DP-camps", or "foreigner camps" – thousands of which existed at that time (Hochmuth and Bispinck, 2014).

To fulfil their function, the camps relied on specific material, architectural and spatial setups. Old barracks structures (often empty concentration camps or army camps), as well as repurposed non-destroyed buildings, were widely drawn upon: perfect for desperately needed ad hoc mass accommodation. Building types and their arrangements constructed the material space, conditioning the possible social relations, mobility within and beyond the camp, and power hierarchies between camp staff and inhabitants.

A genealogy of barracks building types shows how temporariness and mobility are historical core features of camps. In Germany, post-war refugee camps often consisted of RAD barracks (Reichsarbeitsdienst), developed for the Autobahn construction workers in the

Figure 19.1 Nissen hut in camp Friedland.

Source: Copyright Philipp Seuferling

1930s – hence, for mobile workers – and later used for concentration camps. Another type was the so-called Nissen hut, a Canadian model developed during World War One for soldiers, with its characteristic shape of corrugated iron roof (Figure 19.1). In the 1980s and 1990s, the Euro-container emerged, an internationally size-standardised model of temporary accommodation. Mobility and temporariness, a sense of spatial and temporal exception through non-arrival and limbo, all merge and materialise in the architectural form of these refugee camp homes. Moreover, it hardly needs to be explained how such accommodations shape forced socialities, through lack of privacy, crammed spaces and excessive control through administration.

Looking at the design of the entire camp (e.g. in plenty of preserved maps and construction blueprints) shows how the spatial arrangement also included distinctively public spaces:

administration buildings (authorities, hearing rooms, etc.), kitchen and food distribution, churches, workshops and kiosks, but also post offices, community rooms and cinema and theatre barracks. Mobility, sociality, media practices and overall communication, also face-to-face, had to be adapted and fitted into this spatial environment, its public and private areas, and its control. Camp administration, implementing the imaginaries of the refugee regime, had the ultimate power to rule over these spaces, determining which forms of communication and media practices were possible where and when, often defined in official camp rules. For instance, they controlled camp cinemas through ordering the mobile film services, choosing the films and determining screening times.[8] They also gave out licences for shops, cultural events, religious services or own production, e.g. of newspapers[9] or handicrafts. Furthermore, authorities decided upon access to telephones, telegraph or mail services, including the rationing of paper. In reaction to that, camp inhabitants developed ways of voicing their concerns and participating in the spatial construction of the camp. Records of camp assemblies, refugee councils and elected representatives of inhabitants give insight into negotiation processes, where camp inhabitants complained, protested or formulated requests, on topics like availability and access to media (newspapers, cinema, mail), as well as shops, construction of new buildings, etc.[10] These opportunities of participation, however, varied across specific refugee statuses, German citizens having usually more of a say than foreigners. In general, however, the imaginary of the refugee victim, the "speechless emissary" (Malkki, 1996) to be governed by a humanitarian regime, finds its origin in these early camps, where, in the name of welfare, administrations totally ruled over spaces.[11]

What these examples show is how socio-material, spatial construction forms an environment for media practices and communication spaces. Architecture and design matter: where oral or mediated communication and socialities can take place and where not (cf. Mattern, 2017, on histories of urban media infrastructures), and which bodily movements around the camp space this incorporates. Hence, socio-material setups of camps become media practices themselves. The distinctive physical othering of the space, through its secluded-ness, inbuilt encapsulation and control over mobility, forms a specific dimension of camp-based media practices as well as mobile socialities: the social immobilisation of inhabitants becomes embodied and materialised – which shapes which and how media practices are possible.

Media-technological repertoires

Once registered or sent to a camp, refugees find themselves in a new everyday of definite temporariness, of waiting for this limbo to end, either through onward migration, arrival into an own home, repatriation or deportation. In navigating this contradiction of mobility and immobility, media technologies of storing and disseminating content form a media repertoire for camp inhabitants.

> The whole pride of Mr Brod was his radio. "In Russia it would have cost more than 1000 Rubles. And would have made more noise than music." . . . Full of pride his wife showed us what they had bought already. The money they had received in Friedland got spent on the most necessary things only: beds, stove, essential furniture, an old but good sewing machine, a bicycle – very necessary in a remote village – and the radio. . . . "My husband just cannot cope with the world being so contradictive. Suffering and misery over there. Abundance here. He wrote to his brother in Russia

> today. Then everything became vivid again. Reading is hard for him. He needs the radio, so he doesn't drown in contemplation".[12]

Mr Brod's radio reveals the set-up of camp-based media practices: listening to the radio became part of Mr Brod's experience of rupture, soothing him, while writing letters and reading newspapers, according to his wife, only stir him up. It is an affective technology, part of social practices of connectivity, as well as hope. The archival records reveal a whole set of media technologies present in the camp, through which media practices could shape mobile socialities. Radios – specifically mobile, portable transistor radios – were highly popular: they not only enabled listening to international broadcasts, but also could be domesticated perfectly into the mobile camp space, moving with their owners around rooms or even different camps. Media practices in the camp, such as connecting to the world through news (it was of vital importance to the refugees to know about their further destiny), were often communal experiences in the narrow, shared space, as a pastime and a shared social practice, as these two photos from DP-camps (Figure 19.2) show. Communal media practices could even force inescapable mediated socialities upon the inhabitants, as the case of a "camp broadcasting system" shows, installed in the DP-camp Wentorf in 1946, where "200 loudspeakers, carefully placed to cover the whole camp, allowed the benefits of outside as well as internal news and entertainment program", as an administrational report by YMCA noted.[13]

Also newspapers, including self-produced camp journals, as paper-based, mobile media, were incorporated into practices of both news consumption for connecting to the outside world, as well as mediating socialities within camp communities. Figure 19.3 shows a Yiddish newspaper from a DP-camp, commemorating the third anniversary of the Warsaw ghetto uprising – a community-building mnemonic media practice for Jewish Holocaust survivors, who ended up as deported, forced migrants in a DP-camp. As I have elaborated elsewhere (Seuferling, 2020), media technologies are and were crucial for mediated remembering practices in refugee camps – connecting to and upkeeping cultural and personal memory of the lost homeland, documenting the immediate present and imagining a potential future. Photography was employed as a media technology for these motives. Pictures of family and hometowns, which refugees brought along and decorated their rooms with (see Figure 19.2), as well as photos taken with cameras documenting camp life, were mobile carriers and mediators of memory, enabling media practices of negotiating individual and communal past, present and future, as a social practice in the camp community. A photo album, made by inhabitants of DP-camp Cornberg in 1946, impressively demonstrates the role of photography in mediating camp socialities: documenting and storing identities and memories of their camp life.[14]

These examples of media practices all rely on media technologies, which are embedded into the space of the camp and its mobility regime. The lack of permanence, materialised within the crammed and controlled spatial set-up of the camp, created an environment for media practices, whereby media technologies were made part of mobile socialities: radios, newspapers and photography, but also mobile cinemas (mobile projector and film services visiting the camps) and paper-based communication, such as letters or paper notes pinned to the walls of buildings. Media practices around these technological repertoires are sites of mobile socialities, reproducing both immobility and mobility of the impermanent camp space. Forced migrants brought, purchased and produced media (e.g. photos, camp papers, radios, diaries, letters) often just as mobile as them, adapted to their mobile sociality of being thrown out of time and space. Simultaneously, the camp provided media (cinemas, mail service, newspapers), affecting and controlling mobile socialities and media practices through access and availability of media technologies.

Figure 19.2a and 19.2b UNRRA photographs from life in DP-camps, used for public relations at the time. *Source*: UN Archives, "Germany Mission – Photographs #001–120", S-1058-0001-01, 1943–1948

PESACH-NUMER

20 ZAJTN

No. 8

A HEIM

ANTHALT 20 ZAJTN

LEIPHEIM

D. P. LAGER

HOME

DONERSZTIK

11

April 1946

CUM 3-tn JORTOG

FUN OJFSZTAND IN WARSZEWER GETO

17-ter NISAN

5703 JOR

Aszrej hagafrur,
szenisraf
wbecil lhawol

Aszrej halhawa
szebaara bsitrej
lwawot.

Aszrej holwawol
aszer jadun
ladidol bkawod.

Aszrej hagafrur
szenisraf
wbecit lwawol.

CHANA SENESZ

Wojl iz dem szwebele,
wos iz farbrent geworn
un ongecunden flamen.

Wojl iz dem flam,
wos iz farbrent geworn
in di farborgene hercer.

Wojl iz di hercer,
welche hobn gewust,
ojscudojern mit kowed.

Wojl iz dem szwebele,
wos iz farbrent geworn
un ongecinden hercer.

CHANA SENESZ

Briw fun Mordechaj Anilewicz cu zajn
fartreter ojf der Ariszer zajt

Figure 19.3 Yiddish newspaper in DP-camp, "On the 3rd anniversary of the uprising in the Warsaw ghetto".

Source: UN Archive," Area Team 1062 – Augsburg – Copies of Newspaper A Heim Published in Leipheim Camps", S-0436-0009-01, 1946

Mobilising communication

The examples of the preceding section not only reveal the role that available technologies play for media practices, but also hint at social practices of communication, which camp life entails. The archival records reveal communication practices among refugees, which show the symbolic dimension of mediated mobile socialities: media practices of mobilisation, of activism, resistance and establishing communication channels beyond the fences of the camp. What we can observe here are not only *mobile* media technologies long before cell phones and smartphones, but also how media technologies were used for *mobilisation*.

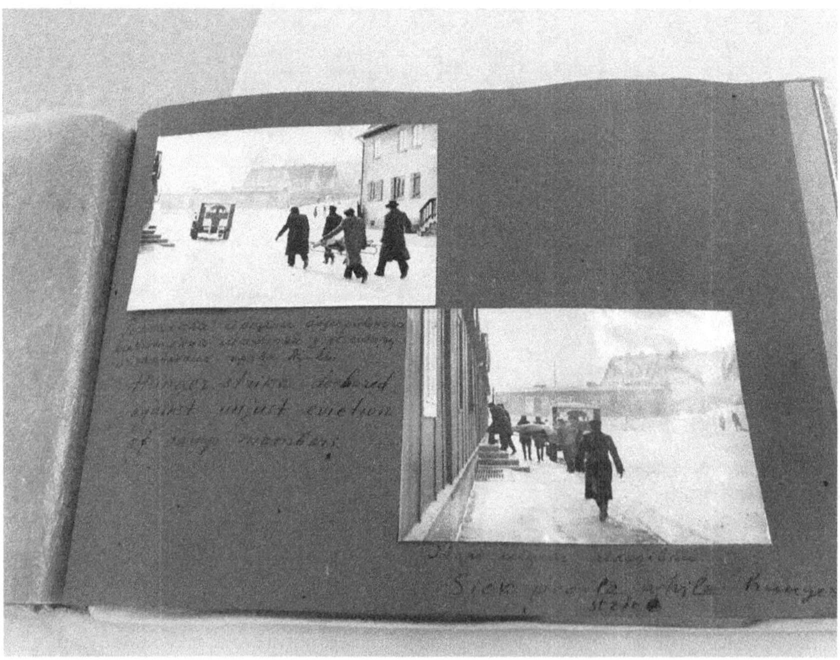

Figure 19.4a and 19.4b Photo album pages, left: "Protest against forcible repatriation"; right: "Hunger strike declared against unjust eviction of camp members".

Sources: UN Archives, "Area Team 1023-DP Camp 566-Cornburg-Photographs and Copy of Newspaper-"For Liberty"-Ukrainian Camp Weekly", S-0436–0004–07

Media repertoires in the camp, as outlined in the previous sections, could hardly been considered a given. Media often needed to be fought for, requested or self-produced. Records of camp assemblies (often called for after escalating conflicts), letters addressed to the authorities and documentations of camp protests reveal agency, vigour and articulated communication needs among camp residents: Hungarian refugees in a Hamburg camp in 1956 asked for "German newspapers and magazines";[15] German refugees in a Bavarian camp "expect cultural institutions" such as "continuous cinema operations",[16] and in a letter requested a permit for a "photo atelier"[17] (which was denied). As the administration had the ultimate control over communication flows within and outside the camp by having to agree on such media-technological investments, inventive media practices by refugees navigated this situation. Reports of improvised notice boards on walls of camp buildings or public buildings in cities (e.g. on the German museum in Munich; or in camp Friedland, see Seuferling, 2019: 212) show how refugees self-initiated communication platforms: paper notes were used for information exchange, mainly meant for finding lost family members and friends again. This paper-based practice was later on picked up by the radio, run by the German Red Cross collaborating with broadcasting stations, reading out names through the ether on a daily basis. Similarly, the mentioned camp newspapers were such self-developed media practices. Communication flows were taken into refugees' own hands, enabling them to mediate socialities of their own purpose and will, within and beyond the camp. Through such activist media practices, camp residents mobilised communication out of their own experience of immobility, and established new communication channels, within and beyond the camp.

Part of this resistance were camp protests, which inhabitants often documented and mediated themselves, e.g. through photography. Protests against specific camp and refugee policies, against concrete living conditions (e.g. "exorbitant shop prices in Friedland"[18]), or against impending deportation or repatriation. The photos in Figure 19.4 stem from a self-produced photo album on the history of DP-camp Cornburg. They show a protest and hunger strike against "forcible repatriation" and "unjust eviction of camp members". The Allies had mainly envisioned for DPs to return to their home countries (often in Eastern Europe), whereas many of them preferred onwards migration to countries like the United States, Palestine/Israel or Canada – a source of conflict.

Self-initiated communication, as well as mediated activism and protest practices, can be seen as further dimensions of mobile socialities in refugee camps: social practices, which emerge out of power dynamics in the camp and subjugation to the refugee regime, and become enacted in and through media practices. Specific communication needs, like tracing missing family, but also a craving for cultural activities, entertainment, or overall connectivity, were articulated and fought for. These are practices of voicing and documenting one's concerns out of the secluded, immobilised position of the refugee camp – symbolically mobilising oneself out of the fixated position of the speechless victim through practices of communication.

Immobile socialities?

Seventy years later, refugee camps have anything but vanished. While the COVID-19 pandemic reshuffled most people's lives, the infectious disease and its mitigation affects refugee camps and asylum-seeker shelters in particular ways. Not only does the complete isolation and practical quarantining of camps, and lack of access to valid health care (e.g. on the Greek islands and in German refugee centres), further immobilise and make precarious camp residents, but also, international agreements for refugee resettlement and reception, like the EU's Dublin regulations, have practically been halted. Regularly, accommodations are being quarantined and shut off after COVID-19-outbreaks. Of course, inside the centres, "social distancing" is practically impossible and large-scale infections are unavoidable. In spring 2020, a group of refugees in Bremen protested against these conditions – and was promptly reported to the police for breaking the hygiene and distancing rules during their demonstration (Presseportal, 2020, March 27) – rules which the state forces them to break on a daily basis in their very homes.

Walter Benjamin would call this situation a "moment of danger" (Benjamin, 1940 [2006]: 391): a moment in time, where an image of history "flashes" at us. In Benjamin's circular understanding of history, memories of the past become visible and graspable, lying in front of us, through a construction in the now. Mobile socialities and media practices in post-war German camps, explored in this chapter, easily lead us into contemporary spaces and experiences of forced migration, and vice versa. Historicising research takes on itself this perspective, in recognising the relevance of historical trajectories of refugee camps, mobility regimes, their mediation, and the subjugation of their inhabitants. Unpacking media practices in refugee camps post-World War Two reveals historical imaginaries and practices around mobility and its mediation in the context of refugee regimes. This period was a critical juncture of the modern refugee regime, when its major practices, institutions and discourses were developed (e.g. Geneva Convention and UNHCR). As we can see, media and communication were always part of establishing, maintaining and controlling the camp space, as well as of experiencing and navigating the socialities it produces. Long before digital media or mobile phones, media practices for disciplining, enabling and shaping mobility were incorporated in refugee individuals' socialities. Looking at trajectories of media-technological repertoires in camps even shows that post-war German refugee camps usually had a relatively well-equipped media environment, when compared to the situation during later critical periods, like the 1980s and 1990s, and recent years. Authorities valued connectivity less and less, and consequently, media-technological equipment, e.g. wi-fi, is not a standard service anymore. Yet, what the applied historical perspective highlights is how media practices were historically incorporated in imagining, designing and experiencing mobilisation and immobilisation in refugee camps.

Apart from counteracting a rhetoric of newness often found in accounts of mobile media and digital communication, historicisations of media practices in camps lead us to preceding articulations of socialities and anti-socialities, where mobility and immobility have been negotiated and imagined – in an extreme counter-space at the margins of society, yet reflecting its utopian ideas of belonging and disbelonging. The various (non-exhaustive) dimensions of mobile socialities I flagged in the previous section demonstrate how media practices are and were sites of social and physical mobilisation and immobilisation: through architecture, spatial arrangement and media technologies for connectivity; as well as for socialities and anti-socialities: forced collectiveness, community building and sharing experiences and identities, or organising protest. This demonstrates that socialities and media practices have to be scrutinised along a "spectrum of mobility and immobility", as Kevin Smets (2019: 3) recently suggested, capturing differentiated experiences of physical, as well as "symbolic" and "affective", immobilities in relation to media technologies and representations. The refugee camp defines itself through the contradiction of mobility and immobility, where people are stuck on the move – and therefore arguably through a continuous spectrum of *mobile and immobile socialities*. Mobility and immobility, as well as socialities and anti-socialities (like protests), overlap in the camp experience. The open-ended media practice lens reveals complex, interwoven mediations of movement and stillness, as well as social ties and encapsulation. Hence, instead of romanticising protest and activist practices as initiatives from below, we can ultimately see them as expressions of deep precarities and products of oppression through social and corporeal immobilisation and experiences of stuck-ness – while at the same time, protests are often aimed at immobility, namely of not getting deported. Media practices and technologies are sites of this spectrum of mobile and immobile socialities. They do not always fulfil clear functions and utilitarian purposes. Much rather, a perspective on mobile and immobile socialities from below and from the margins – which is simultaneously mindful of violent structures of oppression – can shed light on how we as societies construct mobility and immobility regimes and imaginaries through socialities and anti-socialities with their outcasts.

Notes

1 All non-English quotes have been translated by the author.
2 Ré Soupault (1901–1996), born as Meta Erna Niemeyer, was a German French artist. Educated at Bauhaus in Weimar, she worked as a photographer, journalist, fashion designer and radio essayist. She and her husband Philippe Soupault are associated with the Avant-garde and Dada movements.
3 This material stems from my ongoing PhD project on media practices in refugee camps in Germany between 1945 and 2000, charting and historicising media technologies in the camp space and refugee communication.
4 These include: refugee camp, refugee accommodation, asylum-seeker home, central contact point, common accommodation, camp of first reception, emergency reception camp, transit camp, asylum-seeker accommodation, asylum-seeker home.
5 These include: media, communication, radio, newspaper, magazine, television, phone, telecommunication, cinema, film, mail, letter.
6 The UNRRA administrated camps for millions of Displaced Persons (DPs) in post-war Germany. DPs were liberated prisoners and concentration camp inmates who had been forcefully displaced by Nazi Germany.
7 Germany's and central Europe's forced migration events in the aftermath of World War Two can be seen as a root for the contemporary global refugee regime, based in the Geneva Convention of 1951. Imaginaries, practices, definitions and institutions of refugee administration were developed in this context, and form the global handling of forced migration until today (see Gatrell, 2013).
8 Staatsarchiv Nürnberg, "Gewerbebetriebe im Sammellager für Ausländer und Kantinen Allgemein" (Reg. v. Mfr. Abg. 1978, Nr. 19856).
9 UN Archives, "Displaced Persons – Newspapers", S-0402-0001-12, 1946–1957.
10 Landesarchiv Niedersachsen, NDS380 Acc. 32/97 Nr. 8"Beschwerden über das Durchgangslager Friedland", 1955–1956.
11 See e.g. the critical "Harrison report" about DP-camps from 1945.
12 Bundesarchiv Koblenz, 150/3304: "Spendenaufkommen für die Friedlandhilfe", 1958; translated by author.
13 UN Archives, "Z-181 – Story Material Used", S-0405-0019-04, 1945–1947.
14 UN Archives, "Area Team 1023-DP Camp 566-Cornburg-Photographs and Copy of Newspaper-"For Liberty"-Ukrainian Camp Weekly", S-0436-0004-07, 1946.
15 Staatsarchiv Hamburg, 442–7_219, "Finkenwerder Ungarnflüchtlinge", 1956–1960.
16 Staatsarchiv Nürnberg, "Flüchtlingslager im Bereich des Landbauamts Ansbach, Allgemeiner Schriftverkehr und Pläne", 1948–1951 (Reg. v. Mfr. Abg. 1978 Nr.19764).
17 Staatsarchiv Nürnberg,"Gewerbebetriebe im Sammellager für Ausländer und Kantinen Allgemein" (Reg. v. Mfr. Abg. 1978, Nr. 19856).
18 Landesarchiv Niedersachsen, NDS 380 Acc. 32/97 Nr. 8"Beschwerden über das Durchgangslager Friedland", 1955–1956.

References

Agamben, G. (2005) *State of Exception*. Chicago: University of Chicago Press.
Alencar, A., Kondova, K. and Ribbens, W. (2019) "The smartphone as a lifeline: An exploration of refugees' use of mobile communication technologies during their flight", *Media, Culture & Society*, 41(6): 1–17.
Arendt, H. (1956 [2017]) *The Origins of Totalitarianism*. London: Penguin.
Arnold, S. and Görland, S. (2019) "Participatory logistics from below: The role of smartphones for Syrian refugees", in T. Thomas, M. M. Kruse and M. Stehling (eds.) *Media and Participation in Post-Migrant Societies*. Lanham: Rowman & Littlefield, 101–116.

Athique, A. (2017) *Transnational Audiences: Media Reception on a Global Scale*. Cambridge: Polity Press.

Awad, I. and Tossell, J. (2019) "Is the smartphone always a smart choice? Against the utilitarian view of the 'connected migrant'", *Information, Communication & Society*. doi:10.1080/1369118X.2019.1668456

Benjamin, W. (1940 [2006]) "On the concept of history", in H. Eiland and M. W. Jennings (eds.) and H. Zohn (trans.) *Walter Benjamin: Selected Writings. Volume 4, 1938–1940*. Cambridge: The Belknap Press of Harvard University Press, 389–400.

Biselli, A. and Beckmann, L. (2020) *Invading Refugees' Phones: Digital Forms of Migration Control in Germany and Europe*. Berlin: Gesellschaft für Freiheitsrechte e.V. Available at: https://freiheitsrechte.org/home/wp-content/uploads/2020/02/Study_Invading-Refugees-Phones_Digital-Forms-of-Migration-Control.pdf.

Bishop, S. C. (2017) "(Un)documented immigrant media makers and the search for connection online", *Critical Studies in Media Communication*, 34(5): 415–431.

Bolin, G. (1998) *Filmbytare: Videovåld, kulturell produktion & unga män. Doktorsavhandlingar från JMK, 7*. Umeå: Boréa.

Bromley, R. (2012) "Undesirable and placeless: Finding a political space for the displaced in a cinema of destitution", *Interventions*, 14(3): 341–360.

Couldry, N. (2004) "Theorising media as practice", *Social Semiotics*, 14(2): 115–132.

Couldry, N. (2010) "Theorising media as practice", in B. Bräuchler and J. Postill (eds.) *Theorising Media and Practice*. New York and Oxford: Berghahn Books, 35–54.

Diminescu, D. (2008) "The connected migrant: An epistemological manifesto", *Social Science Information*, 47(4): 565–579.

Drotner, K. (1993) "Medieetnografiske problemstillinger – en oversigt", *MedieKultur: Journal of Media and Communication Research*, 9(21): 5–22.

Gatrell, P. (2013) *The Making of the Modern Refugee*. Oxford: Oxford University Press.

Georgiou, M. (2018) "Does the subaltern speak? Migrant voices in digital Europe", *Popular Communication*, 16(1): 45–57.

Gießmann, S. (2018) "Elemente einer Praxistheorie der Medien", *Zeitschrift für Medienwissenschaften*, 19(2): 95–109.

Gitelman, L. (2014) *Paper Knowledge: Toward a Media History of Documents*. Durham, NC: Duke University Press.

Hepp, A., Bozdag, C. and Suna, L. (2011) *Mediale Migranten. Mediatisierung und die kommunikative Vernetzung der Diaspora*. Wiesbaden: Springer VS Verlag für Sozialwissenschaften.

Hochmuth, K. and Bispinck, H. (eds.) (2014) *Flüchtlingslager im Nachkriegsdeutschland. Migration, Politik, Erinnerung*. Berlin: Christoph Links Verlag.

Kaun, A. (2016) *Crisis and Critique: A Brief History of Media Participation*. London: Zed.

Keightley, E. and Reading, A. (2014) "Mediated mobilities", *Media, Culture and Society*, 36(3): 285–301.

Leurs, K. (2017) "Communication rights from the margins: Politicising young refugees' smartphone pocket archives", *International Communication Gazette*, 79(6–7): 674–698.

Leurs, K. and Smets, K. (2018) "Five questions for digital migration studies: Learning from digital connectivity and forced migration in(to) Europe", *Social Media + Society*, 4(1): 1–16.

Madianou, M. and Miller, D. (2012) *Migration and New Media: Transnational Families and Polymedia*. London: Routledge.

Malkki, L. H. (1996) "Speechless emissaries: Refugees, humanitarianism, and dehistoricization", *Cultural Anthropology*, 11(3): 377–404.

Martín-Barbero, J. (1993) *Communication, Culture and Hegemony: From the Media to the Mediations*. Newbury Park, NJ: Sage.

Marvin, C. (1990) *When Old Technologies Were New: Thinking About Electric Communication in the Late Nineteenth Century*. Oxford: Oxford University Press.

Massey, D. (2005) *For Space*. London: Sage.

Mattern, S. (2017) *Code and Clay, Data and Dirt: Five Thousand Years of Urban Media*. Minneapolis: University of Minnesota Press.

Metcalfe, P. and Dencik, L. (2019) "The politics of big borders: Data (in)justice and the governance of refugees", *First Monday*, 24(4): 1–15.

Metzner, M. (ed.) (2016) *Ré Soupault: Katakomben der Seele. Eine Reportage über Westdeutschlands Vertriebenen- und Flüchtlingsproblem 1950*. Heidelberg: Wunderhorn.

Morley, D. (2017) *Communications and Mobility: The Migrant, the Mobile Phone, and the Container Box*. Hoboken, NJ: Wiley-Blackwell.

Presseportal (2020) "POL-HB: Nr.: 0202 Verstöße gegen das Infektionsschutzgesetz bei Kundgebung", March 27. Available at: www.presseportal.de/blaulicht/pm/35235/4558695.

Schießl, S. (2019) *Das Tor zur Freiheit: Kriegsfolgen, Erinnerungspolitik und humanitärer Anspruch im Lager Friedland (1945–1970).* Göttingen: Wallstein.

Seuferling, P. (2019) "'We demand better ways to communicate': Pre-digital media practices in refugee camps", *Media and Communication*, 7(2): 207–217.

Seuferling, P. (2020) "Hopeful and obligatory remembering: Mediated memory in refugee camps in post-war Germany", *Mediální Studia*, 14(1): 13–33.

Shove, E., Pantzar, M. and Watson, M. (2012) *The Dynamics of Social Practice: Everyday Life and How It Changes*. London: Sage.

Smets, K. (2019) "Media and immobility: The affective and symbolic immobility of forced migrants", *European Journal of Communication*, 34(6): 650–660.

Smets, K., Leurs, K., Georgiou, M., Witteborn, S. and Gajjala, R. (eds.) (2020) *The Sage Handbook of Media and Migration*. Los Angeles: Sage.

Stephansen, H. C. and Treré, E. (2020) "Practice what you preach? Currents, connections, and challenges in theorizing citizen media and practice", in H. C. Stephansen and E. Treré (eds.) *Citizen Media and Practice: Currents, Connections, Challenges*. London: Routledge, 1–34.

Urry, J. (2007) *Mobilities*. Cambridge: Polity.

Wall, M., Otis Campbell, M. and Janbek, D. (2017) "Syrian refugees and information precarity", *New Media and Society*, 19(2): 240–254.

Witteborn, S. (2014) "Forced migrants, emotive practice and digital heterotopia", *Crossings: Journal of Migration & Culture*, 5(1): 73–85.

Wittel, A. (2001) "Toward a network sociality", *Theory, Culture and Society*, 18(6): 51–76.

20

SOCIALITIES OF PRACTICE

Stuckedness, accountability and mobile imaginaries among Kenyan migrant fisherpeople descendants

Anna Lisa Ramella

Situating migration and mobility at Lake Baringo[1]

"This is just a place for business", Opondo tells me when I ask him about his house and workshop in Kambi Ya Samaki, an old fishing village at Lake Baringo in northern Kenya. Opondo belongs to the Luo community here, a migrant community from the shores of Lake Victoria who arrived as early as the 1940s to Lake Baringo and other lakes of the Great Rift Valley in search of better fishing opportunities and employment. The colonial government had endeavoured to make this long-neglected, arid stretch of land (Little, 2019: 134ff.) economically viable, initiating several attempts to build an economy around fishing between the 1940s and 1960s, usually involving British entrepreneurs or colonial wardens settled in the area (cf. Rift Valley Province Annual Reports, courtesy of Kenya National Archives). The local community, consisting in their majority of Ilchamus, an ethnic group that is culturally proximate to the Maasai, though they catch and eat fish, had until that point only fished for their own consumption, although such practices date back at least a century or two (Petek, 2015: 105). Luo fishermen from Lake Victoria in western Kenya (former Nyanza province) were also employed in Baringo to teach the local community at Kambi Ya Samaki how to turn fish into an economic asset. This included mostly the introduction of new fishing methods with nets and lines, as well as manoeuvring the concomitant materials and tools, all imported from Lake Victoria, where Luos had already been fishing economically for many years. The fishing method of the Ilchamus, which is still practised today, was mostly viable for quantities of fish sufficient just for personal consumption, however; it did not bring in enough of a haul for selling. More than a decade after the first Luos had arrived in Baringo, a fish processing factory with proper freezing rooms was built at the lake's shore in 1956, and rebuilt in 1965 after a flood. This brought to Lake Baringo more Luo fishermen who had been searching for new pastures (or waters) given the ever-decreasing fishing opportunities at the time in Lake Victoria, and served to make fishing a welcomed opportunity for the local Ilchamus and Tugen, whose dominant activity remains to be cattle pastoralism (Anderson and Bollig, 2016: 4).

However, some Luo had already discovered the fishing opportunities at Lake Baringo on their own. While men fished the lake, women were mostly engaged in the processing of the fish – gutting, smoking, drying, packing – and in establishing economic networks, initially with the nearby town of Nakuru, and later on all the way to Nairobi's markets.

Some descendants of those first fishermen and -women now still live in Kambi Ya Samaki, as do others who have arrived after them. However, after downturns in the fishing economy in the 1990s due to lower water levels and the lake's temporary closing by the fisheries department to curtail overfishing, many of the fishermen have left and returned to their lands in western Kenya. In this chapter, I would like to share stories of some of those second-generation Luo descendants who stayed, and who now engage in different activities around and apart from fish and fish trading. All of them are adamant that they are not at home here, and many narrate experiences of feeling 'stuck' by the possibilities of work Baringo offers with regards to their homeland.

The perceived necessity of return to western Kenya is narratively linked to Luo customary law, which demands final burial on one's own land in the region of one's descent (Dundes Renteln, 2004: 176). Even though moving back 'home' is often stated as the ultimate aim of most of these descendants and current inhabitants, the attempts of a number of them have failed due to the absence of social networks, or have not even been possible in the first place due to lack of funds. The established socialities at Baringo, however, are understood to be temporary and mobile, and typically connoted with a sense of "stuckedness" (Khan, 2016; Hage, 2009) in this place that interlocutors have discussed as being a "place for business".

By looking at the practical and social implications of this perception of one's own abode as being both here and not here, I will discuss the implicit practices that account for senses of perceived mobility, temporariness and provisionality. I wish to explore the concept of 'mobile socialities' as a means of situating the lifeworlds of this lakeside Luo community within their current state of (perceived) mobility, as well as towards imaginations of an anticipated future in their home region. A look at practices here reveals the complex relations that drive and are driven by social practices of accountability. Through my broader research on Kenya, I hope to offer a new perspective on the way migration and mobility play out in terms of social bonds that are neither mobile nor in any one place. Rather, these trajectories and the imaginations that accompany them offer an opportunity to understand the various ethnic, familial and political relations as being embedded within diverse socialities.

Socialities of practice: accounting for im/mobility

The ethnographic material this chapter draws on was produced during fieldwork in the Kenyan Rift Valley in 2019 and 2020. In Baringo, I have worked closely with a research assistant from the Ilchamus community, and have regularly accompanied fishermen on their boats, as well as fish traders at their trading beach. Opondo and Okwany, the two Luo migrant descendants whose stories I will share in this chapter, have been my close contacts with whom I have spent many days – in Opondo's workshop and house, or by Okwany's boat. I have conducted audiovisual methods by filming practices and focusing on how people form and engage in communities of practice. Through the framing interviews of observing and filming the practices of fishing, repairing, building and not least furnishing and inhabiting, I have caught the notions with which these migrant descendants are conceptualising their stay at Baringo. Without particularly asking about 'home', this issue came up immediately with reference to the western Kenyan home region when talking about their activities in the Rift Valley, which were framed as temporary, and serve the goal of finally realising the move back. I have then observed how this particular narrative is mirrored in practices of temporariness, and/or of a shortly lived means to gain capital over the long term. Via the analysis of these observations and conversations, I have come to understand that many of the practices serve to communicate such temporariness, or are connected therewith.

The feeling of stuckedness within a perceived mobility, as reported by migrant descendants in the Kenyan Rift Valley, offers to sketch out the various levels of tension across diverse temporalities at play here. While forming sustainable social bonds in Baringo as the place where one is established – even temporarily – promotes a sense of home, practices of unsteadiness, such as building temporary housing, account for the provisionality and temporariness that is to be articulated towards socialities at 'home'. Thereby, an imagined mobility (being transient, not at home here) meets an imagined immobilisation in the future (to settle in one's home region), and both are accounted for by practices manifesting socialities across geographical landscapes. At the same time, as will become evident in my ethnographic material, the social bonds held at home may suffice for returning as an elder, but not to conduct business there. It is therefore directed at a temporally distant, future mobility.

'Mobile socialities' as a concept lends itself to such boundary work (as developed by Suchman, 2002), as it can capture the tensions that arise from such determination, yet inability to 'make it' at home, while being established in the place of one's migratory past. It is characterised by persons' "boundary crossings" (Suchman, 2002: 94) on a social and local scale in a situated manner.

In this sense, this chapter aims to offer a new look at migration, which despite its apparent physical immobility comports perceived traces and experiences of mobility, visible via the various socialities that can be discerned in im/mobility. In this context, I address the way in which mobility here is defined and engendered, and how an understanding of practices that shape and stabilise socialities can be understood as "media [practices] of accountability" (Thielmann, 2012) that determine and justify one's sense of belonging. Thielmann, while referring to social media platforms, draws on Garfinkel's ethnomethodology work from the 1950s (cf. Rawls, 2002), in which he relates social practices to practices of documentation. Taking these thoughts a bit further, I will sketch out how certain practices can account for e.g. a sense of stuckedness as they document a will to leave; and in the same way, practices that manifest a strong connection to home, document and communicate to others an imagined mobile scenario in which 'going back' seems realistic.

Departing from a mobilities perspective that perceives both movement and standstill as potentially 'mobile' circumstances, I want to look at the practices that evolve through the establishment of these socialities as a means of differentiating and understanding my interlocutors' notions of home through their own understanding of homing practices. By engaging with the stories of two Luo migrant descendants at Lake Baringo, I will conceptualise how practices manifest their socialities across distances and in situ, claiming that im/mobilities cannot be equated with movement or being in place, but rather must be seen through the practices that account for either. By way of conclusion, though offering no formal concluding remark, I will try to look forward, first by describing the circumstances of lockdown I encountered during my fieldwork in Baringo due to the spread of COVID-19, then asking how futures and future-making – including the ongoing quest of (not) being able to go home – can, at a moment's notice, be put on hold.

Im/mobilities and sociality in imaginary and practice

While migration figures centrally into the origin of the Luo in Kenya – Western Nilotic people who have moved from Sudan slowly towards the shores of Lake Victoria since pre-colonial times – it has been an asset in the future-making practices of contemporary Luo during and after British colonial rule. Due to their experience in farming and fishing, Luo people have for decades relocated on their own terms for farming jobs or as part of development projects in areas of pastoralist inhabitation (cf. Rift Valley Province Reports, courtesy of Kenya National Archives); others had found their way to the Rift Valley lakes by means of engagement in trading, as fishermen or as middlemen for fish trade. However, rather than explicitly calling it 'migration', the narratives of my interlocutors suggest that their movements are better understood through terms such as 'commuting', despite the fact that a back-and-forth movement associated with such is largely not even the case. In discussing their lives, people evoke images of work-related places of abode rather than of making – or trying to make – a home elsewhere. This is often justified in relation to Luo customary law, which expects a man to build a home close to the homestead he has been raised in, and to be buried on the grounds of his home upon death (cf. Shiino, 1997). Many may not achieve the capital they have aspired to raise in order to go back home, and as their plans are postponed ever again, they feel like they are 'stuck in commute' – always, already, but somehow never actually, moving.

It goes without saying that Luo migration is situated within a complex field of political, cultural and economic factors that determine not only movement, but also ways of immobilising it – as in other contexts of migration, as well. Specific for Luo migration in Kenya is a history of deteriorating industries in western Kenya, along with a cultural narrative which impedes a long-term migration in theory. The lack of wage labour in the home region is part of why it has become difficult for many to raise capital back home, as the education officer in Baringo county, who is also a Luo, has outlined the situation. Other factors such as the distribution of land have produced a scarcity that brought about an increase in land prices, and thereby deteriorated opportunities of small-scale agricultural production.

Migration to Baringo from western Kenya has been largely connected to attempts at economically exploiting the lake for its fish resources, and the redux of migrants going back has been mostly inspired by either changes in lake levels or decreased fish populations, leading to a lack of viable employment or the possibility to actually fish.

Yet people's success at Lake Baringo does not only depend on economic and ecological factors, but is also deeply connected to the ways in which Luos are involved and embedded in the local communities, organised in diverse settings of e.g. self-help or self-organised savings groups, or by means of economic ties to their larger networks that reach back 'home' or towards the markets of Nakuru and Nairobi.

Given these factors, I argue that the migratory setting of western Kenyan migrants at Lake Baringo depends heavily on the juggling of diverse socialities that are related to a number of factors ranging from cultural narrative, social ties, and the socio-political and economic situation. These come to the fore, as I will show, by means of accountability practices that stabilise and establish their sense of both mobility *and* sociality. I will explore understanding this setting through the lens of *mobile socialities*, a perspective that enables tying together aspects that feed off of both, and one that lets me speak directly to this cultural narrative that is often evoked in conversations about home, but at the same time connotes the distress encountered by living elsewhere and most likely staying or being stuck there.

Mobilities and immobilities

Academic understandings of mobility have undergone quite a few shifts in recent decades, as Salazar and Smart (2011: 1) point out, moving from a sedentarist (Malkki, 1992; Hannam et al., 2006) towards a nomadic metaphyiscs (Cresswell, 2006), before returning attention to the points in time in which mobility comes to a standstill (Adey, 2006; cf. Ramella and Norum, 2021). Given this tension, *mobile socialities* is a promising concept, as it lets us more comprehensively understand the ways in which people form social bonds, both across localities and temporalities.

While mobilities approaches question assumptions of interrelatedness of identity, culture and place (cf. Cangià and Zittoun, 2020: 2) and therefore serve to dismantle the oft-recounted culturalistic perspective on Luo cultural identity, thinking about socialities leaves room for considering both cultural narratives and social embeddedness. I understand socialities as a concept that can include both 'home' and 'here', without juxtaposing them. Luos in Baringo are involved in the translocal social relations they hold in their home region, as well as many other diverse local socialities, such as neighbourhoods, fishermen communities and farming groups.

As we will see in the stories of Okwany and Opondo, two Luo descendants of the first fishermen at Baringo, these socialities cannot be distinguished by the mere locative presence of physical encounter; sometimes they exist through a network of phone calls, or the acting out of kinship relations via remittances, or even the mere employment of traditional practices from the home region. Whether referring to communication or other forms of media, I am interested in how specific practices account for a belonging to various socialities at play. In this sense, media here are understood in the broadest sense, and include practices of planting or of building as 'media practices' that account for a certain articulation of movement or stability. Putting these practices into perspective with a focus on imagination, thereby adds to im/mobilities and social worlds that are not just lived and experienced, but also dreamed of, feared or aspired to.

Why socialities and not networks?

A lot of ink has been spilled during recent decades to understand (or misunderstand) the seemingly more rapidly moving world, particularly by paying attention to its moments of stillness and immobility (e.g. Hage, 2009).

Larsen et al. (2006) introduce the focus on networks with a section on "mobile societies" (Larsen et al., 2006: 1), which they characterise as people being more on the move *and* at the same time more connected through communicative media. Rather than densely populated "little boxes" (Larsen et al., 2006: 1), societies seem to be more and more understood through terms of the network. By employing the concept of socialities rather than network, however, I aim to circumvent the notion that seems to put geographical factors such as distance and location – as well as factors of time such as past, future or present – into relation with the way people form social bonds, thereby implying a differentiation of mobility and immobility that is based on place, time and movement (or the lack thereof). Rather, from the perspective of socialities, a myriad of factors that do not necessarily counteract each other might be involved in sensing a belonging – a settled practice can emote a mobile sociality elsewhere, or mobility can be perceived even though no movement is performed. Thereby, as we will see, immobilisation in one place through practices of settledness can evoke forming part of a sociality 'in commute'. In the same way, a sociality in commute may mean the stability of place, regardless of time.

In terms of research methodology, the 'ethnographic place' itself has undergone shifts in the way that it has been described, conceptualised and (un-)bounded. As Vered Amit (2000) challenged in her book from two decades ago, our ethnographic fields can no longer be seen as bounded to a locality. Sarah Pink and John Postill later developed the idea that ethnographic places are neither bounded territories nor groups or communities, but are rather "clusters or intensities of things of which *both localities and socialities* are elements" (Postill and Pink, 2012: 123; emphasis mine), and which Rapport and Amit call "hybrid forms of sociality through which the ethnographer and her research participants gain variously mediated senses of contextual fellowship" (Rapport and Amit, 2002 cited in Postill and Pink, 2012: 10ff.). In the same way as these authors describe the relationship between researcher and research participants, *socialities* can serve to think through the bonds that the participants of my research might have with their distant relatives, and their 'new' neighbours, as ones of contextuality.

The authors here emphasise employing the perspective of 'socialities' rather than of 'networks' as a means of dissolving the "network thinking" that departs from disparate points to their connections, evoking an Ingoldian understanding of the mesh (Ingold, 2007: 80ff.), where not the knots but the lines connecting them enter into focus. This approach, they argue, attends to the qualities of social relationships rather than their being part of a 'community' (see Postill, 2008: 5). The concept is helpful in considering the various forms of sociality encountered in Luo migrants' lives – socially, locally, culturally – as well as the various forms these can be mediated, whether via physical encounter or a networked device, thereby materialising the relationships into practice.

Imagining home, mobilising sociality

It is difficult to describe and conceptualise the idea of home that is being portrayed here without putting it into relation with imaginaries. In this section, I want to briefly sketch out some theoretical underpinnings that help me to further think through mobile socialities in my research field. Ida Wentzel Winther (2006) has distinguished four different ways for understanding home: home as a place, home as an idea, feeling at home, and homing as a practice to establish such home-feeling (as has been summarised by Arp Fallov et al., 2013: 477). She has developed this differentiation in order to describe how home and mobility can be thought as part and parcel of the same configuration. Homing, she describes, is a way to relate oneself to everyday life and to obtain the necessary impression of 'coping' (with e.g. mobility): " 'Homing oneself' is not connected to one place, but is a nomadism – a capability of being in the world. It is a strategy and kind of tactic, as Michel de Certeau (1984) puts it; a way to use the space, not to *arrive* in a place" (Winther, 2009: 66; emphasis in original). Now, in the context of the Kenyan migrant fisherpeople descendants, these categories seem to be distorted: while they are imagining home as a place in the present – the geographical locus of their ancestors' homeland – and their own homestead as the idea of a material infrastructure to signify this home in the future, they are performing present homing practices in the migratory place of their abode. These practices, however, point to diverse imaginaries: planting crops in their *shambas*, for example – a seemingly 'sedentarist' practice that differentiates many Luos in Baringo from the local pastoralist community of the Ilchamus – is closely tied to their 'home imaginary' and the fertile land of their descent. Apparent homing practices, such as building structures to live in, are built in a manner that is temporary and provisional, so as to materialise their sense of belonging elsewhere. Forming bonds to secure social and economic stability, on the other hand, positions them socially in their migratory place.

Charles Taylor describes the "social imaginary" as

> a common understanding that makes possible common practices and a widely shared sense of legitimacy . . . it incorporates a sense of the normal expectations that we have of one another, the kind of common understanding which enables us to carry out the collective practices that make up our social life.
>
> (Taylor, 2002: 106–107)

Roger Norum has used this notion to describe the "hypersocialities" of expatriates in Nepal, which involves a particular "relationship between individual and collective imaginaries. . . [and] offers a nuanced means of expressing meaning across various social categories" (Norum, 2015).

Planting crops for Luos in Baringo speaks to those various levels of sociality as it 'homes' them in the sense of a common and known practice which evokes their region of origin, serves to counter economic instabilities by means of self-subsistence, and at the same time places them in a new sociality with the local communities who are eager to engage in farming and learn from them.

Flavia Cangià and Tania Zittoun, in their editorial "Exploring the Interplay between (Im) mobility and Imagination", put the possibilities to move physically into relation with imagining and aspiring future lives and selves (Cangià and Zittoun, 2020: 1). They describe how imagination is part of what makes one move or hinders one's movements. With reference to Schuetz (1945), they write: "[One] can go physically and imaginatively back home, while psychologically or socially never be able to return to the place that was before" (Cangià and Zittoun, 2020: 2). What this points to is that there is potentially also a sense of imaginary in the stuckedness, not only in mobility. Ghassan Hage elaborates on exactly this point, arguing "that just as there is an imaginary existential mobility, there is an imagined existential stuckedness. . ., [highlighting] the fact that social and existential mobility are not the same thing" (Hage, 2009: 99). In the case of Baringo, this is an important point to keep in mind, as the sense of stuckedness that people report is not due to the fact that they cannot reach a place where they can work; rather, they are stuck in the place *due to* the possibilities it offers, and are thereby hindered to move back 'home'.

While what makes them want to go home is a sense of belonging elsewhere, they emplace themselves (Grasseni, 2009: 8) through practice in their local sociality.

Grasseni, who has written about the development of skill based on Lave's (1991) notion of communities of practice, suggests addressing the question of identity and emplacement through an emphasis on practice, landscape and contexts, rather than from the point of view of social, political and ethnic identity boundaries (Grasseni, 2009: 8). She notes:

> From an anthropological point of view, then, 'communities of practice' (Lave, 1991), in their broadest sense, offer concrete scopes and targets of observation to highlight the dynamics of cognition, skill and social roles in their making. Cultural belonging would be the result of 'enskilment', not of enculturation (Ingold, 1993), of 'resonance' rather than of interpretation (Wikan, 1993), and understanding cultures a problem of learning rather than one of translation (Pálsson, 1993).
>
> (Grasseni, 2009: 12)

It is from here that I derive my concept of *socialities of practice*, which is aimed to consolidate notions of Lave's "communities of practice" in that they are involved in mutual learning and enskilment processes, with the notion of 'mobile socialities', which serves to understand the implications of the imaginary on the practices performed to account for one's self-perception and the stabilisation of socialities with others. This shall become more clear once I detail ethnographically how people materialise imagination and sociality in their practices, and serves as a lens to bridge to the analysis via the following ethnographic section.

Born here, but not at home here? The stories of Okwany and Opondo

"We Luos follow lakes and rivers", Okwany tells me in one of our first conversations, by way of explanation why it is no wonder that Luo migrants such as his father have come here decades ago. "However", he adds, "we must always go back to our people". Okwany is one of the few Luos here at Lake Baringo who still works as a fisherman. He owns a small timber boat that he inherited from a fishermen's association which he used to work for. Every morning at 8am, he enters the lake with his boat to check on his fishing nets. The nets remain in the lake all night and the fisherperson goes to collect his fish in the morning. They must hurry because according to the fisheries department, in order to ensure freshness, no fish are to be 'landed' (brought out from the water onto land) after 2pm. Okwany is married to a Maasai woman, but has not yet paid the full dowry to her family, which explains why he cannot bring her from her home village and take her with him. Once he has paid his debt, he wants to make enough money to build a house – either 'here' in Baringo, renting it out to someone else and then finding another place to stay, or right away 'there' – in western Kenyan, that is. "I must go there", he says, "because where my people are is there. Home is home".

Okwany is the oldest of his siblings. When his mother passed away, his father was urged to go back to his home village in Kisumu – several hundred kilometres west of Baringo – to bury her and, as custom has it, to stay there in order to be in his own homestead. Because there was no work in Kisumu, Okwany stayed in Baringo and took care of his siblings, while he worked as a fisherman to send remittances home to his father. Now that his own four children have grown, Okwany still sends money home to his father. "I would like to go back," he says. "Now, I'm working, if I could get some [sufficient] amount [of money], I could go when I'm still strong, so that I organise and get used to life there".

As long as he does not have that sort of money, he says, he has to stay here, in Baringo. Because back home, people do not know him – after all, he has never really lived in Kisumu – he cannot count on a network of support there. In Baringo, however, his neighbours and friends – a community made up both of locals, as well as Luo – share and offer support to anyone in need.

Opondo, just like Okwany, is also a descendant of a Luo fisherman. His father, who had first come to Baringo in 1938 to engage in donkey trading back to Kisumu in western Kenya, started out by building a small camp under a tree and, as Opondo's sister Jennifer tells me, decided to quit his donkey business. When he found the fish population to be lucrative enough, he returned to Kisumu to bring two of his brothers back with him to Baringo, and soon they had their first customers in a nearby town, where some settlements connected to the sisal estates in the region had other Luo inhabitants. He built a few timber boats and settled in the place today known as Kambi Ya Samaki (Swahili for 'Fish Camp'), reportedly before anyone had lived here.

In front of a former road-side shop which used to sell necessities to tourists sits Opondo in the shade of a tree, fixing a car part with his welding machine.

Tourism had arrived at Lake Baringo much later than fishing. It was long after the collapse of the fish filleting company in the 1960s that tourism began to bloom in this area, a stretch of land off the popular circuit. Drawn to the Baringo basin by the large number of pelicans at Lake Bogoria, a smaller lake close to Lake Baringo, tourists would pay a visit to Lake Baringo's lodges, book a boat trip to the island and roam across the lake in search of hippo and bird viewings. In this way, tourism became a lucrative alternative to fishing, particularly for those who owned a boat. It was in this realm that Opondo learned the necessary skills of woodworking and metal construction. While his sister Jennifer continues to process and sell fish, as she did for their father when only a schoolgirl, Opondo found work with the family that owned both the freezing room and a tourist camp, and who supported him to obtain qualifications such as welding and mechanics, as well as becoming a licensed professional driver. Equipped with a number of skills, Opondo decided to leave Baringo to go back to his home in Kisumu and start his mechanics business there. However, his small workshop there did not go very well, as he reflects, due to the lacking social bonds. While he had stayed in touch via phone, made visits to see his 13 brothers and sisters, and even remained in contact with his extended family, he was not known in his home region as a regular visitor and had too few connections to enable him to establish a solid business there.

He gave himself one year before he returned to Baringo, where people knew and trusted him.

"I should not be in my father's compound now that my hair is grey", Opondo says. However, he cannot yet afford to build his own homestead back in Nyanza. "What about the house that you built here?" I ask. He laughs. "That's just for business. Don't you see it's made of iron sheets? This is not the house I will build back home", he replies, and adds: "No one will be buried here". His children will inherit it, in case they want to pursue further business here. Since the land in Nyanza is scarce, many sons have to go and buy land elsewhere. At some point, however, they must return home and own a compound on which their own sons can build their house, even though they might not ever live there themselves. In fact, empty houses feature greatly in homesteads across western Kenya; a material sign of migration as a common strategy and practice, and one that highlights the complexities and subtleties that persist within many homing and migratory practices.

Mediating socialities and accountabilities across distance and in situ

Opondo's and Okwany's stories show how Luo migrant descendants at Lake Baringo manoeuvre manifold socialities. Drawing on this, I want to develop how these socialities play out via (media) practices that serve to negotiate both, a local emplacement and the imagined belonging to 'home'. I will do so by considering these as 'media practices of accountability', situating them within their complex setting between narratives, imaginations and justification practices towards oneself and others.

While some concepts have bridged these different perspectives convincingly – such as Winther's concept of home (2006) – I want to suggest that the notion of mobile socialities, too, can offer us cross-cutting insights that enable understanding seemingly juxtaposed aspects of everyday life. I will demonstrate this by taking into account the ways in which such socialities are manifested – that is, brought forth – and stabilised by concrete practices that serve to account for what is here understood as mobility/provisionality and immobility/stuckedness.

Following a praxeological approach to media studies as features in Nick Couldry's work (Couldry, 2010), and which also relates to a "non-media-centric media approach" as Shaun Moores, David Morley and others have employed long before they came to call it such (Krajina et al., 2014), I define media through the notion of what "people do with [them]". From this angle, media cannot be understood as an entity per se, but rather only come into being via the practices that shape them and through which they are being shaped. In the context of descendants of migrant fishermen at Lake Baringo, I wish to take into account various "social media practices' as instances that manifest the diverse socialities in situ and across distance: these do not refer necessarily to 'communication media', but include landscapes, materialisations, buildings. Via Shannon Mattern's definition of media, I understand such as inscribed with, transmitting and embodying information (Mattern, 2017: xii, citing McCullough [2013]) – articulating a certain positioning which I analyse through practices. I understand 'social' media, then, in the sense that these media are practiced to manifest a sociality or social bond. Following Thielmann's (2012) work on "media of accountability" in which he further theorises Garfinkel's (1967) notion of "social accounts", these practices serve to accumulate social capital (in terms of trust, skill, networks of support) in both the mobile sociality which can be located at the place of abode (Baringo) and the mobile sociality brought about from 'home' via social media practices. One such practice is sending remittances, understanding money itself as a social medium (Passmann, 2018: 16) that negotiates social relationships and interactions. A social bond to home in, for example, Opondo's case, then, is primarily practised by remitting, visiting, and keeping in touch via communication media. By these means, he stabilises and continuously reaffirms the relationship to his genealogical community, thereby *accounting for* his belonging to such sociality.

On the other hand, in Baringo, where Opondo lives, the very materiality of his house serves to visualise and prove the provisionality which he ascribes to his living here: the iron sheets serve to account for Opondo's specific perception of mobility, and his sense of stuckedness, towards himself and towards others. For Okwany, similarly, his plot in Baringo at most serves for rentals, in order to take the money 'home' with him. A narrative of belonging elsewhere is evoked and determined by (media) practices of accountability to substantiate and stabilise the idea of not being at home here.

Mobile socialities, then, are established in a two-fold way: One requires to grasp the understanding of mobility that underlies Opondo's actions: as he perceives himself as not settled, but always already prepared to strike camp and upend for elsewhere, his socialities in Baringo are based on a normative principle of mobility. On the other hand, those social bonds he maintains back 'home', which establish his sense of belonging elsewhere and in a way solidify his possibility to at any point move back – despite his several failed attempts to do so thus far – are based on his 'mobility' or rather his mobile status in the Rift Valley, as the practices that establish them have come into being only *because* he is away.

In a sense, then, Opondo is paying into different social 'accounts', in the Garfinkelian sense as elaborated in Thielmann (2012), by accumulating both at Baringo and at 'home'; however, this only works through the specific narrative of mobility that is here being employed. My argument here is that through these practices, both mobility and sociality are being framed, explained, defined and demarcated: How one makes one's practices accountable, and what one accounts for 'mobile' and 'social', is defined by the means of how mobility and sociality are established.

In this sense, what I am describing is a sort of reverse practice theory: Taking the example of the iron sheet house, the specific way of *making it accountable* is what defines and determines the practice; the practice, then, is what defines what being mobile or being social is. Thereby, the idea is defined by practice – not the other way around. Daily practices such as provisional arrangements account for one's own perceived transitoriness and further the question of the social person per se – a thought that deserves further reflection beyond the scope of this chapter. Suffice it to say that through traversing static forms of genealogical depth that is inscribed in the landscape, the social person here refers to both living and dead; it forms itself through both the practice and the imagination of migration.

In order to understand the complex field in which mobile and social are defined, we must consider that the socialities at play here come into being by a sense of being stuck in one's own future-making possibilities; that is, the interminable accomplishment of one's own future fixates oneself and ascribes one to this place. If social mobility means that the social defines what is mobile, then it would suggest that mobile sociality means that it is the mobile that defines the social. But how is mobility understood in this? Opondo and Okwany feel mobile because they identify *only* with the place that is left behind; their definitions of both mobility and of place define their emplacement practices. Their sense of belonging is tied to a place defined by genealogy – Lake Victoria as genealogical landscape, thereby making all other places landscapes of the commute or of mobility. However, this gets us into an ontological quandry regarding how mobility and emplacement are defined: even seemingly sedentary practices, such as having a *shamba* and growing maize, are not understood as a sign for having settled, but can just as much be seen as a way to merely produce food while here, temporarily at first, but then – by virtue of always ready to but, in actuality, never really departing – more permanently later on.

Tristan Thielmann sharpens Harold Garfinkel's groundwork in ethnomethodology to consider the theoretical background of the question "in which situations which media become media through which use" (Thielmann, 2012: 87; my translation). He refers to Jörg Bergmann, who has found a number of paraphrases for Garfinkel's notion of "accountable", such as "observable, reportable, recordable, detectable, countable, comparable, picturable, tellable, storyable, tell-a-story-aboutable, analysable, representable, visible, testable, available to observation and report, available to inventory, cursory representation, anecdote, enumeration of professional psychological assessment" (Bergmann cf. Thielmann, 2012: 92; Garfinkel, 1967: vii). He analyses that Garfinkel has thereby brought to the fore how reporting and self-reporting activities are integral markers of social practices and of the same social order which is subject of the reporting (cf. Garfinkel, 1967: 191). If we were to consider a broad sense of media practices through the lens of this analysis, one could say that those practices that lay at the root of social bonds formed and stabilised through them, such as showing each other how one is connected through a mutual experience of stuckedness, by way of building provisional houses; keeping bonds to distant relatives and organising groups to gain capital, in order to leave again, can be seen as what Garfinkel called "indexical media" (Garfinkel, 1967: 206; cf. Thielmann, 2012: 97). As Thielmann notes, these "documented representations" serve to understand what actors expect of themselves and others (Thielmann, 2012: 97). They become the visualised and materialised elements of what establishes specifically mobile socialities.

Combined then with the practices that form and solidify social bonds in situ, such as fishing together, building together and planting together, but also organising, sharing and supporting each other, these practices shape and manifest what I understand to be socialities of practice. I take from Jean Lave's (1991) "communities of practice" the focus on situated social practice, which "emphasizes the relational interdependency of agent and world, . . . the inherently socially negotiated quality of meaning and the interested, concerned character of the thought and action of persons engaged in activity".

The perspective on 'socialities of practice' suggests, furthermore, a non-localised and mobile setting, one which is grounded on social practice and imaginaries of belonging. Socialities of practice can thereby be conceptualised as the co-operation of diverse, but not exclusive, emplacements and belongings that individuals account for by performing and enacting situated social media practices.

The quest of (not) going home: an outlook on futures at Lake Baringo

In my conversation with Opondo from the beginning of this chapter, he makes quite clear that this place, and his house, are nothing to be thought of as a "home". He engages images similar to those recounted to me by his sister about their father, building a shed under some trees just to sleep somewhere while he fishes.

From the narrations of these Luo descendants, it would appear as though the settlement of Kambi Ya Samaki is itself a mobile place. Indeed, when fish became fewer in number, some Luos actually settled on the other shores of the lake, in Sokotei, Kiserian and surrounding beaches, where the fish populations are still intact because the breeding areas have moved. Still, Kambi Ya Samaki – the 'fish camp' – remains the number one trading spot on the lake. Yet it also seems to have remained what had been imagined by the first Luos who arrived and who were involved in the foundation of this village: a camp, a place for business, a place for fish.

In recent months, however, as the risk of COVID-19 has impeded much traffic to Nairobi and the markets, Baringo's fishing families have been struck by further economic fragilities.

Importantly, however, this has neither driven people out of the region nor forced them to return to their 'homes', but has rather even more poignantly manifested their stuckedness. Travels back to western Kenya have become even less affordable due to the increased transport prices, while business revenues are decreasing. As the consequences of the pandemic are experienced all over Kenya, some inhabitants of Baringo have unwittingly and suddenly become the sole breadwinners of their extended families, a fact which compels them to continue making ends meet. This makes visible another aspect of being stuck, one that "by definition [is] a situation where a person suffers from both the absence of choices or alternatives to the situation they are in and an inability to grab such alternatives even if they present themselves" (Hage, 2009: 100).

In this sense, processes and practices of future-making continue to be governed by past and present socialities, which in their own terms are driven by imaginings of mobility and return. As actors engaged in practices situated across concurrent, concomitant social worlds, these imaginings account for people's experiences of both mobility and stuckedness, as well as their different emplacements.

Note

1 I wish to thank my research assistant Wilson Tiren in Baringo, without whom this project would have been hardly possible, and my interlocutors Opondo and Okwany for sharing their stories with me, as well as the inhabitants of fisherpeople and fish mongers at Lake Baringo who have shared their lives with me for a period of time.

The research was funded by the German Research Foundation (DFG) through the collaborative research center "Future Rural Africa", funding code TRR 228/1. I thank my PI Martin Zillinger for his valuable input in shaping this chapter, and my colleague Mario Schmidt for bringing me to Lake Baringo, as well as David Anderson for his generosity in sharing with me his deep knowledge of the Baringo area. I also owe my thanks to Alfred Anangwe at the Kenyan National Archive in Nairobi and Lucy McCann at the Bodleian Library in Oxford for their help with the archival research.

References

Adey, P. (2006) "If mobility is everything then it is nothing: Towards a relational politics of (im)mobilities", *Mobilities*, 1(1): 75–94.

Amit, V. (2000) *Constructing the Field: Ethnographic Fieldwork in the Contemporary World*. London: Routledge.

Anderson, D. M. and Bollig, M. (2016) "Resilience and collapse: Histories, ecologies, conflicts and identities in the Baringo-Bogoria basin, Kenya", *Journal of Eastern African Studies*, 10(1): 1–20. doi:10.1080/17531055.2016.1150240.

Arp Fallov, M., Jørgensen, A. and Knudsen, L. B. (2013) "Mobile forms of belonging", *Mobilities*, 8(4): 467–486. doi:10.1080/17450101.2013.769722.

Cangià, F. and Zittoun, T. (2020) "Exploring the interplay between (im)mobility and imagination", *Culture & Psychology*, 1–13. doi:10.1177/1354067X19899063.

Certeau, M. de (1984) *The Practice of Everyday Life*. Berkeley/Los Angeles/London: University of California Press.

Couldry, N. (2010) "Theorising media as practice", in B. Bräuchler and J. Postill (eds.) *Theorising Media and Practice*. Oxford: Berghahn Books, 33–54.

Cresswell, T. (2006) *On the Move: Mobility in the Modern Western World*. New York: Routledge.

Dundes Renteln, A. (2004) *The Cultural Defense*. New York: Oxford University Press.

Garfinkel, H. (1967) *Studies in Ethnomethodology*. Englewood Cliffs: Prentice-Hall.

Grasseni, C. (2009) *Developing Skill, Developing Vision: Practices of Locality at the Foot of the Alps*. New York: Berghahn Books.

Hage, G. (2009) "Waiting out the crisis: On stuckedness and governmentality", in G. Hage (ed.) *Waiting*. Melbourne: Melbourne University Press, 97–106.

Hannam, K., Sheller, M. and Urry, J. (2006) "Editorial: Mobilities, immobilities and moorings", *Mobilities*, 1(1): 1–22.

Ingold, T. (1993) "The art of translation in a continuous world", in G. Pálsson (ed.) *Beyond Boundaries: Understanding, Translation and Anthropological Discourse*. London: Berg, 210–230.

Ingold, T. (2007) *Lines: A Brief History*. London: Routledge.

Khan, N. (2016) "Immobility", in N. B. Salazar and K. Jayaram (eds.) *Keywords of Mobility: Critical Engagements*. New York: Berghahn Books, 93–112.

Krajina, Z., Moores, S. and Morley, D. (2014) "Non-media-centric media studies: A cross-generational conversation", *European Journal of Cultural Studies*, 17(6): 682–700.

Larsen, J., Urry, J. and Axhausen, K. W. (eds.) (2006) *Mobilities, Networks, Geographies*. Aldershot: Ashgate.

Lave, J. (1991) "Situating learning in communities of practice", in L. Resnick, J. M. Levine and S. D. Teasley (eds.) *Perspectives on Socially Shared Cognition*. Washington, DC: American Psychological Association. doi:10.1037/10096-000.

Little, P. D. (2019) "When green equals thorny and mean: The politics and costs of an environmental experiment in East Africa", *African Studies Review*, 62(3): 132–163.

Malkki, L. H. (1992) "National geographic: The rooting of peoples and the territorialisation of national identity among scholars and refugees", *Cultural Anthropology*, 7: 24–44.

Mattern, S. (2017) *Code and Clay, Data and Dirt: Five Thousand Years of Urban Media*. Minneapolis: Minnesota Press.

McCullough, M. (2013) *Ambient Commons: Attention in the Age of Embodied Information*. Cambridge: MIT Press.

Norum, R. (2015) "The hypersocial: Transience, privilege and the neo-colonial imaginary in Expatria, Kathmandu", unpublished D.Phil Thesis, University of Oxford, Institute of Social and Cultural Anthropology, Oxford.

Pálsson, G. (ed.) (1993) *Beyond Boundaries: Understanding, Translation and Anthropological Discourse*. London: Berg.

Passmann, J. (2018) *Die soziale Logik des Likes: Eine Twitter-Ethnografie*. Frankfurt: Campus.

Petek, N. (2015) "An archaeological survey of the lake Baringo Lowlands 2014: Preliminary results", *Nyame Akuma*, 83, 100–111.

Postill, J. (2008) "Localising the Internet beyond communities and networks", *New Media and Society*, 10(3): 413–431.

Postill, J. and Pink, S. (2012) "Social media ethnography: The digital researcher in a messy web", *Media International Australia*, 145(1): 123–134.

Ramella, A. L. and Norum, R. (2021) "Mediengeographie", in T. Thielmann and M. Kanderske (eds.) *Handbuch Mediengeographie*. Baden-Baden: Nomos (in print).

Rapport, N. and Amit, V. (2002) *The Trouble with Community*. London: Pluto.

Rawls, A. W. (ed.) (2002) *Ethnomethodology's Program. Working Out Durkheim's Aphorism*. Lanham: Rowman & Littlefield.

Salazar, N. B. and Smart, A. (2011) "Anthropological takes on (im)mobility", *Identities*, 18(6): i–ix. doi:10.1080/1070289X.2012.683674.

Schuetz, A. (1945) "The homecomer", *American Journal of Sociology*, 50(5): 369–376.

Shiino, W. (1997) "Death and rituals among the Luo in South Nyanza", *African Study Monographs*, 18(3–4): 213–228.

Suchman, L. (2002) "Located accountabilities in technology production", *Scandinavian Journal of Information Systems*, 14(2): 91–105.

Taylor, C. (2002) "Modern social imaginaries", *Public Culture*, 14(1): 91–124.

Thielmann, T. (2012) "Taking into account: Harold Garfinkels Beitrag für eine Theorie sozialer Medien", *Zeitschrift für Medienwissenschaft*, 4(1): 85–102. doi:10.25969/mediarep/2680.

Wikan, U. (1993) "Beyond the words. The power of resonance", in G. Pálsson (ed.) *Beyond Boundaries: Understanding, Translation and Anthropological Discourse*. London: Berg, 184–209.

Winther, I. W. (2006) *Hjemlighed – Kulturfænomenologiske studier*. København: Danmarks Pædagogiske Universitets Forlag.

Winther, I. W. (2009) "'Homing oneself': Home as a practice", *Haecceity Papers*, 4(2): 49–83.

INDEX

www.ingramcontent.com/pod-product-compliance
Lightning Source LLC
LaVergne TN
LVHW080806011026
849160LV00015B/1552
* 9 7 8 0 3 6 7 5 4 6 1 7 5 *